# CATASTROPHES AND EARTH HISTORY

# CATASTROPHES
# AND EARTH HISTORY
The New Uniformitarianism

W. A. BERGGREN AND
JOHN A. VAN COUVERING, EDITORS

1984

PRINCETON UNIVERSITY PRESS

PRINCETON, NEW JERSEY

Copyright © 1984 by Princeton University Press
Published by Princeton University Press,
41 William Street, Princeton, New Jersey 08540
In the United Kingdom: Princeton University Press,
Guilford, Surrey

All Rights Reserved
Library of Congress Cataloging in Publication Data
will be found on the last printed page of this book

ISBN 0-691-08328-2
ISBN 0-691-08329-0 (pbk.)

This book has been composed in CRT Times Roman
Clothbound editions of Princeton University Press books
are printed on acid-free paper, and binding materials
are chosen for strength and durability.
Paperbacks, although satisfactory for personal collections,
are not usually suitable for library rebinding.

Printed in the United States of America by
Princeton University Press, Princeton, New Jersey

# CONTENTS

# FOREWORD

W. A. BERGGREN

Department of Geology and Geophysics
Woods Hole Oceanographic Institution

In June of 1974, the Graduate Education Program of the Woods Hole Oceanographic Institution sponsored a week-long symposium entitled *Organisms and Continents Through Time.* The symposium was primarily organized for the sake of acquainting the graduate students at Woods Hole with current research in this area, and it was run in an informal manner so as to allow a maximum of exchange between speakers and audience. Some thirty invited speakers, supported by their published papers on related topics (a compendium was distributed to everyone in attendance) and by the guidance of moderator John A. Van Couvering, combined to make the symposium a resounding success.

At the suggestion of Robert W. Morse and A. Lawrence Peirson (dean and assistant dean, respectively, of the Woods Hole Oceanographic Institution Education Program), I was asked early in 1977 to organize a second symposium along lines similar to those of the first. John Van Couvering and I discussed several subjects before deciding that an highly appropriate theme would be the current state of uniformitarianism as a scientific subject.

The second symposium took place in the Redfield Auditorium of the Woods Hole Oceanographic Institution from June 6 to 10, 1977. Some twenty-five speakers presented lectures on various aspects of the concepts of uniformitarianism and catastrophism as working hypotheses in the earth sciences. Many of the talks that were presented at this symposium are included in this volume. In addition, we have also included several oral papers from a symposium held in connection with the Second North American Paleontological Convention in August of 1977 at Lawrence, Kansas. Since this symposium dealt with the problems of the Cretaceous-Tertiary boundary event, these papers have been gathered together—through the efforts of Norman Newell, who organized the meeting—for presentation in this volume, where they appropriately fit with our theme.

We would like to thank Robert Morse and Lawrence Peirson for providing help in the organization of this symposium, lovely facilities, and financial assistance through the Woods Hole Oceanographic Institution Education Program.

# CATASTROPHES AND EARTH HISTORY

# INTRODUCTION

John A. Van Couvering

American Museum of Natural History

The papers in this volume are intended to involve the reader in the current reappraisal of uniformitarianism that is radicalizing geology. We have not attempted to review the dialogue in every subdiscipline of earth history, because such a review would necessarily be little more than a piecemeal study of the momentum (some might say "inertia") of specialist literature and tradition; nor have we sought for a unifying treatise on modern geohistorical philosophy, since such books as David B. Kitts' *The Structure of Geology*—which could be read as both an introduction and a conclusion to our book—already exist. Finally, we believe that the symposium format (despite the good intentions and hopeful claims of many a convener and publisher) is not really conducive to a smoothly integrated overview of either type.

When we organized the Woods Hole symposium, from which most of our chapters are derived, we felt that it was necessary—and would be most interesting—to adopt a "case history" approach and to consider only those topics for which we had both the best data and available speakers. Furthermore, since our meeting was sponsored on behalf of resident and visiting graduate students, we felt that they could most fruitfully absorb the underlying concept if it were actualized in a methodological narrative. Nevertheless, to emphasize and summarize our intent, we have included in this volume several other papers that treat the basic philosophy. These papers comprise the first part of the book and introduce the liberated uniformitarianism that enables us to deal with the large and small "catastrophes" described in the case studies which then follow.

What exactly is being challenged in the "new uniformitarianism," and what are the limits of the question? It seems to us that the battleground between the old and the new lies in the realm of those events which cannot be related to human experiences and for which "the present" is not, therefore, intuitively the key to "the past." Uniformitarian analysis, no

3

matter how sophisticated or successful, is never anything other than a mere analogy with the present and is, therefore, only an opinion about the meaning of rock features. I am fond of saying that a geologist writes like a person overcoming very grave reservations. This is because no geologist can operate as an earth historian without continuously doubting such opinions—regardless of the accuracy of the observations upon which they are based. The realization of our uncertainty makes us uneasy, as does knowing that our explanations of the past are not more true, but only more plausible, than the stories told by creationists, extraterrestrialists, and other seers. This degree of plausibility, in turn, arises in our own minds in direct relationship to the strictness with which we adhere to the experienced world as a model for past worlds and neither add nor subtract according to our human desires: in a word, by uniformitarianism.

It is no wonder, then, that geologists are hostile to any explanation appealing to a state of affairs that has not been observed to exist. However, the merely uniform and linear extrapolation of the human experience of nature—gradualism, in the strict sense—does not account for many of the surprises that, through interpretation of unusual features in the geological record, have already been incorporated into our view of the world. In this regard, earth historians began with explanations of the most obvious non-experiential "catastrophes" of the prehistoric past: extinct volcanism in central France, continental "ice ages" in Scotland and Massachusetts, and ice-dam floods in the channeled scablands of Washington state. This initial phase of trans-uniformitarian deduction was later followed by the recognition of astroblemes and submarine landslides as geological agents. We are currently in the process of allowing uniformitarian status to unwitnessed—and, therefore, nonanalogous—marvels such as global eustatic sea-level change, ocean basin dessication, megavolcanic blasts, and geomagnetic polarity reversals. On the other hand, the formulation of the principles of plate tectonics—however revolutionary in geological thinking—seems to lead less to an explanation of remarkable and unpredictable events and more to a simple and sensible description of the present world, as well as to an expanded understanding of the nature of processes that can be extrapolated without much variation into the past. Thus, it might be said that plate tectonics gave new life to gradualism; it might also be said that, in speaking of such unexpected discoveries (not to mention the still speculative events discussed in this book), we are actually talking, not about modifying uniformitarianism to include catastrophes, but rather about redefining gradualism.

Gradualism, strictly applied, is the refusal to apply uniformitarian logic to events that are ouside of the human frame of reference; as such, it is a failure to consider scale or time. Nearly everyone who has thought about the great extinction events—such as that of the Cretaceous-Tertiary, which is discussed at length in this book—has confronted a genuine mystery. Rather than admitting to the existence of that mystery, some have said that these events are nonexistent; others, impatient for an answer, have summoned up explanations that work in a very *ad hoc* manner. Still others, however, have approached the subject by applying uniformitarian logic with a greater care and a more proper dubiety. In such cases, the reader will be interested to note how these tactics can slowly and progressively resolve both the shape and the character of such perfectly unimaginable events. Uniformitarianists must always be wary of the fact that the model of the world is not yet completely described, but gradualists must accept that human experience now includes the discovery in the rock record of unpredicted events—paleomagnetic reversals being salutary examples.

The point of this book is, therefore, that geology has a metahistorical capability which should be formally acknowledged. Like astronomers, geologists can recreate and investigate events of a magnitude, or subtlety, so extreme as to be imperceptible to real-time instruments—let alone to human eyes—if they were happening today (and possibly, some of them are). Only the rock record is capable of registering certain natural expressions of nuclear, molecular, and thermodynamic processes; some of these have remained for historical geologists to discover, simply because they lie beyond the limits of human experience. Since we have yet to explain every feature of the rock record, it seems obvious that some of the explanations that are yet to come will contain further surprises. Uniformitarian logic is still the only tool with which we can read the rocks, isolate the evidence, and describe inexplicable "catastrophes." Insofar as these events and their causes transcend our present knowledge of the world, the "new uniformitarianism" differs from the old in that we now also realize that the past may be the key to the present.

# THE CONCEPT OF CATASTROPHE
# AS A NATURAL AGENT

Chapter 1

# TOWARD THE VINDICATION OF
# PUNCTUATIONAL CHANGE

STEPHEN JAY GOULD

Museum of Comparative Zoology, Harvard University

*PART I*   LYELL'S VISION AND RHETORIC

The sanctified writings of a profession are often among the most misunderstood, largely because so few people read them. Sir Charles Lyell's *Principles of Geology* (1830–1833) rests prominently among such works. Most geologists revere it as a painstaking, scrupulously objective, empirical catalogue that established their calling by demonstrating the power of modern causes to explain all past results. In fact, it is a lawyer's brief, ingeniously constructed to push a point by all means, most fair, but some foul. And it was written by a lawyer, for Lyell trained and qualified in the profession banned from Utopia by Saint Thomas More (1516): "They have no lawyers among them, for they consider them as a sort of people whose profession it is to disguise matters."

Lyell had a vision of the earth and its history. He wished to transfer the timeless majesty of Newton's cosmos to an earth that most of his colleagues viewed as progressing in definite and limited directions, powered by occasional, devastating paroxysms. As the planets circle continuously, always in steady motion, always returning to previous positions, so would the earth move, as a changing yet timeless planet, through its own history—in modern jargon: a dynamic steady state. Land and sea would change places as the products of continents slowly eroded to fill up oceans, but land and sea would always exist in roughly constant amounts. Species would die and new ones would arise, but the mean complexity of life would not alter and its basic designs, created at the beginning, would endure to the end of time. In short, Lyell's vision of earth history and geological process rested upon two cardinal tenets: *gradualism* and *nondirectionalism.*

Lyell called upon a "principle of uniformity" as his primary intellec-

9

tual weapon against catastrophism and directionalism. As a primary tactic in his belief, Lyell argued that the procedures of science itself required a belief in uniformity, since the alternate claims of "catastrophism" precluded rational inquiry and resolution:

> The student, instead of being encouraged with the hope of interpreting the enigmas presented to him in the earth's structure,—instead of being prompted to undertake laborious inquiries into the natural history of the organic world, and the complicated effects of the igneous and aqueous causes now in operation, was taught to despond from the first. Geology, it was affirmed, could never rise to the rank of an exact science,—the greater number of phenomena must forever remain inexplicable, or only be partially elucidated by ingenious conjectures (Lyell, 1842, vol. 1, p. 325).[1]

Yet, what is unscientific about the hypothesis that catastrophic events shape a planet's surface? (Look at Mercury and our moon.) Or, is it unscientific to hypothesize that the history of terrestrial vertebrates includes progressive trends? (Indeed, it does.) "Uniformity" had a host of distinct meanings for Lyell. Some are (as he stated) methodological prerequisites for doing science. Others are specific, substantive claims about the empirical world: they may be true, false, or somewhere in between without threatening the enterprises of science. Lyell was strongly committed to the substantive claims, and he supported them (perhaps unconsciously) with a consummate trick of argument: he gave the same name—uniformity—to the methodological presuppositions and to the substantive claims, and then he maintained that scientists must accept the substantive claims because the practice of their profession rests upon the methodological presuppositions.

## THE MULTIPLE MEANINGS OF UNIFORMITY

During the 1960s, several geologists and historians independently recognized the hybrid nature of "uniformitarianism"[2] (Hooykaas, 1963;

---

[1] I cite from the sixth English edition (1842), which differs in no important way from the first of 1830–1833.

[2] Lyell did not invent the term "uniformitarianism." The philosopher William Whewell coined it in his review of the second volume of Lyell's *Principles of Geology*. An incisive analytical thinker, Whewell did not follow Lyell's tactic of including several concepts under one name; rather, he confined his use of the term to the theory of gradualism (the third uniformity—see text). Later writers then extended its usage to include all of Lyell's meanings for "uniformity." It is no accident, I think, that the

Gould, 1965; and Simpson, 1963). Since then, several historians, particularly Rudwick (1972) and Porter (1976), have forcefully asserted this "revisionist" interpretation of the history of geology. (See also: Lawrence, 1973; and the Lyell issue—July 1976—of the *British Journal for the History of Science*.) Lyell the Cardboard Hero, the white knight of science against lingering supernaturalism, is being replaced by Lyell the Passionate Believer, who pits his system of a balanced and stately earth against equally passionate (and equally scientific) beliefs in definite directions and catastrophic changes. A much more interesting man, and one much truer to the original.

Rudwick (1972) has teased apart four different meanings of uniformitarianism that I arrange as follows into two distinct subcatagories (see Gould, 1965).

METHODOLOGICAL PRESUPPOSITIONS ACCEPTED BY ALL SCIENTISTS

*1) The Uniformity of law*   Natural laws are invariant in space and time. John Stuart Mill (1881) argued that such a postulate of uniformity must be invoked if we are to have any confidence in the validity of inductive inference; for if laws change, then an hypothesis about cause and effect gains no support from repeated observations—the law may alter the next time and yield a different result. We cannot "prove" the assumption of invariant laws; we cannot even venture forth into the world to gather empirical evidence for it. It is an *a priori* methodological assumption made in order to practice science; it is a warrant for inductive inference (Gould, 1965).

*2) The uniformity of process (actualism[3])*   Whenever possible, explain past results as the outcome of causes that are still operating on the earth. Do not invent causes with no modern analogues when present causes can render the observed results. Philosopher Nelson Goodman (1967) has recognized that an assertion of actualism represents a particular defense of the general principle of simplicity. As such, it is another *a priori* methodological assumption shared by all scientists and not a statement about the empirical world.

---

multiple recognition of uniformity as a hybrid (see Merton, 1963) came in the early days of a new historiography emphasizing the power of theories as comprehensive visions with sources and support far more complex and varied than the empirical data of a profession (see Kuhn, 1962, as the primary document).

[3] Actualism, a European term little used in the United States, makes good sense in most continental languages; for the French, *actuel* meansd "at the present time" and, unlike the English "actual," does not imply "real."

SUBSTANTIVE ASSERTIONS ABOUT THE WORLD

*3) Uniformity of rate (gradualism)*    Friedrich Engels once expressed his frustration about the course of reform in his adopted country with these words: "It moves, as all things in England, with slow and measured pace." Lyell's vision of the world lay within this stately tradition. His earth was surely in constant flux, but at a pace so leisurely that a human observer during a lifetime of watching might proclaim it static. Yet, the immensity of time guaranteed enormous results in accumulation. All the great events of earth history, from the cutting of the Grand Canyon to the rise of the Himalayas, must be interpreted as being the outcome of or-dinary causes advancing step by innumerable step—millimeters of downcutting or uprising per century, accumulated slowly and steadily through eons. In 1829, Lyell wrote to Murchison about his grand vision of an earth governed by causes that "never acted with different degrees of energy from that which they now exert."

Unlike the first two uniformities, gradualism is not a presupposition of method. It is a definite empirical claim about the world. It may be true or false. It must be tested, not assumed.

*4) Uniformity of conditions (non-directionalism, dynamic steady state)* Lyell's earth was a place of constant change, yet permanent aspect. Like the wind and rivers of Ecclesiastes, it cycled endlessly[4] with no direction. Lyell was quite confident that Paleozoic mammals would soon be found; after all, the Mesozoic had just fallen with the discovery of Jurassic mammals at Stonesfield. Moreover, the extinction of large groups implied no direction to the history of life, for they would reappear when the "great year" made its full cycle:

> Then might those genera of animals return, of which the memorials are preserved in the ancient rocks of our continents. The huge iguanodon might reappear in the woods, and the ichthyosaur in the sea, while the pterodactyle [sic] might flit again through the umbrageous groves of tree-ferns (Lyell, 1842, vol. 1, p. 193).

Again, uniformity of conditions is not a necessary assumption, but a testable claim about the real world. Indeed, it is quite incorrect in Lyell's strict formulation. We do not expect to encounter a Silurian rat, much less a post-Holocene dinosaur.

---

[4] Unlike Hutton, with his famous dictum of "no vestige of a beginning, no prospect of an end," Lyell didn't deny that the earth must have an ordained beginning and end. But he believed that decisions about ultimate origin and final fate did not lie in the domain of science.

## MODERN GEOLOGY AS A BLEND OF LYELL
## AND THE CATASTROPHISTS

What, then, was Lyell's battle with the "catastrophists" all about? We are taught, in the conventional textbooks of geology, that Lyell routed a group of theological apologists and established geology as a modern science. This homily supposes that the catastrophists directly denied science by rejecting the first two uniformities in favor of an earth ruled directly by a god who capriciously changed his own laws.

The catastrophists did no such thing. They held as staunchly as Lyell to the uniformity of law, for they matched him in their commitment to science (Rudwick, 1972). Even the Reverend Thomas Burnet, author of the most fantastic scriptural geology (1681), invoked Lyell's favorite metaphor of the Gordian knot,[5] and for the same reason—to attack a belief in miracles (suspension of natural law) as a denial of science itself: "They say in short, that God Almighty created waters on purpose to make the Deluge. . . . And this, in a few words, is the whole account of the business. This is to cut the knot when we cannot loose it" (Burnet, 1691 ed., p. 33).

Moreover, catastrophists supported with equal fervor the second methodological uniformity (actualism). One must, they argued, begin with the observable modern causes, gauge their effects, and estimate their explanatory power in yielding past results. If modern causes suffice, then no others should be invoked. In any case, the only good argument for a cataclysmic cause is a result that modern causes cannot produce. Thus, Alcide d'Orbigny, a firm believer in frequent universal cataclysm, wrote:

> Natural causes now in action have always existed. . . . To have a satisfactory explanation of all past phenomena, the study of present phenomena is indispensable. . . . Science owes much to Mr. Lyell for the development of this system, supported by copious research as wise as it is ingenious (d'Orbigny, 1849–1852, vol. 1, p. 71).

Catastrophists and uniformitarians disagreed on matters of substance: the third and fourth meanings of uniformity. Catastrophists denied a uniformity of rate since they attributed much of the earth's structure and topography to intermittent paroxysms, often on a worldwide scale, rather than to the simple accumulation of tiny changes. In addition, and nearly to a man, they adopted a directional view of earth history. The nebular hypothesis, with its consequence of a continually cooling earth, provided the dynamics both for direction and for intermittent global catastrophe.

---

[5] "We see the ancient spirit of speculation revived, and a desire manifestly shown to cut, rather than patiently to untie, the Gordian Knot" (Lyell, 1842, vol. 1, pp. 325–326).

As the earth cooled, it contracted, drawing the more fluid interior away from the rigid crust. Instability built up until the crust collapsed upon the constricted interior, linear mountain belts were thrown up along its major cracks, and in the process, much of life was wiped out. New creations had to cope with less hospitable, cooler climates; life had to become progressively more complex in order to survive in more rigorous environments.

The textbook tale of uniformitarian goodies versus catastrophist baddies is a bit of self-serving, historically inaccurate rhetoric (see Porter, 1976, on Lyell's creative use of history). Lyell was a brilliant man, one of the great scientists of the nineteenth—or, indeed, of any—century. But he was not using the weapon of uniformitarianism specifically to uphold science against a group of aging theological apologists, who wished to retain the earth as a domain of miracles. The catastrophists were as "scientific" as the uniformitarians. Everyone upheld the two methodological uniformities as part of the definition of science. The two schools had major substantive disagreements about rates and direction. Moreover, in advocating paroxysmal change, the catastrophists were not letting a reverence for God blind them to the facts of earth history. If anything, as they saw it, they were the empirical literalists, who read their story directly, without interpolation. Local and regional geological sections usually record an episodic history with profound faunal and stratigraphic breaks. Lyell was the interpolationist, the non-literalist. To preserve his belief in gradualism, he argued that appearances are misleading and that gaps in the record can explain away nearly all supposed catastrophes—if a record preserves only one step in a thousand, then truly gradual changes will appear to be abrupt.

Modern geology is a fairly even mixture of what Lyell espoused and what the catastrophists believed. Both schools accepted the first two uniformities. Lyell largely triumphed with his third uniformity of rate, while our current ideas on the history of earth and life lie closer to the directionalism of catastrophists than to Lyell's steady state. Yet, modern geology bears Lyell's name. We are all educated to call ourselves uniformitarians and to enshrine Lyell's doctrine as "the greatest single contribution geologists have made to scientific thought" (Longwell and Flint, 1955, p. 385).

In short, Lyell won with rhetoric what he could not carry with data. Late in his life, he dropped his rigid insistence on an earth in steady state, when the evidence of progression in life's history inspired his conversion to evolutionary theory (Wilson, 1970; Gould, 1970). As for the substantive uniformity of rate (gradualism), he continued to assert its necessity by conflating it with the methodological uniformities of law and process. In the following famous passage, for example, Lyell first argues that catastrophism is unscientific on methodological grounds, and then he asserts that a substantive belief in gradualism can make geology a science:

> Never was there a dogma more calculated to foster indolence, and to blunt the keen edge of curiosity, than this assumption of the discordance between the ancient and existing causes of change. . . . Geology, it was affirmed, could never rise to the rank of an exact science . . . the greater number of phenomena must forever remain inexplicable [second uniformity]. . . . The course directly opposed to this method of philosophizing consists in an earnest and patient inquiry how far geological appearances are reconcilable with the effect of changes now in progress. . . . For this reason all theories are rejected which involve the assumption of sudden and violent catastrophes and revolutions of the whole earth, and its inhabitants [third uniformity] . . . theories which are restrained by no reference to existing analogies, and in which a desire is manifested to cut, rather than patiently to untie, the Gordian knot (Lyell, 1842, vol. 1, pp. 324–326).

Lyell's conflation had two major effects which have resounded throughout the subsequent history of geology. The first has led to an annoying, but basically harmless, confusion about the meaning of uniformitarianism. The second, however, has had a profoundly negative impact by stifling hypotheses and by closing the minds of a profession toward reasonable empirical alternatives to the dogma of gradualism.

*1) Confusion*    The history of debate on the meaning of uniformitarianism has been dominated by confusion. Lyell mixed methodological and substantive issues under the same name, and geologists have been engaged in pseudodebates ever since. Paul Krynine (1956), for example, dubbed uniformitarianism a "dangerous doctrine," because it held too strictly to a constancy of rate and conditions (substantive meanings 3 and 4). Zangerl and Richardson (1963, p. 4) labelled Krynine's opposition as "rather pointless," and then they defined uniformitarianism as "the conclusion that nature's laws are unchanging" (methodological meaning 1). Yet, since science seems to proceed satisfactorily in the face of philosophical speculation from its practitioners, these debates have wasted some time and paper, but they have otherwise not impeded the practice of geology.

*2) Restriction*    Presuppositions become more serious when they restrict the range of hypotheses that scientists are willing to entertain in explaining phenomena. The restriction is particularly unfortunate when it rests upon a doctrine enshrined as unquestioned and obviously true and not recognized as only one possible bias among many. Gradualism, the third uniformity of rate, has held this status for most geologists, largely because Lyell succeeded with his rhetoric—he managed to conflate the substantive (and possibly false) uniformity of rate with those necessary presuppositions of method (induction and simplicity) that he also called "uniformity." Thus, many geologists have felt that a belief in slow and steady change defines the necessary practice of their success. Hawkes, for example, rejected, as inadmissable *a priori,* the hypothesis of rapid polar wandering: "The idea of polar wandering in a series of relatively short spurts from one stable position to another—the 'random walk of the poles'—is heady wine to place before the paleogeographer! It constitutes a departure from the doctrine of uniformitarianism" (Hawkes, 1958, p. 405). But maybe the poles do move rapidly. The claim should at least be investigated, rather than being dismissed without a test.

## PART 2 THE VALIDITY OF PUNCTUATIONAL CHANGE

I come, then, to the second part of this manifesto. In the first, just concluded, I argued that Lyell won support for the substantive doctrine of gradualism by conflating it with methodological presuppositions accepted by all scientists. In the second part, I wish to argue the following:

1) Gradualism has operated for the past one hundred and fifty years as a serious constraining bias in the history of geology.

2) Gradualism was never "proved from the rocks" by Lyell and Darwin, but was rather imposed as a bias upon nature. The roots of this doctrine lie as much in general culture and ideology as in inferences from natural phenomena.

3) Although some geological results fit the conventional model of gradualism, many others do not. We need a more pluralistic view of the nature of change.

4) Many important changes in the earth's physical and biological history proceed by "punctuational change,"[6] that is, by relatively rapid flips

[6] I reject the term "catastrophic" for this style of discontinuous change, because the geological theory of catastrophism generally referred to rapid change of a global scale (as in Elie de Beaumont's theory of crustal collapse). Modern proponents of catastrophism, particulary Velikovsky (1950, 1951) and his allies, have also emphasized the worldwide extent of the sudden changes that they advocate. My emphasis is upon rate, not upon extent—whether geographic or subjective (in the sense of profundity or magnitude). I envisage a world composed of quasi-stable systems that resist stress to a breaking point and then flip rapidly to a new equilibrium (for example, the boiling of water or the fracturing of rocks). I refer to this mode as "punctuational change" to emphasize both the stability of systems and the concentration of change in short epi-

between fairly stable equilibria. Systems often absorb stress and resist change until the stresses accumulate past a breaking point; systems then flip to a new stable state.

5) Punctuational change has become popular in discipline after discipline during the past decade. In defending the reality of discontinuous change, catastrophists, with their adherence to empirical literalism, had grasped an essential part of nature.

My defense of punctuational change shall be largely anecdotal. Indeed, I do not assert any dogma, for my complaint is that gradualism has itself operated as a restrictive dogma for more than a century. The world is too multifarious to permit any philosophy of change to hold exclusive sway. I make a simple plea for pluralism in guiding metaphors about natural change, although I do not hide a personal preference for punctuational tempos (if only because they have been so unfairly neglected, and because I am enough of a Pythagorean to believe that certain forms have a stability far beyond the persistence of the objects that temporarily assume them).

With my primary goal being to increase the number of styles of change deemed admissable in geological inquiry, I will have done my job if I can convince readers that gradualism has acted as a restrictive dogma and that punctuational change is a reality, not an artifact of an imperfect record. I present, therefore, two examples: one, a specific case in which a (presumably correct) punctuational hypothesis was rejected—and even denigrated—because it conflicted with gradualist dogma; and the second, a general statement about the dilemmas of an entire profession wedded to gradualism and about a punctuational release from the impasse.

## THE CHANNELED SCABLANDS OF EASTERN WASHINGTON

In 1923, J. Harlen Bretz proposed a striking and unorthodox explanation for the channeled scablands of Eastern Washington. This peculiar topography is developed within a series of elongate basins called coulees. The coulees are subparallel and have a continuous gradient. They are enlargements of pre-existing drainage lines; all are developed on the Columbia basalts and cut through a thick overlying deposit of loess. They are traceable up gradient to the southern extent of the last glaciation and down gradient to the Snake or Columbia rivers. Now, the peculiar features: the

---

sodes that break old equilibria and quickly reestablish new ones. I can't help but think that some very deep metaphors about the nature of things is embedded in this debate over gradual versus punctuational change. Shall we see flux as fundamental and objects as ephemeral? Or, is the world ordered in a hierarchy of bounded structures?

subparallel channels anastomose extensively across the old divides, as though they were once flooded over; numerous hanging valleys line the channel sides, suggesting that the channels were once filled with water to a great height; the deep gouging of basalt within the channels, where hills of older loessial topography are arranged as if they represented the uncovered areas of a gigantic braided stream, does not look like the ordinary work of rivers; and finally, the coulee floors are covered with discontinuous deposits of basaltic stream gravel containing rock that is foreign to the area.

Bretz concluded, therefore, that the coulees had been carved by a single gigantic flood that had filled them to a depth of more than 1,000 feet, had cut through hundreds of feet of basalt in places, and had ended in a matter of days. He envisioned the scope of the event as follows:

> Fully 3,000 square miles of the Columbia plateau were swept by the glacial flood, and the loess and silt cover removed. More than 2,000 square miles of this area were left as bare, eroded, rock-cut channel floors, now the scablands, and nearly 1,000 square miles carry gravel deposits derived form the eroded basalt. It *was* a debacle which swept the Columbia Plateau (Bretz, 1923, p. 649).

Bretz's hypothesis evoked from the geological establishment a flood of commentary, nearly all of it negative. The common theme running through all of this criticism was the rejection of his ideas in favor of gradualistic explanations, often on *a priori* grounds. In 1927, for example, the Washington Academy of Sciences heard a lecture by Bretz and, subsequently, published commentary upon it by several leading geologists, most of whom were in the employ of the United States Geological Survey. I cite statements from the various disputants:

W. C. Alden admitted that he had never seen the Columbia plateau, but he ventured, nonetheless, an opinion about the "main difficulty":

> the idea that all the channels must have been developed simultaneously in a very short time. . . . The problem would be easier if less water was required and if longer time and repeated floods could be allotted to do the work (in Bretz, 1927, p. 203).

James Gilluly clearly expressed an *a priori* preference for gradualism by arguing that it was not necessary to explicitly discredit Bretz's data and that an equally plausible case for slow and steady change was reason enough to reject catastrophic flooding: "That the actual floods involved at any given time were of the order of magnitude of the present Columbia's or at most a few times as large, seems by no means excluded by any evidence as yet presented" (in Bretz, 1927, p. 205).

E. T. McKnight proclaimed the scabland features as "normal channels deposits of the Columbia during its eastward shift over the area in pre-glacial, glacial, and post-glacial times" (in Bretz, 1927, p. 206).

G. R. Mansfield presented yet another bit of gradualistic guesswork:

> The scablands seem to me better explained as the effects of persistent ponding and overflow of marginal glacial waters, which changed their position or their places of outlet from time to time through a somewhat protracted period (in Bretz, 1927, p. 207).

O. E. Meinzer, who had studied the scablands, admitted that "the erosion features of the region are so large and bizarre that they defy description." Then, he continued:

> However, the Columbia River, is a very large stream, especially in its flood stages, and it was doubtless still larger in the Pleistocene epoch. Its erosive work in the Grand Coulee and Quincy Valley, impressive though it is, appears to me about what would be expected from a stream of such size when diverted from its valley and poured for a long time over a surface of considerable relief that was wholly unadjusted to it (in Bretz, 1927, p. 207).

Finally—and more honestly and explicitly than most—Meinzer stated his *a priori* preference for gradualism: "Before a theory that requires a seemingly impossible quantity of water is fully accepted, every effort should be made to account for the existing features without employing so violent an assumption" (in Bretz, 1927, p. 208).

In autobiographical notes that have been informally distributed among students and colleagues in recent years, Bretz has characterized his own adversaries:

> Uniformitarian-minded geologists were shocked and all asked for longer lived glacial streams of more moderate volumes. . . . For nearly thirty years I published successive papers on my theme and had rebuttals verbal and in print from time to time, all arguing for more time and for much smaller rivers.

But in the 1950s, Bretz triumphed, as new information vindicated his hypothesis. First, a source for his water emerged in the catastrophic emptying of ice-damned glacial Lake Missoula; and then, giant ripples 60 to 425 feet in chord length and 15 to 22 feet high were found both in the area of Lake Missoula and on the gravel bars of the scablands themselves. Baker (1973) argued that a flood powerful enough to construct such ripples might have discharged as much water as $752 \times 10^6$ cfs into the flood channels and might have carried boulders up to 36 feet in diameter! First identified in aerial photographs, these ripples had eluded Bretz at ground level—a good illustration of the effects of scale upon observations; Bretz had walked right over them and missed them "under a cover of sage brush" (Bretz 1969, p. 509).

I do not wish to overemphasize the simplistic moralistic character of this tale, although it does exemplify my contention that gradualism stifles valid hypotheses. I neither regard Bretz's opponents as fools or blinded dogmatists, nor Bretz himself as a patient exemplary hero. Bretz's own style of science virtually guaranteed that he would not triumph with the data he had in the 1920s. He did not have a complete story, only a set of peculiar results in the coulees themselves. He did not postulate a reasonable source for the prodigious amount of water he required, but rather offered only a few apologetic suggestions about volcanic melting of ice and dramatic climatic amelioration (Bretz, 1972). Moreover, he seemed singularly unconcerned about this fundamental missing chapter in his tale. As a rigid empiricist of the naive indictivist school, he apparently felt that the sheer length of his list of evidence from the coulees would overwhelm the opposition. As he put it: "I believe that my interpretation of channeled scabland should stand or fall on the scabland phenomena themselves" (Bretz, 1927, p. 209). But orthodox theories are almost never overthrown by a simple list of contrary observations (for an extensive biological example see Gould, 1977). Orthodox theories are replaced by alternate ones. And, without a source for his waters, Bretz did not have a cohesive theory, only a set of odd observations. Who can blame his opponents for seeking along conventional lines for an *ad hoc* explanation of these observations? With Lake Missoula as a source, Bretz's observations finally cohered into a comprehensive story. Moreover, Bretz's opponents cited evidence for more than one flood, and they were right. Bretz himself now speaks of eight, each equally catastrophic (Bretz, 1969).

This story has a happy ending, for Bretz, a feisty man well into his 90s, has lived to enjoy his vindication. He closed his latest article with these words:

The International Association for Quaternary Research held its 1965 meeting in the United States. Among the many field excursions it organized was one in the northern Rockies and the Columbia Plateau in Washington. . . . The party crossed the head portions of the Cheney-Palouse and Telford-Crab Creek tracts, traversed the full length of the Grand Coulee, part of the Quincy basin and much of the Palouse-Snake scabland divide, and the great flood gravel deposits in the Snake Canyon. The writer, unable to attend, received the next day a telegram of "greetings and salutations" which closed with the sentence, "we are now all catastrophists" (Bretz, 1969, p. 541).

Moreover, Bretz's hypothesis has had utility well beyond the resolution of a local dilemma. Other lakes emptied catastrophically, and their results have been recognized in the light of Bretz's work. (See Malde, 1968, on Lake Bonneville.) In addition, Bretz has become the darling of planetary geologists, who find in the fluvial erosion of Mars many phenomena first recognized by him in the channeled scablands (Masursky et al., 1977).

## PUNCTUATED EQUILIBRIA AND THE TEMPO OF EVOLUTIONARY CHANGE

Darwin, writing for once without false modesty, claimed that half his work had come out of Lyell's brain. Among the key elements of Darwin's thought, an abiding faith in gradualism ranks as prominently as a deep belief in the power of natural selection. But the fossil record displays very few, if any, gradual transitions; and this fact, more than any other, bothered Darwin. As he wrote the *Origin of Species:*

> Why then is not every geological formation and every stratum full of such intermediate links? Geology assuredly does not reveal any such finely graduated organic chain; and this, perhaps, is the greatest objection which can be urged against my theory (Darwin, 1859, p. 280).

Darwin's answer was the same as Lyell's, and it reflected the same style of argument—a denial of empirical literalism with an interpolation of unobserved states dictated by theory. In short, Darwin argued that a woefully imperfect fossil record had left but one step in thousands, thereby giving a false punctuational character to the distribution of life in local selections:

> The geological record is extremely imperfect and this fact will to a large extent explain why we do not find interminable varieties, connecting together all the extinct and existing forms of life by the finest graduated steps. He who rejects these views on the nature of the geological record, will rightly reject my whole theory (Darwin, 1859, p. 342).

But Thomas Henry Huxley was not convinced. On the day before publication of the *Origin* (he had read an advance copy), he wrote a famous letter to Darwin. In it (see L. Huxley, 1901, p. 189), he offers to "go to the stake, if requisite" in Darwin's support. These lines are often quoted, but few authors add that Huxley hedged his offer of immolation by reserving it for certain sections of the book. For others, he offered his criticism in the same letter. As his major objection, he wrote (in L. Huxley, 1901, p. 189): "You have loaded yourself with an unnecessary difficulty in adopting *Natura non facit saltum* so unreservedly."

In citing Linnaeus' old aphorism—"nature does not proceed by leaps"—Huxley urged Darwin not to tie the theory of natural selection to an unnecessary (and false) belief in gradualism; the theory of natural selection was in for enough trouble on its own merits. Natural selection is not, fundamentally, a theory about rates. It may demand that new species not arise all at once, lest selection lose its creative cumulative role and operate only as a headsman to eliminate the unfit. But nothing in the theory of natural selection precludes a notion of rapid change during very short periods relative to the average duration of a species.

Nevertheless, Darwin persisted; for the new science of evolutionary paleontology, he set a task: find and document the preciously few cases of gradual change in stratigraphic sequences spared somehow from the ravages of metamorphism, tectonism, and simple non-deposition. Ever since, the history of this field has been dominated by Darwin's task. Most, if not all, of the "classic" examples are simply wrong (Hallam, 1968; and Gould, 1972, on the coiled oyster *Gryphaea*), either proclaimed on evidence insufficient in principle for deciding the issue[7] or, in some cases, asserted on the basis of no evidence at all (see Gould, 1974, on the "Irish Elk"). Moreover, the search for gradualism and the paucity of results it has yielded have perpetuated an unfortunate gap in communications between evolutionary theorists in paleontology and the practical scientists—the biostratigraphers—who do nearly all the work. Biostratigraphers know perfectly well that species, in general, do not change within their stratigraphic range, and this fact is reflected in the scientists' schemes of correlation (for example, the principle of overlapping range zones—see Eldredge and Gould, 1977). The assertion that evolution *means* gradual change has led the biostratigraphers to pay little attention to evolutionary paleontology, a discipline speaking so poorly to their own

---

[7] Rowe's (1899) famous *Micraster* collections cannot be sorted temporally within zones. Hence, we can only construct a mean morphology for a zone; we cannot tell whether variation around it is geographic or temporal; and, if it is temporal, we cannot order it. Indeed, some "trends" in mean points do more in the same direction through three zones; however, equally many others—ones not discussed by Rowe—do not. But what can one assert on the basis of three points? We may as well have two punctuational events of speciation, as opposed to having a gradual trend through the sequence. Moreover, it now appears that both the phylogeny and the stratigraphic distribution of *Micraster* are far more complex than Rowe imagined (Stokes, 1977).

observations. In the meantime, evolutionary paleontologists have paid a supremely high price for their continued adherence to gradualism. Since they also know that the vast majority of species do not exhibit continuous trends, they have had to seek refuge in Darwin's classic argument on the imperfection of the fossil record. The argument works, but at what a cost: the admission that they never (or hardly ever) see evidence of evolution, the very phenomenon they wish most to study.

There is a way, though an unconventional one, out of this impasse: we may return to the empirical literalism of the catastrophists. Two characteristic features of fossil species inspired Darwin's recourse to imperfections.

*1) Stasis:*   most species do not change appreciably during their tenure on earth, although they may fluctuate mildly in morphology as environments alter. (As an exception to this statement, increase in size seems to be fairly common; but simple increase in size without other modifications—beyond allometric compensation—does not go a long way towards producing the major features of evolution.)

*2) Sudden appearance:*   in local sections, old species are replaced abruptly by new ones; at this first appearance, the new ones look pretty much the same as they will throughout their duration.

But perhaps stasis and sudden appearance *are* the proper predictions for how the fossil record should render evolution as we understand it. Perhaps the problem is not the nature of the fossil record itself, but a false equation: gradualism equals evolution.

In 1972, Niles Eldredge and I constructed a defense of this literalist argument by proposing a model of "punctuated equilibria." This work has aroused, to say the least, a fair amount of passionate commentary. (See Gould and Eldredge, 1977, for a review with a point of view.)

Evolutionary change may occur in either of two modes: 1) an entire population may transform itself from one state to another—phyletic evolution; or 2) a subpopulation of a parental species may attain reproductive isolation and branch off to form a new species—speciation.[8] Most evolutionists maintain that speciation is both the primary style of evolu-

---

[8] Here we encounter an unfortunate ambiguity of terminology. Phyletic evolution also leads to the formation of new species (A transforms to B), but only branching is technically called "speciation." Speciation is the only process that augments organic diversity. (Ancestors tend to persist after their descendants branch off.) Hence, in a world dominated by inevitable extinction, speciation is wholly responsible for the ultimate continuity of life. (All phyletic lines ultimately die out, regardless of how much they transform; diversity is restored only in branching.)

tionary change and the mode of origin for the vast majority of contemporary species (Mayr, 1963). We would also add the argument that characteristic rates of phyletic change are far too slow to account for much of importance in evolution (Gould and Eldredge, 1977).

If speciation is the dominant mode of evolution, then we must inquire about its usual form and rate. If ancestral populations commonly split into even parts and subsequently diverge slowly at rates characteristic for phyletic evolution, then Lyellian gradualism would be strongly affirmed. But this rarely happens. Instead, new species usually arise from very small subpopulations that are geographically isolated at the periphery of their parental range.[9] And they arise rapidly, usually in hundreds of thousands of years. (Such a rate would not impress a human observer as particularly rapid; indeed, during a lifetime of observation, he might not notice it at all. But a few hundred years translates to a bedding plane in most geological sections. In any case, since most established species persist for a few million years, an origin in a few thousand represents a fraction of one percent of their tenure on earth.) After a new species establishes itself as a successful central population in its own right, it is unlikely to change very much; the inertia of large numbers is a powerful impediment to the spread of favorable variants.

In summary, most evolutionary change is concentrated in events of speciation; speciation tends to occur rapidly in very small subpopulations isolated at the periphery of their ancestor's range. How, then, should speciation translate its effects into the fossil record? The answer would seem to be: in the exact way that we see—namely, by stasis and by sudden appearance. Most fossils are the records of large, established central populations, and elsewhere (that is, away from the central poulations preserved as fossils). We do not expect to find the transition itself. We first meet the new species when it invades the ancestral range as an established, successful form in its own right—sudden appearance. Evolution is not the slow and steady transformation of ancestors into modified descendants, but is rather a series of rare punctuations that disrupt persistent and stable systems—punctuated equilibria.

Punctuated equilibrium is a model for the level of speciation alone. But I believe—again contrary to established gradualistic expectations—that

[9] Evolutionists refer to this view as the theory of allopatric speciation ("allopatric" means "in another place"). Until a few years ago, the allopatric theory reigned supreme. Recently, it has been strongly challenged by a reassertion of earlier beliefs in sympatric speciation (in the same place as ancestors)—although I think that most supporters of sympatric speciation would still admit that allopatry is the more common mode. The model of punctuated equilibria does not require that speciation be allopatric. It does demand that speciation occur quickly in small groups; sympatric models tend to outdo allopatric schemes in their insistence upon rapid transition in small subpoulations (Bush, 1975).

punctuational change also characterizes several other levels of evolution. I did not challenge conventional gradualism at the lower level of speciation within the time frame of the event itself—hundreds or thousands of years. Yet, even at this lowest level of evolutionary change, punctuational rumblings are now being heard. For example, Hampton Carson, a distinguished population geneticist and student of Hawaiian *Drosophila,* writes in an explicit questioning of gradualistic biases:

> The classical view of speciation holds that it is a gradual microevolutionary process. Thus, the genetic events which lead toward speciation are considered to be individually trivial or simple. Changes, such as a shift in gene frequency, may accumulate slowly in a population. . . . Most theories of speciation are thus wedded to gradualism, using the mode of origin of intraspecific adaptations as a model. . . . I would nevertheless like to propose that. . .speciational events may be set in motion and important genetic saltations toward species formation accomplished by a series of catastrophic, stochastic genetic events (Carson, 1975, pp. 87–88).

At the higher level of transition between major organic designs (vertebrate from invertebrate, jawed from jawless fishes), gradualism has always been in trouble, although it has stonewalled with commendable tenacity. No one has ever solved Mivart's old (1871) dilemma of "incipient stages of useful structures." The finished jaw is an engineer's delight; the same bones worked equally well to support the gill arch of a jawless ancestor. But can you really construct a graded series of workable intermediates? (Needless to say, we have no intermediate fossils.) What good is a series of bones detached from the gills, but still too far back to function as a mouth? Did they move forward, millimeter by millimeter, until they finally assumed a coordinated position surrounding the mouth?

Darwin took Mivart's objection very seriously and even added an entire chapter to later editions of the *Origin* in an attempt to refute it. Basically, he answered that a graded series of intermediates had to exist and that our failure to specify their function merely expressed our lack of imagination. In a characteristically frank comment, he wrote: "If it could be demonstrated that any complex organ existed, which could not possibly have been formed by numerous, successive, slight modifications, my theory would absolutely break down" (Darwin, 1859, p. 189).

But, again, Huxley was right; Darwin had conflated gradualism with natural selection. Natural selection does require transitional forms,[10] but

[10] If new designs arose all at once, fully formed, then selection could only play the negative role of removing the unfit (the untransformed ancestors). The essence of Darwin's theory lies in the creative role it assigns to natural selection—selection produces the fit by accumulation.

it does not demand an insensibly graded series of intermediates. Why not move the bones forward all at once as the result of a small genetic change with major morphological results through an alteration in the timing of early development? Small effects in early embryos often accumulate through ontogeny and lead to profound modification of the subsequent adult. Early maturation, for example, may make an adult with many features of its ancestor's larva. This phenomenon, called progenesis, has been invoked to explain the rapid origin of many major groups (Gould, 1977). Lovtrup (1974) and Frazzetta (1975) have recently written books to defend a punctuational origin of major groups by small genetic changes with major effects upon ontogeny.

At the highest level of patterns in the history of life, gradualistic biases again dictated previous preferences. By way of example, as a student I learned that the prebiotic evolution of life must have been an enormously slow and gradual event—step by improbable step through at least a billion years being required for the conversion of the extremely improbable into the virtually certain (Wald, 1954, for example). But now we know that life emerged very shortly after the earth cooled down sufficiently to support it. (The oldest rocks are 3.8 billion years old; the oldest rocks that could preserve evidence of life are 3.4 billion years old, and they do contain abundant microorganisms—see Knoll and Barghoorn, 1977; Gould, 1978.) Likewise, the most popular theories of my student days (the mid 60s, not so long ago) tried to deny—or, at least, to dilute substantially—the Cambrian explosion and the Permian extinction by labelling them "preservational artifacts." But the more these events are studied, the more they resist attempts to spread them out into gradualistic oblivion. And we now have reasonable boundary conditions and theories that render them as probable events in the punctuational mode: the Cambrian explosion may be little more than the expected logarithmic phase for exponential growth in diversity within a system of superabundant resources (Sepkoski, in press); and the Permian extinction emerges as a predictable consequence of the coalescence of Pangaea (Schopf, 1974; Simberloff, 1974). The history of marine invertebrate diversity is set primarily by its rare punctuations. The geological time scale is a mirror of these punctuations. No unit is without its "boundary problems," but the units are reasonably objective packages, not the arbitrary divisions of a gradual and uniform flow of life. Strength of punctuation established the hierarchy: the largest mark eras (Cambrian explosion, as well as Permian and Cretaceous extinctions); the smaller eras, periods.

*PART 3*   THE CULTURAL CONTEXT OF GRADUALISM

If gradualism stands up so poorly as a universal dogma when subjected to detailed examination, then why did it maintain its hegemony for so long? This question has no simple resolution, but I am certain of one thing: the popularity of gradualism did not arise from nature. Lyell and Darwin did not "see" gradualism in the rocks and thus cast their generalization as a simple induction from the facts of geology. Nature is multifarious; she speaks ambiguously on any issue as broad as *the* nature of change.

Nature had an input in the formulation of gradualism—some of her processes surely work slowly and cumulatively. But the doctrine of Lyell's third uniformity has more complex roots. I am convinced that the cultural and political context of European society had an input equal to, or greater than, nature herself. In saying this, I do not criticize Lyell and Darwin for letting "extra-scientific" influences cloud a supposedly objective judgment; nor do I claim that Lyell and Darwin were explicitly aware of this influence. The notion that science operates apart from culture by a universal method that yields truth according to canons of observation and experimentation is a myth that has been carefully nurtured by scientists themselves. Science operates, as does all creative thought, within a cultural context that influences all practitioners in various subtle and unacknowledged ways.

When scientific theories of a static world order collapsed towards the end of the eighteenth century, a new ideology rose to justify social stability within a world now dominated by ceaseless change. If change is intrinsic and fundamental, what could be better than a notion that it must proceed with excruciating slowness, move from one system to another through countless intermediary stages, and always be weighted down by an inheritance from the past?

Gradualism became the quintessential doctrine of liberalism as it faced a world increasingly engulfed by demands for revolutionary change. Anecdotal evidence for this use of gradualism abounds. In 1891, the noted British anthropologist General Augustus Lane-Fox Pitt Rivers wrote for the provincial museum he had established at Farnham, Dorset, a prospectus that argued:

> For good or for evil. . .we have thought proper to place power in the hands of the masses. The masses are ignorant, and knowledge is swamped by ignorance. . . . The knowledge they lack is the knowledge of history. This lays them open to the designs of demagogues and agitators, who strive to make them break with the past, and seek the remedies for existing evils, or the means of future progress, in drastic

changes that have not the sanction of experience. It is by a knowledge of history only that such experience can be supplied. . . . The law that Nature makes no jumps, can be taught by the history of mechanical contrivances, in such way as at least to make men cautious how they listen to scatter-brained revolutionary suggestions (Pitt Rivers, 1891, pp. 115–116).

Closer to home, and rather more painfully, note Booker T. Washington on the necessary path of black progress—and understand why he was the darling of whites while he lived, yet is so widely rejected by blacks today:

Finally, reduced to its last analysis, there are but two questions that constitute the problem of this country so far as the black and white races are concerned. The answer to one rests with my people, the other with the white race. For my race, one of its dangers is that it may grow impatient and feel that it can get upon its feet by artificial and superficial efforts rather than by the slower but surer process which means one step at a time through all the constructive grades of industrial, mental, moral, and social development which all races had to follow that have become independent and strong (Washington, 1904, p. 245).

Gradualism is not the only available philosophy of change. Alternative systems have a long history in Western philosophy. Hegel's dialectic, as reformulated by Marx and Engels, has become an official state philosophy in many nations. The so-called dialectical laws are explicitly punctuational; particularly so is the doctrine of transformation of quantity into quality. Putting aside the mumbo-jumbo of jargon, this principle simply holds that change occurs by leaps (new qualities) after accumulating inputs (quantities) strain systems to the breaking point. Heat water, and eventually it boils. Bend a beam, and eventually it breaks. Oppress the workers more and more, and eventually they revolt. The official Soviet handbook of Marxism-Leninism proclaims:

The transition of a thing, through the accumulation of quantitative modifications, from one qualitative state to a different, new state, is a leap in development. . . . It is the transition to a new quality and signalizes a sharp turn, a radical change in development. . . . The evolutionary development of society is inevitably consummated by leap-like qualitative transformation, by revolutions (Anon., pp. 88–89).

Now no reader of this work will have any trouble identifying, quite correctly, the ideological component in this philosophy of change. Yet, we tend to view our own beliefs as objective renderings of nature. I merely ask that we also recognize the ideological roots of gradualism.

Several late-nineteenth-century Marxist intellectuals argued that differing social conditions favored different philosophies of change. Karl Kautsky, for example, had this to say on the origins of gradualism:

> The bourgeoisie . . . must seek more effective arguments with which to stigmatize the revolution, and these are found in the newly-arising natural science with its accompanying mental attitude. . . . Once the capitalist revolution was ended, the place of the catastrophic theory was taken by the hypothesis of a gradual imperceptible development, proceeding by the accumulation of countless little advances and adjustments in a competitive struggle. . . . Everyone is involuntarily influenced by the mental attitude of the class amid which he lives and carries something from it into his scientific conceptions (Kautsky, 1902, pp. 12–13; see also, Plekhanov, 1908).

If we should ask what lessons a scientist might draw from the recognition of ideological components in philosophies of change, I would cite Kautsky's subsequent statement:

> The fact that an idea emanates from any particular class, or accords with their interests, of course proves nothing as to its truth or falsity. But its historical influence does depend upon just those things (Kautsky, 1902, p. 13).

My demonstration that gradualism arose as much from politics and culture as from nature does not render it false. The inadequacy of gradualism as an exclusive philosophy of change derives from its failure to account for much of the world, not from its status and origin as a political doctrine. I emphasize the cultural roots of this dogma becuase I hope to convince scientists that their unquestioned adherence to it arises largely (and unconsciously) from this extra-scientific source. It is not a fact of nature, and if it does not work, it may be abandoned—despite Lyell's claim that geology without it is virtually unimaginable.

## THE GROWING POPULARITY OF PUNCTUATIONAL CHANGE

The *zeitgeist* is now in for an overhaul. Punctutional change is attracting attention and excitement in a host of diverse disciplines. In 1962, Thomas Kuhn published one of the century's most influential books: *The Structure of Scientific Revolutions.* In it, he sought an alternative to the standard gradualistic account of the history of science—progress to truth by the steady accumulation of new facts. He viewed the history of science as a domain of persistent stasis in explanatory schemes, or ruling paradigms (normal science). Anomalies accumulate within paradigms and are either explained away by *ad hoc* rationalizations or ignored with conspiracies of silence. New paradigms, usually generated by scientists working outside the tradition, replace old ones in rare and brief revolutionary episodes. Scientists are forced to reconstruct their world from time to time; they are not continuously reevaluating it and rebuilding it laboriously, piece by piece.[11] More recently, the noted French philosopher and historian of ideas Michel Foucault has attracted great attention with his punctuational theory of history. In a review of his work, C. Geertz writes:

> Foucault believes that history is not a continuity, one thing growing organically out of the last and into the next, like the chapters in some 19th century romance. It is a series of radical discontinuities, ruptures, breaks, each of which involves a wholly novel mutation in the possibilities for human observation, thought, and action (Geertz, 1978, p. 3).

David Warsh, writing in Forbes Magazine (of all places), recently blamed Newton and Lyell for imposing a restrictive uniformitarian bias upon economic thought:

> Now prepare for a bit of a shock; Karl Marx was more right than Adam Smith. . . . The idea of catastrophic development has become quite familiar—of a system sitting quietly in equilibrium while its underlying forces are slowly changing until a point is suddenly reached where equilibrium breaks down and the system snaps to a new equilibrium (Warsh, 1977, pp. 173–176).

And Robert May has demonstrated how gradualistic biases have long prevented ecologists from understanding some events of great practical importance:

> Continuous variation in a control variable can produce discontinuous effects. Thus smooth changes in stocking rates can cause discontinuous changes in the grazed vegetation; continuous changes in harvesting rates can cause discontinuous collapse in fisheries; continuous changes in environmental parameters or foliage growth or predation rates can lead to discontinuous outbreaks of insect pests; continuous changes in snail or dipteran population densities can cause discontinuous appearance or disappearance of helminthic infections (May, 1977, p. 447).

[11] Laborious building does occur all the time, but it only operates within paradigms and in their interest. It contributes little, if anything, to the replacement of one paradigm by another. Eldredge and I view the relationship between phyletic gradualism and speciation as operating in a very similar manner. Minor gradualistic change may occur within a species, but these alterations will have little effect upon the history of major groups if that history is propelled by more profound changes during rapid events of speciation.

Much of this new focus on punctuational change has centered upon René Thom's attempt to construct a mathematics for discontinuous alteration—catastrophe theory, as he calls it (Thom, 1975). (The New York Times of November 19, 1977, gave his work front page coverage—a first, perhaps, for a topic in pure mathematics.) Whatever one thinks of the swirling controversy now surrounding Thom's particular formulation and of the excess zeal of some of his votaries (see Kolata, 1977), the excitement that it has attracted certainly records an increasing recognition of punctuational change as an important part of our world.

I do not know how much of this new fascination for punctuational change resides in the stresses of our general culture. Information theory and general systems theory—with their concepts of equilibrium, steady state, homeostasis, feedback loops, and positive feedback leading to rapid autocatalytic change—certainly evoke a set of punctuational metaphors. Our uncertain world of nuclear armaments and deteriorating environments must also encourage a departure from gradualism.

Yet, I believe that much of this current advocacy for punctuational change represents nature reasserting herself against the blinders of our previous gradualistic prejudice. Geology, more than any other discipline, has been mired in this prejudice. Lyell put us there for a host of complex reasons that had little to do with the empirical world as he observed it. We should acknowledge his influence by recognizing that some processes do tick along in the gradualistic tempo. But we should reject gradualism as a restrictive dogma. Punctuational change, with rapid flips between stable states, may characterize more of our world. As geologists, we should pay special attention to the punctuations in our record, for we have systematically ignored them and may well have missed the dominant tempo of natural change.

# LITERATURE CITED

Anon., n.d. *Fundamentals of Marxism-Leninism Manual* (Moscow: Foreign Language Publishing House), 891 pp.

Baker, V. R., 1973. Paleohydrology and sedimentology of Lake Missoula flooding in Eastern Washington. *Geol. Soc. Amer. Spec. Paper*, 144, 79 pp.

Bretz, J. H., 1923. The channeled scablands of the Columbia plateau. *Jour. Geol.*, 31:617–649.

————, 1927. Channeled scabland and the Spokane flood. *Washington Acad. Sci. Jour.*, 17:200–211.

————, 1969. The Lake Missoula floods and the channeled scabland. *Jour. Geol.*, 77:505–543.

Burnet, T., 1691. *Telluris Theoria Sacra* [English ed.: *The Sacred Theory of the Earth* (London: W. Kettilby, 1690)].

Bush, G. L., 1975. Modes of animal speciation. *Rev. Ecol. Syst. Ann.*, 6:339–364.

Carson, H. L., 1975. The genetics of speciation at the diploid level. *Amer. Nat.*, 109:83–92.

Darwin, C., 1859. *The Origin of Species* (London: John Murray), 490 pp.

Eldredge, N., and S. J. Gould., 1972. Punctuated equilibria: an alternative to phyletic gradualism. In: T. J. M. Schopf (ed.), *Models in Paleobiology* (San Francisco, Calif.: Freeman, Cooper & Co.), pp. 82–115.

―――, 1977. Evolutionary models and biostratigraphic strategies. In: E. G. Kauffman and J. E. Hazel (eds.), *Concepts and Methods of Biostratigraphy* (Stroudsburg, Penn.: Dowden, Hutchinson & Ross, Inc.), pp. 25–40.

Frazzetta, T. H., 1975. *Complex Adaptations in Evolving Populations* (Sunderland, Mass.: Sinauer Associates), 267 pp.

Geertz, C., 1978. Stir Crazy. *N. Y. Rev. Books*, Jan. 26, pp. 3–6.

Goodman, N., 1967. Uniformity and simplicity. *Geol. Soc. Amer. Spec. Paper*, 89:93–99.

Gould, S. J. 1965. Is uniformitarianism necessary? *Amer. Jour. Sci.*, 263:223–228.

―――, 1970. Private thoughts of Lyell on progression and evolution. *Science*, 169:663–664.

―――, 1972. Allometric fallacies and the evolution of *Gryphaea*. *Evol. Biol.*, 6:91–119.

―――, 1974. The evolutionary significance of "bizarre" structures: Antler size and skull size in the "Irish Elk," *Megaloceros giganteus*. *Evolution*, 28:191–220.

―――, 1977. *Ontogeny and phylogeny* (Cambridge, Mass.: Harvard Univ. Press), 501 pp.

―――, 1978. This view of life: An early start. *Nat. Hist.*, LXXXVII (2):10–24.

Gould, S. J., and N. Eldredge, 1977. Punctuated equilibria: the tempo and mode of evolution reconsidered. *Paleobiology*, 3 (2):115–151.

Hallam, A., 1968. Morphology, palaeoecology and evolution of the genus *Gryphaea* in the British Lias. *Roy. Soc. London Phil. Trans.*, B 254:91–128.

Hawkes, L., 1958. Some aspects of the progress in geology in the last 50 years. *Geol. Soc. London Quart. Jour.*, 114:395–410.

Hooykaas, R., 1963. *The Principle of Uniformity in Geology, Biology and Theology* (Leiden: E. J. Brill), 237 pp.

Huxley, L., 1901. *The Life and Letters of Thomas H. Huxley*, 2 vols. (New York: D. Appleton), 539 pp. & 541 pp.

Kautsky, K., 1902. *The Social Revolution* (Chicago, Ill.: Charles H. Kerr & Co.), 189 pp.

Knoll, A. H., and E. S., Barghoorn, 1977. Archean microfossils showing cell division from the Swaziland System of South Africa. *Science*, 198:396:398.

Kolata, G. B., 1977. Catastrophe theory: the emperor has no clothes. *Science*, 196:287.

Krynine, P. D., 1956. Uniformitarianism is a dangerous doctrine. *Jour. Pal.*, 30:1003–1004.

Kuhn, T. S., 1962. *The Structure of Scientific Revolutions* (Chicago, Ill.: Univ. Chicago Press), 172 pp.

Laurence, P., 1973. *The Central Heat of the Earth.* Harvard Univ., Dept. History of Science, Ph.D. thesis.

Longwell, C. R., and R. F. Flint, 1955. *Introduction to Physical Geology* (New York: John Wiley), 432 pp.

Lovtrup, S., 1974. *Epigenetics* (New York: John Wiley), 547 pp.

Lyell, C., 1842. *Principles of Geology,* 6th ed., 3 vols. (Boston: Hilliard, Gray & Co.).

Malde, H. E., 1968. The catastrophic Lake Pleistocene Bonneville flood in the Snake River Plain, Idaho. *U.S. Geol. Surv. Prof. Paper,* 596, 52 pp.

Masursky, H., et al., 1977. Classification and time of formation of Martian channels based on Viking data. *Jour. Geophys. Res.,* 82 (28):4016–4038.

May, R. M., 1977. Thresholds and breakpoints in ecosystems with a multiplicity of stable states. *Nature,* 269:471–477.

Mayr, E., 1963. *Animal Species and Evolution* (Cambridge, Mass.: Harvard Univ. Press), 797 pp.

Merton, R. K., 1963. Resistance to the systematic study of multiple discoveries in science. *European Jour. Sci.,* 4:237–282.

Mill, J. S., 1881. A system of logic. Reprinted in: J. S. Mill, *Philosophy of Scientific Method,* 8th ed. (New York: Hafner, 1950), 461 pp.

Mivart, St. G., 1871. *On the Genesis of Species* (New York: D. Appleton), 314 pp.

More, T., 1516. *Utopia.* (Oxford: Clarendon Press, 1895), 347 pp.

Orbigny, A. d', 1849–1852. Cours élémentaire de paléntologie et de géologie stratigraphiques (Paris: V. Masson), tome 2.

Pitt Rivers, A. L. F., 1891. Typological museums, as exemplified by the Pitt-Rivers Museum at Oxford, and his provincial museum at Farnham, Dorset. *Jour. Soc. Arts,* 40:115–122.

Plekhanov, G. V., 1908. *Fundamental Problems of Marxism,* 1969 ed. (New York: International Publishers), 145 pp.

Porter, R., 1976. Charles Lyell and the principles of the history of geology. *Brit. Jour. Hist. Sci.,* 9:91–103.

Rudwick, M. J. S., 1972. *The Meaning of Fossils* (London: MacDonald), 287 pp.

Rowe, A. W., 1899. An analysis of the genus *Micraster,* as determined by rigid zonal collecting from the zone of *Rhynchonella Cuvieri* to that of *Micraster cor-anguinum. Geol. Soc. London Quart. Jour.,* 55:494–547.

Schopf, T. J. M., 1974. Permo-Triassic extinctions: relation to sea-floor spreading. *Jour. Geol.* 82:129–143.

Sepkoski, J. J., Jr., 1978. A kinetic model of Phanerozoic taxonomic diversity, 1. Analysis of marine orders. *Paleobiology.* 4:223–251.

Simberloff, D. S., 1974. Permo-Triassic extinctions: effects of area on biotic equilibrium. *Jour. Geol.,* 82:267–274.

Simpson, G. G., 1963. Historical science. In: C. C. Albritton, Jr. (ed). *The Fabric of Geology* (Reading, Mass.: Addison-Wesley), pp. 24–28.

Stokes, R. B., 1977. The echinoids *Micraster* and *Epiaster* from the Turonian and Senonian chalk of England. *Palaeontology,* 20:805–821.

Thom, R., 1975. *Structural Stabiity and Morphogensis* (Reading, Mass.: W. A. Benjamin), 348 pp.

Velikovsky, I., 1950. *Worlds in Collision* (New York: Doubleday), 401 pp.

———, 1955. *Earth in Upheaval* (New York: Doubleday), 301 pp.

Wald, G.D, 1954. The origin of life. *Sci. Amer.,* 191 (12):44–53.

Warsh, D., 1977. The great hamburger paradox. *Forbes,* 120:166–167.

Washington, B. T., 1904. *Working With the Hands* (New York: Doubleday, Page & Co.), 246 pp.

Wilson, L., 1970. *Sir Charles Lyell's Scientific Journals on the Species Question* (New Haven, Conn.: Yale Univ. Press).

Zangerl, R., and E. S. Richardson, Jr., 1963. The paleoecological history of two Pennsylvanian black shales. *Chicago Nat. Hist. Mus., Fieldiana Geol. Mem.,* 4, 352 pp.

Chapter 2

# PERFECTION, CONTINUITY, AND COMMON SENSE IN HISTORICAL GEOLOGY

RICHARD H. BENSON

Smithsonian Institution

"Until we habituate ourselves to contemplate the possibility of an indefinite lapse of time having been comprised within each of the modern periods of earth's history, we shall be in danger of forming most erroneous and partial views of geology."                    (Charles Lyell, 1872, vol. 2, p. 43).

## INTRODUCTION

This paper is about how most geologists—and paleontologists who are trained basically as geologists—regard the passage of events recorded in historical geology. It examines the difference between the analytical consequences of those who follow the style of the pragmatic essentialists (individuals such as George Cuvier, who sought the intrinsic order in nature by attempting to define its perfection and thereby concluded that there were interruptions in its historical succession) and the analytical consequences of those who follow the style of the gradualists (persons such as Charles Lyell, who believed in processes of development and in a balance between natural forces that resulted in continuous change and who thereby concluded that geologic history is generally a Heraclitian flow).

Neither of these views is exclusively correct; nor were the past advocates of either position ever unaware of the other point of view. Fashions in science change, however, as individual practitioners become dissatisfied with the rate of discovery of one or another view. We are today in the middle of such a change: from the acceptance of continuity as a necessary and sufficient procedural assumption to a concurrence that interrupted stasis is an obvious condition of development. As we observe the common sense of one time evolving into a new synthesis, it is well to reflect that these are not new or absolute positions. The old geometers, who tried to describe the universe in terms of ideal space, gave to us the rigor and discipline of model construction, with all of its illogical inadequacies. The empiricists, who saw the obvious movement in the real world, helped us to see the necessity of proceeding even when logic fails.

Will Durant (1954) said, "Most history is guessing, and the rest is prejudice." To this I would add: "as seen through the constant refinement and partial rejection of prevailing common sense." Attitudes about uniformitarianism and about crises in the process of geologic processes and events have become less rigid in recent years. They were possibly never as rigid as student memory and some of the more extreme statements in the literature would suggest. We know, of course, that the image of controversy usually attracts interest to a theoretical problem and that clear issues are not common. Still, it is helpful to dissect our views and to understand that the position of the observer may be as important as that of the observed. Most geologists are too busy either describing the phenomena of the geologic record or solving problems to be concerned with assumptions that border on the philosophical. Yet, they often use terms that carry the weight of theory without suspecting that those very terms support that theory.

When I began to write this essay, I was tempted to support the contention that there are in historical geology special elements of historiography that make this science unique. This is a common belief, sometimes referred to as the "historical aspect." In elementary texts, uniformitarianism is often cited as a universal principle of historical geology, the implication being either that it is somehow unique to geology or that it has predominant importance. Uniformitarianism is important, but so too are other principles. M. K. Hubbert (1967) suggested that the "Principle of Uniformity" had conceivably lost its usefulness and that we should consider "the logical essentials in the deciphering of history, not geologic history, but any kind of history." Perhaps this is so; I have concluded, however, that there is nothing unique about the historiography of historical geology. In fact, outside of philosophy, there is nothing unique about historiography except for its subject matter, and that is all of the difference necessary.

Clarence King, revered as a leader in the description of American geology in the early days, once said (1877; in Albritton, 1963): "Men are born either catastrophists or uniformitarians. You may divide the human race into imaginative people who believe in all sorts of impending crises . . . and others who anchor their very souls to the *status quo*." It is easy to dismiss this statement as an extreme characterization of human nature. Such a dismissal, however, would steal some of our interest in the study that follows. I hope to treat both Cuvier and Lyell not only objectively, but also as people in their own times.

To start on an even footing, I ask the reader to consider the following meanings for some terms that are often used elsewhere, but may have special connotations in this paper.

*Structure:* a configuration of form or activity that serves, as a pathway of reaction, to convey stress from one point to another. In a changing world, where characteristics are altered, the appearance of a new structural configuration is an event.

*Similitude* (the principle of): the concurrence between structure and geometry of form that allows casual inference between reality and a model.

*System:* a set of interrelated structures that may consist of an hierarchic arrangement of subsets or subsystems.

*Essentialism:* a "realistic" philosophical position that dates back to the idealism of Plato and Aristotle, was then placed into religious scholastic argument by Thomas Aquinas, and finally put into modern methodological form by Karl Popper (1957). Essentialism is concerned with understanding the potential, or universal, meaning of the descriptors of a class or set; it takes an opposite view of that of nominalism, which believes that such descriptors are only labels of characteristics. The concept of essentialism, as first proposed, argued that those properties which are common to the many individuals of a set, a class, a group, or, perhaps, a taxon have a reality exceeding that of their mere occurence as parts of the individual members of such sets, classes, and so on. This position, then, allows one to search within natural groupings for common plans, designs, and patterns, with the expectancy that such patterns will be understandable by analogy. In recent years, essentialism has been considered somewhat sterile, at least in the natural sciences, because it focused on universals of meaningless generality. Plato, on the other hand, used this position to search within nature for order, regularity, and perfection.

*Historicism:* the approach to science—particularly to science that is historical, as is geology in large part—which assumes that historical prediction (or retrodiction) is a principal aim, attainable by discovering the patterns, laws, and trends that underlie the evolution of history (Popper, 1957).

*Crisis:* an event that occurs in the history of a system, when stress is sufficient to cause the imminent alteration of the system's principal structures, but, through the absorption of this stress into its subsystems, the system survives.

*Catastrophe:* an event that occurs in the history of a system, when stress is sufficient to cause the imminent alteration of the system's principal structures, and the subsystems fail to absorb all of the stress but survive, although the system fails. In such cases, a new and modified system is then formed to take the place of the failed system.[1]

[1] Stephen Jay Gould, as noted in the previous chapter, prefers not to use this term, possibly because of stigmas attached to the concept of "catastrophism." Should the term "essential" be abandoned because of the connotations of "essentialism"? Let the reader decide.

*Cataclysm:* an event that occurs in the history of a system, when stress is sufficient to cause the imminent alteration of the system's principal structures, and both the system and its subsystems fail.

In each of the three events just described, the source of the stress is left undefined; but, for the present, it can be inferred to be external. Crises occur often, catastrophes happen less often, and cataclysms rarely occur on a grand scale.

Several more terms, because they are often used with their meanings implied, require some comments.

*Common sense:* that sense which some say (when they want to win an argument) that God gave to us; or, as Locke might have said, "a self-evident truth"; or, the sense based on those assumptions of childhood that are later challenged by increasingly abstract theory. The abstractions of yesterday, if they survive cultural processes and become habit, may become the common sense of today.

*Romanticism:* an expression of longing for the return of older, simple analogies and a prior common sense. This longing usually results in one exaggerating the importance of attributes or forces by relying too heavily on intuition or enthusiasm.[2] Romanticism is a rebelling against the advance of that reason which outdistances common sense.

*Novelty:* a term often used to describe the sudden appearance of something new in the geologic record. Novelty is a characterization of a disjunct condition in the progress of change. Since the term is loaded with secondary implications, a few comments are warranted to provoke some thought about its usage. Julian Huxley (1960) described evolution as a one-way, irreversible process that generates novelty, diversity, and higher degrees of organization. Eldridge (1979) suggested that a principle problem in describing the course of evolution is determining the patterns of shared novelty.

A thought central to the subject of this paper is that the use of the term "novelty" can be an indication of an incomplete thought. The existence of novelty implies the inability to see morphologic connection (similitude) in structural terms. As soon as similitude is seen, novelty either disappears or is resolved by analogy. The "discovery" of novelty—that is, its identification—implies that its origin is disconnected from the "view" of the discoverer.

The terms "uniformitarianism" and "catastrophism" will be left undefined for the present. Both terms have been used in quite different ways and often (unfairly) in opposition to each other. To attempt to define them here would only anticipate some of the discussion that follows.

---

[2] Enthusiasm originally referred to an especially strong revelation, usually religious; but, of course, all insights into the origin of things were once either considered religious or took a religious context.

One last note about language: I believe that the word "catastrophe" in French is far more commonly used and carries less emotional baggage than it does in English. One can make his own judgment as to the possible reasons.

In the following section, the discussion will focus on George Cuvier, a platonic perfectionist and the author of the geological-historical concept of world revolutions in nature; on some of his methods; and on why he chose a catastrophic point of view. The section which then follows will broaden the scope of investigation from France to England at the beginning of the nineteenth century and will discuss how divergent common sense affected both English catastrophism and uniformitarianism. Next, Charles Lyell will be seen in the light of my own experience with Mediterranean geology, and some guesses will be made about how his procedures were affected by his legal, as well as his class, experiences.

From this point, we will look at the general problem of thinking about continuity and discontinuity in history; and in the subsequent section, we will examine how concepts of structure and systems can be used to imagine that large scale, sudden events can happen in nature. Finally, there will be some comments on evolution and sudden events.

The following conclusions that will be reached are presented here to assist the reader in marshalling his own common sense and prejudices well in advance of the text.

1) In the course of development of geological knowledge, the common sense used empirically to describe new observations is always out of step with advancing theory in direct proportion to the rate of learning.

2) Although the more forceful and impatient among us may devise new theories that arise from the inadequacy of common sense, these new theories must be closely tied to that common sense. Too much abstraction, too soon, breeds discontent.

3) Models are useful, because they can fail, and the modeler can still survive. This relationship is not always understood in the conflict between abstraction and common sense.

4) Cuvier was a careful, logical inductionist, who was forced to create an historical model because he was dissatisfied with the incomplete way in which Hutton's position explained observed fact. Hutton's time frame was too brief, his rates and forces were exaggerated. There was no mechanism to absorb stress in the system, because he focused primarily on the perfection of its design. A part had to be replaced each time it failed. Revolution was part of his common sense. He advocated, as did Plato, the study of perfection in nature as means of perfecting the observer.

5) Lyell, as the spokesman for uniformity in nature, did not want to believe that natural systems changed fundamentally, although he recog-

nized change as a fact. In his earlier years, he believed in a "steady state" physical world without biological progress. After microcatastrophic extinction, species were replaced from outside. He had no mechanism, until Darwin, by which biological subsystems might absorb stress before systems were threatened; he was torn, therefore, between assuming a continuous steady state and concluding that all systems must fail. He also saw a linkage between nature and morality: as a barrister, he saw the imperfections of man and urged reasonable respect for a natural law that allowed others to advance to his own level of perfection.

6) Some historians, given the "facts" of history, see few trends that either indicate inevitable development or justify historicism (Popper, 1957). Others see no reason to study history other than to look for such trends.

Since we cannot consider our view of nature and its present structure to be complete, it is illogical to conclude that we can with any greater degree of certainty reconstruct a fragmentary record of the past. Logic, however, plays a limited role—albeit a useful one—in reconstructing natural history. Perhaps invention and improvement of common sense play major parts. History is the science—or, if you will, the art—of learning from unique events. Their uniqueness should not be a stumbling block.

7) Lastly, this chapter is a plea for a better understanding of the engineering of nature as an expression of history: that is, of how reactions can bring about sudden change by selecting alternate—if not, to us, unexpected—pathways in parts of systems, thereby causing the reorganization of the whole. We have too often assumed that the evidence of the mechanism of change was removed from our experience by erosion, by accident of geography, or by the origin of things in nature being rare and random accident. There is a reasonable compromise somewhere between the search for design and the observation that change has brought it about.

## THE ORIGINS OF CATASTROPHISM

Baron Georges Leopold Chretian Frederick Dagobert Cuvier (1769–1832) was born of a poor Lutheran military family in Wurttemburg, now a part of Burgundy, France. Educated at Caroline University near Stuttgart, he became a geologist and, later, a paleontologist. He lived through turbulent times: the fall of the nobility, the French Revolution, the reign of Napoleon, the return of the nobility, the fall of the Church, and the resurgence of its influence. He was a "survivor," who died rich, famous, and powerful. His vanity was boundless, as was his hunger for honors and praise. He was said to have had an exceptional memory and to have known the contents of all 19,000 books in his library (Bourdier, 1971).

Cuvier was hostile to theory, be it scientific, philosophic, or social. He believed in facts and description. He was a chameleon-like individual, whose opinions seemed to vary with his political surroundings and whose true motivations and convictions will probably never be known. He has been accorded the respectful title of "Father of Comparative Anatomy" and is usually considered the founder of vertebrate paleontology. He was the originator of the geological concept of catastrophism and the coauthor of the term "Tertiary." Throughout most of his professional career, he held the post in Anatomy at the Museum National d'Histoire Naturelle in Paris. A man of exceptional ability, he became most influential in the early part of the nineteenth century. He has not, however, been well known among American authors of geology texts, some of whom imply that he was either sinister or misguided.

Cuvier seldom went into the field except around Paris with his friend and colleague Alexandre Brongniart (1770–1847), Professor of Mineralogy at the Museum. Brongniart described the cyclic alteration of marine and fresh-water formations in the Paris Basin. The sharp breaks in the sedimentary record of this sequence was used in part by Cuvier to support his theory of sudden change causing the extinction of terrestrial faunas. Meanwhile, Brongniart's detailed biostratigraphy proved useful, even before William "Strata" Smith's Jurassic studies were published, for correlating across lithologic changes over extensive areas in order to demonstrate that fossils could form a basis for stratigraphy and historical succession.

Cuvier believed, as did most scholars of his time, that all of the events of earth history had taken place in a brief time span of about 75,000 years, that the earth had begun in a molten state, and that it was cooling down. Although some rough estimates of the age of the earth and some speculations on the nature of its origin had been made from observations of temperatures in deep mine shafts in Auvergne, this figure of 75,000 years probably came from the estimates of Buffon (1802). Cuvier himself (as well as Brongniart) was primarily concerned with the strata and the vertebrate fossils of the Paris Basin.

The time span he accepted for the development of the dinosaurs in Thuringia, the plesiosaurs of the chalks, and the mammals, as well as for the structural displacement of strata in the surrounding mountainous regions of France, would have been only five times longer than human history. Human history was, in his time, closely associated with the studies of the savants who accompanied Napoleon to Egypt.

The disruption of strata with traceable unconformities caused him great concern. Their succession led to his conception of revolutions in crustal history forming "islands" in seas that were continuous but fluctuated in sea level. These disruptions coincided with major changes in the fossil record. They were numerous and sudden. We can credit Cuvier with anticipating the concept of regional mountain building as basis for stratigraphic classification that would later appear in such terms as the "Appalachian Revolution," "Laramide Disturbance," and so on, as well as in Chamberlin's (1910) conclusion that diastrophism is the ultimate basis of correlation.

Cuvier's respect for the power of geologic processes of wind, water, ice, and volcanism should not be overlooked. In "Discours sur les Revolutions de la Surface du Globe," chapter of 174 pages in his *Ossemens Fossiles* (Cuvier, 1825 ed.), he devotes considerable effort to the description of the geological processes of erosion and sedimentation, to geomorphic profiles of equilibrium or disequilibrium, to coastal changes and dune formation, and to historical evidence of submergence and emergence. As Minister of the Interior, he had obtained from civil and military engineers many examples of stream courses and harbor changes. His notice of facies differences due to ecological differences in stratigraphic sections inland from the coast led him to postulate rapid transgressions of marine conditions.

So, it was not ignorance of the power of modern geologic processes—actualism—that led Cuvier to dismiss Hutton's uniformity. He rejected the "eternal" aspects of simple geomorphic forces as being insufficient to explain the structural displacements he had observed. He argued, albeit briefly, that actualism was necessary, but that it was not sufficient to explain why the same marine strata he had seen lying flat and near the sea were also found, disjunct and faulted, high in the mountains. He apparently did not seriously consider Hutton's general explanation of the origins of continents rising from the bottom of the ocean, although he mentioned this in a footnote. Rather than Hutton's isostatic theory, Cuvier obviously preferred eustatic explanations for the continental distribution of Mesozoic and Cenozoic marine strata in the Paris Basin.

Cuvier's prowess as a paleontologist was impressive. It was based on a knowledge of functional analogy and descriptive homology in vertebrate skeletal systems. He saw the same parts repeated in different systems with different adaptive configurations. He developed a set of principles (for example: "existence," that the component parts of each organism must be arranged in a manner that enables the whole animal to function in its surroundings; and "subordination of characters," that all parts mutually interact to eliminate some parts in some animals and to require other parts in other animals). These concepts later made it possible to find evidence of evolution in the fossil record, although Cuvier was never able to see it.

The theory of organic evolution current in France at the end of the eighteenth century was an extension of Buffon's concept of the mutability of species that was being advocated by Cuvier's museum colleague Jean Baptiste de Monet, Chevalier de Lamarck (1744–1829). Lamarck believed that the development of the earth's surface had been gradual and constant and that all life had responded directly in its form to its surrounding environment. This directness resulted in a morphological flux, in which each generation was shaped by external forces that were translated within the animal to create an "inner consciousness," and this process resulted in the modification. The development of a particular organ was determined by its usefulness to the creature that possessed it; if it was useful, it was then transmitted to the next generation.

Lamarck's "development hypothesis," the response to felt needs (Greene, 1959), was soundly rejected by Cuvier, who argued that there existed in the fossil record no evidence for the gradual transformation of one species to another. He cited mummified cats, recently brought from Egypt, as examples of the lack of morphologic change within a species over a significant historical interval.

It has often been stated that Cuvier believed in successive creations of life that were separated by geologic upheavals. This is not true. His acceptance of Biblical history had nothing to do with religious reverence. He simply, and erroneously, went along with the prevailing opinion that there had been only one Creation, probably at about 40,000 years B.P. The sudden appearance of new species in the fossil record of France was due to the immigration from distant—and, as of then, unstudied—lands, possibly from Africa. In fairness to him, he just deferred the question of the origin of species to an earlier time than the "Diluvium" (Pleistocene) of the Paris Basin, the period that concerned him in his studies of the vertebrates of the chalks.

If Cuvier did not discover mass extinction as an historical fact, his work in the Diluvium on elephants and rhinoceros certainly provided evidence of the sudden disappearance of a familiar, exotic, and impressive species. It shook the claims of those who stated that there was a design and a direction to nature and that man was its goal. The discovery of part of a frozen carcass of a mammoth in Siberia brought further evidence of the potential "suddeness" of the extinction event. Of course, the Pleistocene tills of Europe were thought to be water-lain deposits that were, possibly, the historical remains of the Noachian Deluge. Cuvier never would understand why these deposits had reindeer and no monkeys. Why were there no remains of human inhabitants before the great flood?

Cuvier did not invent the deluge concept to explain the Diluvium. He included the Deluge and the Creation as the two datum levels in his general historical scheme. These were the two principal events in world pre-history upon which the scholars of his time seemed to agree. From there on, he spent considerable effort trying to calculate the time required for the various events of human history that were known from legends, the Bible, and Egyptian discoveries just then being made—all of which came to about 6,000 years—as well as the time, rather short, that was necessary for all known geological history. He came to the conclusion that to accommodate the long periods of stability that were punctuated by extinctions and structural discordances, sudden catastrophic changes were required. He proposed, incidently, no marine catastrophes, only terrestrial ones that were caused by marine inundations and floods. It seemed to him that much of the geologic record consisted of cycles of terrestrial and marine deposition that were punctuated, in some places, by convulsions in the crust.

This was only a portion of the work of a man who was a voluminous writer and prodigious thinker, one who would have to be considered such even had he not decided to assume the task of proposing a grand historical synthesis that included revolutions, extinctions, deluges, and catastrophes. John Greene, an historian who read Cuvier's report, concluded that he was very limited in his motivation toward, and in his ability for, theoretical explanation. Greene (1959, p. 171) states that for Cuvier it was sufficient "to name, describe and classify . . . [and that] . . . Cuvier never had a dynamic conception of nature . . . he gave up searching for general laws governing nature's operations . . . to look up patterns upon which nature's production had been modeled." This testimony at least suggests that Cuvier was not given to idle flights of hypothetical inventiveness. Why, then, did he suddenly become motivated to propose a grand theory of history? His respect for inductive reasoning was very strong, as was his avoidance of historical explanation with quasi-religious overtones. He, like the many others who were the intellectual descendants of the "Ency-clopedists" and materialists, considered himself a practical Newtonian. If I am accurately judging Cuvier, he was a man who could not stand by idly without synthesizing what was obvious. He was not one to concede that his was not common sense. The pressure of Hutton's work was stronger than he was likely to admit.

As Greene has suggested, Cuvier was a skilled observer of parts of systems and of how these parts were arranged, but he lacked real knowledge of the mechanics that governed the way in which these parts fit together. He reasoned that groupings of animals he observed reflected a natural order among the parts. The parts were unique, however, and did not change; they were only replaced. Lamarck's model would not allow stress to accumulate, and this no doubt posed questions of extinction, as well as of evidence. Cuvier's model had no means by which stress might be dissipated either in organisms or in tectonics. The response had to be sudden. Strata marked periods in the geologic past when conditions were stable and the preservation of fossil remains was possible; that is, sedimentation and life were constant, as peace is constant between wars, and prosperity between famines or plagues. By further analogy, there must have been violent periods, as could be seen in the angular unconformities and extinctions. The play was repeated, but the actors changed, each new cast of characters coming in from offstage to repeat the tragedy with Moliere-like irony. We will later see how this scenario differs from Lyell's proposed one, in which the actors change one at a time, but the same play keeps going constantly.

To suggest that Cuvier, or Cuvier with Brongniart (1811), stayed close to literal interpretations of their field observations implies that a model of intermittant stability is not as dependent on theory as one of flux. I remember remarks by R. C. Moore (1970), who noted how continents were stable even though lithologic sequences in Europe and North America (especially in the Paleozoic) are extremely similar. In retrospect, with the help of tectonic hindsight, his explanation that each past age was universally characterized by unique conditions seems less natural, usual, or in conformance with current common sense than it did at the time.


## BRITISH CATASTROPHISM AND UNIFORMITARIANISM

It makes a good story (at least for English readers) to say that common sense was brought to France by Voltaire (1694–1778) after his visit to England and his exposure to the empirical philosophy of John Locke (1632–1704). This presumes, of course, that philosophers have any real impact on the way people think and that the French were lacking common sense in the first place.

Regardless, the French enjoyed a marvelous period of rational enlightenment that lasted until the end of the eighteenth century, when revolution and war began to take their tolls. Cuvier and Napoleon both benefited from the respect for practical reasoning that developed during these years.

This was a time when, according to the rationalists, a gentleman could hope to learn, by the time he was forty, all that was worth knowing. The famous Encyclopedists compiled all the practical knowledge of their time into red leather volumes that were to lie in state in the squires' libraries. Reason, especially practical reason, was supposed to make men's lives better and freer (Durant and Durant, 1967).

Unfortunately, when enlightened reason becomes the property of a privileged few and outstrips common sense, fear and distrust arise, especially when the benefits of this enlightened reason are not equitably shared. Institutions that once were intended to serve everyone begin to serve only those that run them. Ultimately, someone points out that the masses are closer to nature and reality, and the seeds of revolt are sewn. People begin to shout, "Enough of this snobbish nonsense!"

Historically, when social or technological change occurs, intellectuals all too often remain isolated—for whatever reason—from the masses. Laws that had been intended to preserve order are then used to enforce it. This was undoubtedly true among the priests of Egypt when Alexander invaded. It was also true in medieval times, when the Church offered, on the one hand, the only promise of a just return for a difficult life and, on the other hand, made reading illegal for everyone except those who could enforce or assure order. Until the nineteenth century, when universal education—that same general education that Charlemagne had promised centuries before—began to become a reality in western society, there existed two kinds of common sense: one was born from work with the hands and from daily life in the world as it was, and the other was derived from the mind's sense of inner order and the power of the world as it ought to be.

The difference between these two kinds of common sense is relevant to the birth of geology. This was especially true in Britain, where compromise had become a necessity, where the skills of the hands were becoming increasingly important to the growth and security of the nation, and where laws—both ecclesiastical and secular—were adjustable (albeit reluctantly). Geology and Natural History were born, perhaps, more from the marriage of these two kinds of common sense than from either the trial and error kitchens of chemistry or the inventions of physics. As we will see from the example of Charles Lyell, both the laws of the world (as some thought it should be) and the laws of nature (that others hoped to find) were implicit in the common sense of the class that sought them.

Jean Jacques Rousseau (1712–1778) was the "flower child" of the eighteenth century. He claimed that culture and establishment reasoning had corrupted the innate goodness in man. He cried for a movement back to nature, with the "noble savage" as his idealized model. Both he and Voltaire died eleven years before the attack on the Bastille. The romanticism that he helped to spread against the failing influence of the Church, as well as the problem that he illuminated, brought down the Bourbon house and put a military hero in its place. These events made it acceptable for some to believe that history did, indeed, progress through revolution.

In Britain, where there had been a bloody civil and religious war in the seventeenth century, there was in the eighteenth century a newfound tolerance for religious differences and a respect for prudence, observation, and individualism (the reasons for American independence notwithstanding). Happiness and property became admired as goals, and compromise was seen as the means to these ends (Russell, 1946). Land reform was taking place in Britain, while feudal estates were beoming larger on the Continent. Empiricism flowered as nationalism increased.

Britain was, to a degree, isolated from the rise of romantic discontent on the Continent both by water and by having a mercantile class that began effectively to develop experimental technologies. Science was still, for the most part, a gentleman's hobby. Far more people still lived in small towns, in villages, and on farms than in cities. Each year in London, more people died than were born. Across the English landscape, one could see from one church steeple to the next; but in the midlands and cities, coal smoke was beginning to blot out the view. Over the years, a slow flux of healthy young people moved from the countryside to the cities and ports, where they died at an early age from disease or pollution; or else, they emigrated. While the anachronistic monarchy postured and the parliament deliberated, there was a need for the clergy and lawyers to keep the lid on a social pot that was beginning to boil, even as the navy was busy keeping the pot intact. Riots and open questioning of class structure were momentarily quieted (for 23 years) by the Napoleonic Wars, during which time, Charles Lyell grew up and began to study law and natural history, while uniformitarianism took root from its origins in Scotland.

James Hutton (1726-1797) was a physician and farmer turned geologist. He was part of—or, at least, near to—the Scottish Enlightenment in Edinburgh. One can see in his thinking about natural history evidence of his having been influenced by Newton and Harvey. He was also, obviously, a protestant and shared those drives and fears. He published his main works in 1789 (the year of the storming of the Bastille) and in 1795 (the year that Napoleon became First Counsul).

In writing about geologic processes, Hutton (1796) states, "[when] we view the connection of those several parts, the whole represents a machine of a peculiar construction which . . . is adopted to a certain end. . .[and]. . .erected in wisdom." He sought to dispel any idea of a capricious tampering with the machine by "any natural supposition of evil and destructive accident of nature or to the agency of any preternatural cause."[3] Lyell will later imply the need to overcome the distaste for the

[3] "Preternatural" means "other than natural." Note that Hutton does not say "supernatural," since this would evolve into a religious argument about whether or not

unnatural that was characteristic of British romantic common sense. It would seem that, deep within every Englishman, there lurks a Celt, who must be controlled by fear, respect, or reason.

Uniformitarianism (see Gould's analysis in Chapter 1) is a mixture of common sense observation, "hands on" experience of modern geologic processes, extrapolation by analogy to past events (in Europe, "actualism"), and the assumption that the Aristotelian models of First Cause and an eternal world are correct.[4] The reverse of the principle assumes that the processes operating today are causal extensions of those of the past. The second view, which is actually easier to use, flows from a longer history of thinking. Whitehead has said:

> the greatest contribution of medievalism to the formation of the scientific movement is. . .the inexpugnable belief that every detailed occurrence can be correlated with its antecedants in a perfectly definite manner, exemplifying general principles (Whitehead, 1925, p. 18 as reprinted in Cloud, 1970).

The concept of uniformity pervades rationalism. We would be hard pressed to make a geological inference without it.

Hutton (1895) asserted that common sense observations of geologic processes and inductive inference were preferable to the kind of cataclysmic thinking about the aqueous origin of trap basalt that was allowed by the Werner school (English translation of Werner, 1788). The mechanisms now seem so different that only the timing seems important. One can only imagine Hutton's dismay had he known—as we do today—that most of the ocean floor is covered by a precipitate (catalyzed and transformed by a biologic intermediary) and is underlain by trap flow basalt. Playfair's (1802) insistence on the need for more time gave birth to gradualism. He did away with the necessity of valleys being formed by the tearing asunder of the crust, but he did not realize that the largest valleys, those between continents, were made in that very way.

As I write this, I remember John Martin's paintings in the Tate Gallery in London; these canvases depict great cracks forming in the earth, and people—presumably sinners—falling into the abyss. Martin was also an

---

God interferes with the natural workings of the world. Using an Aristotelian model, Thomas Aquinas had centuries before effectively argued that this was unnecessary; hence, the birth of uniformitarianism.

[4] This idea—that the universe was designed with all of its potential motions included—is also implicit in Galileo's Principle of Uniformity in Motion, but without Aristotle's essentialism. Averröes, a Moslem in Cordova, Spain, is said to be responsible for translating Aristotle from Arabic during the twelfth century, thus reintroducing his work into Western civilization.

engraver, who (among other tasks) illustrated geologic texts (Pendered, 1924). With such a fervor, romantic naturalism in those days blended with theology.

British Catastrophism combined empirical geological discovery with this romantic historical explanation, in which both certain aspects of wonderment and a blind acceptance of Scripture were satisfied. Historical authority became entangled with moral authority, and the ease with which geologists could speculate was thus impaired. There was a danger that nature, as a lesson in morality, would prove to differ from traditional teachings and that social order would then come unglued. Catastrophism was a form of acceptable terror, one that was typical of some parts of the Old Testament. A concern for morality in nature continued throughout the nineteenth century and, of course, on into our own. More than one early geologist set out to prove in the rocks Christian truth and law.

While the lessons of physical geology were being pondered, new and exciting fossil finds were being reported. This is not to say that the British had been unaware of fossils. For more than a thousand years, they had been seeing them as products of the supernatural. (Echinoids were called "fairy loaves.") But now, for some, fossils were becoming useful: with them, one could make maps of the strata that could be followed when digging canals; later, these maps would be of use for routing railraods; and, later still, they would be used for building roads. For others, however, fossils were becoming a testimony to God's work.

Yet, for the more detached minds, it became apparent that knowledge of life need not be confined to the living world. The vast thicknesses of strata were thought to have been deposited almost within historical times. The fossil species were often amazingly well preserved, but some represented completely unfamiliar forms of life. The search for the oldest—the primordial—fauna began to show that the succession of fossil evidence was both discontinuous and difficult to reassemble from its fragmentary distribution.

From the discontinuity of this record, it was concluded that each assemblage of strata represented a geologic epoch that was characterized by its own life and by a time of quiescence. Each interval of quiescence ended with a cataclysm, in which every living thing was swept away and replaced by brand new creation; there then followed another time of quiescence. The sequence was "like a succession of rubbers of whist." The replacement was *en masse,* and the breaks were complete. Whereas Cuvier had posited relatively few catastrophes, one was now being proposed for every geologic interval, approaching a hundred in all.

Although Robert Jameson's translation (annotated and somewhat altered) of Cuvier's (1817) *Theory of the Earth* may have had some influence on Britain's catastrophism, it was probably only more fuel for an already raging fire. Cuvier particularly avoided religious implications; as stated earlier, he had but one Creation and only terrestrial catastrophes, usually by island flooding. He described no marine catastrophes and no cataclysms. For the British, cataclysms were total and final.

Not until after Darwin's contribution, can it be said—as it was by T.H. Huxley in 1859 (Huxley, from Bibby, 1967, p. 162)—that catastrophism was eliminated from geologic speculation. The belief arose that species were actually replaced one by one, rather than in assemblages and that, if all of the evidence were available, the species "would fade into one another with limits as undefinable as those of distinct yet separable colors of the solar spectrum" (Huxley, in Bibby, 1967).

## CHARLES LYELL

Charles Lyell (1797–1875) was born the year that Hutton died.[5] He was described by Darwin (Barlow, 1958) as notable for his clarity of thought, his caution, candor, sound judgment, and originality. As a theoretical innovator, Lyell was perhaps more conservative than Darwin, who characterized him as going about ". . . correcting and adding up new information in old train"—(see Barlow, 1958, p. 232)—from Darwin's notes. He is said to have had considerable sympathy for the work of others and a keen interest in how his own works lent dignity to mankind. He was a liberal in religious belief—"or disbelief" (see Barlow, 1958, p. 100)—and a strong theist, or at least one who believed that the universe resulted from a primary causation rather than from secondary ones.

Lyell grew up in the south of England in a well-to-do family with a tradition in both the navy and the legal profession. He went to public school, matriculated from Oxford as a "gentleman commoner," and took up the study of law. After some travels in France and Italy at the end of the Napoleonic Wars, he developed his interest and skills in natural history, in between a few years of practice as a barrister.

In 1823, he visited Paris to meet Cuvier, Brongniart, and Constant Prévost. He was greatly impressed by the serious attitude of the French geologists, became very interested in French institutions and culture, and was to return to France and Italy many times to develop the ideas that became some of the foundations of modern geology. In 1831, Lyell abandoned the practice of secular law to take a position as Professor of Geo-

[5] This is a curious parallel to Newton's having been born the year that Galileo died. For a complete biography of Lyell, see L. G. Wilson, 1973.

logy at King's College in London. One might say, however, that he never gave up his interest in law, but merely aspired to plead his case before a more universal court.

To understand Lyell, it is necessary to understand both the times in which he lived and the conflict that was then taking place before the uniformitarian and the catastrophic points of view. Although no man can ever fully understand another, knowledge of certain aspects of Lyell's character and training might help to explain why he was, at first, scientifically ambivalent about this conflict and then radically (albeit reluctantly) changed his mind and the history of geology.

I propose that two factors help to explain Lyell's contributions as more than simply expository science. The first is a lawyer's respect for the values of evidence, analogy, and refutation within the framework of the defense of natural law. Lyell had a lawyer's understanding of the importance of personal dignity. Almost every chapter in his famous *Principles of Geology* ends with a moral. The premise of secular law—that the purpose of a legal system is to assure a "steady state"—led Lyell to plead for uniformity and constancy in the existence of species. The second factor is his scientific assumption of a model of crustal mechanics in which almost all movement was vertical and controlled by erosion, deposition, and volcanism-related heat flow systems. Nearly all of his examples of uniform geologic processes were chosen near the sea. He assumed that sea level was constant.

Lyell had little trouble in understanding that consistency in natural law through time could explain the development of the physical geologic record. We will trace this development in a moment. He ran into difficulty, however, when he began to examine the fossil record. It was easy enough for Lyell to suggest that the physical world was a "perpetual motion machine," "a clock with a self-winding main spring," or one that had been wound up by the Creator; but the discovery of species-extinction posed a problem: what was the mechanism for species-replacement? How could one substitute materially self-adjusting machinery for a Supreme Creative Intelligence without ultimately conceding to a theory that was derogatory to human dignity?

Trained in legal principles, Lyell realized that any proposed system of natural or secular law which encompasses man's conception of himself (thereby implying his relative position in the universe) must include an assumption of self respect. If ideas of order are to have any chance of being accepted, they cannot depart very far from the common sense of their time. Although fear may serve as a temporary deterrent to potential offenders, most law is effective, not because of policing, but because it coincides with inherent concepts of personal dignity. Any proposed natural law that by implication detracted from man's traditional image of himself would have run contrary to Lyell's legal or religious sense of proportion. To suggest that life, with man at the summit, was a succession of mechanical accidents, not only defied common sense, but also implied an egalitarianism with the bestial world that Lyell's British pride could not swallow, at least not until more dignity could be recognized in the state and conduct of that world. Natural theology may have made landscapes seem like reflections of God's grandeur, inherent goodness, and order; but man still had to confront his recurrent memories and the evidence of his savage remission to a less than civilized state.

The legal arguments become involved with abstruse language, as legal arguments are wont to do. Lyell was looking for a suitable theory or body of natural laws that did away with the repeated intervention of the first cause by substituting the regular action of secondary causes. To succeed, a brief must both employ an appeal to common sense and assure that the interests of the client are preserved. The advocate's chain of evidence must be complete, while also demonstrating that the opposition's is not. Lyell's arguments often used this operational language, but his closing statements were about the moral implications of geology.

There is another part to this story, which is, of course, here much abstracted. Lyell had visited Giovanni Risso in Nice and had been told that the local Tertiary deposits had 18 percent living species. When he visited Turin in Italy, Franco Bonelli told him that the local strata there had faunas similar to those in the Bordeaux region, but different from the ones in the northern Apennines. After discussions with Bonelli about these faunas and those of Sicily, Lyell decided that the Tertiary could be divided and classified on the basis of the percentages of living species found among the fossils. Gerard Paul Deshayes in Paris had also arrived at this same conclusion, and Lyell asked him to prepare a catalogue of the shells and formations of the Paris Basin. Later, with the help of William Whewell, who coined the term "uniformitarianism," Lyell would give to the three Tertiary units the names by which we now know them: the Eocene, the Miocene, and the Pliocene. A time scale was thus erected, by means of which modern analogies of geologic change could be extrapolated backward into the past.

I do not wish to undermine anyone's confidence in the foundations of uniformitarianism. After having worked in Italy and Sicily for more years than Lyell had at this stage in his career, I can only admire the patience with which he described detail and his generalized perspective. Considering his acceptance of Hutton's theory of relative isostacy, I would be loath to state that Lyell was unjustified in drawing such universal conclusions about uniformity in change. Yet, the "Father of Modern

Geology" based the power of his text and his arguments on a very limited and erroneous tectonic theory and on an inference about the passive state of the oceans that we now know is not true. Most of his historical inferences by analogy from the field examples he described would not, by today's standards, justify extrapolation past x $10^6$ years B.P.

In Sicily, the last twelve million years of geological history is a jumble of fragments, in which there exists evidence of a major catastrophic event (see Benson, Chapter 17) in the very rocks where Lyell could not find fossils. One can only surmise what damage might have been done to Lyell's confidence—and to the future of geology—had he known about the vast layers of Miocene evaporate in the floor of the Mediterranean, the evidence of great changes in sea level, the olistostromes and other evidence of massive lateral movements, or, above all, about the absence of thick sequences of primary and secondary deposits in the floor of the Mediterranean. He seems to have been relatively unconcerned by the fact that the source areas of the "Pliocene" deposits of southern Italy and Sicily were to the west, within the Tyrhennian Sea, which is now under 12,000 feet of water. According to his isostatic beliefs, the bottom of the sea could be raised to expose sediments that had been eroded from the land, but he did not address the problem of how a portion of the continent could suddenly disappear. Today, the original analogy of continuity in the rates of change is strained by our knowledge of too many details. Nonetheless, Lyell decided, on the basis of the evidence he had observed, that the physical world as a whole was a steady system in so far as geology was concerned. We see this same evidence as part of a pattern of continental subduction that causes periods of regional instability and oceanic catastrophism.

In 1872, Lyell read Lamarck's *Philosophie Zoologique* (1809), and he included an abstract of Lamarck's views in the 1832 edition of the *Principles of Geology*. Many years later, he admitted that he had not done justice to the ideas of Lamarck; but, in the meantime, he had become famous (according to Darwin's autobiography) for attacking Lamarck's positions. Whatever harm he may have done—for example, indirectly delaying Darwin's efforts—he later more than made up for by supporting natural selection.

Lamarck had observed that, as collections of related specimens and species increased, the gaps in difference between them decreased, until the voids were filled, and their determinations became arbitrary—that is to say, if one ignored inconsequential variations in form. He assumed conversely that, where gaps appeared, the collections were incomplete. He also noted that, when individuals changed their environments, their appearances were altered, and their organizations changed. He began to run into trouble when he considered the origin of the forces that brought about this change. He said that they originated internally, and, of course, Darwin later showed that the major selective force was external (he did not, however, resolve the question). Lamarck went on to state that, when the relationship of the changing species came into stable equilibrium with a new environment, the change slowed—or, perhaps, stopped—until the process was repeated, and another link in the chain of development was forged.

Lamarck more broadly concluded that a continuum existed between simple and complex organisms, and, futhermore, that the simpler organisms existed in the fossil record at dates earlier than those of the complex one. This was diametrically opposed to the classical view which held that perfection originated first and that poor quality control was responsible for forms of life departing from the original plans. Implicit in Lamarck's ideas was an historical plan for the progressive improvement of life.

This set of ideas must have greatly disturbed Lyell. Here was a theory that allowed him to show continuity in the process of the development of the organic world; it extended the concept of geologic time in a way which was the same as that needed to justify physical uniformitarianism. Yet, he set out to destroy the concept with a zeal similar to Cuvier's.[6]

Implicit in Lamarck's theory of transmutation of species was a greater concept: that all of nature was a delegated power, an instrument, a mechanism acting by necessity, a simple extension of the Grand Architect, and not the results of his discrete judgments. Nature created by degrees, because it did not have the power to create directly. Yet, all of life's diversity and complexity existed; it had been formed apparently without guidance. This concept ran contrary to Lyell's sense of religious proportion that justified man's—and, perhaps, his own—position in the structure of things; it departed too much from the common sense of his time. He might win his day in court on the issue of physical uniformitarianism, but he would not defend Lamarck. With human dignity at stake, no scientific jury in Britain would uphold such a broad and sporadically documented theory on the basis of the evidence available. So, Lyell joined with Cuvier, his conceptual adversary, and stuck with empirical observations and the idea that species were immutable. He did, however, alter the catastrophic model so that it would allow for species to be replaced by the Creator gradually, when habitats were selectively vacated by extinction.

[6] Cuvier died in 1832. Lyell's attack on Lamarck appeared in the 1832 edition of the *Principles of Geology*, which was released just after Darwin had departed on the Beagle. Darwin received this edition in Brazil, yet there is some question (Smith, 1960) as to whether or not Darwin ever thoroughly read Lyell's remarks.

In this way, Lyell was able to have his physical uniformity and, at the same time, to stay with the establishment thinking of his peers.

Lyell effectively refuted Lamarck's theory in Britain, not because he could prove that the internal causation model was inadequate, but because Lamarck would not cite a single observable example of an actual gradual transition of an old and previously known species into a new one. There was no instance of the substitution of some entirely new sense, faculty or organ in an old animal.[7] This criticism pointed to "a chasm in the chain of evidence" (Lyell, 1872, vol. 2, p. 253). Lyell triumphed—or so his personal success suggests—even though Lamarck had strongly stated that, in order for the evidence to be judged properly, more time was needed than human experience could provide. In fairness to Lyell, he made a good point: in man's experience, no one had reported the special creation of a new species either. This was because it would have probably been missed, since in Europe it would likely occur only once in 40,000 years.

In 1856, Lyell suddenly converted to organic gradualism—so the story goes—when Darwin convinced him that the mechanism of natural selection had sufficient explanatory power to overcome his previous objections. He saw in Darwin's explanation a logical elegance approaching that of the special creation model; perhaps, he recognized a shift in the attitudes of his prospective jurors; or, perhaps, he realized that in the 1850s, if man's image of proximity to the Creator did not imply such great dependence, man's dignity might be less affronted than it would have been in the 1920s. After all, Britain was now much further into the age of industrialization and materialism; the country had survived and prospered, when most every other major power had failed.

Mayr (1972) has argued that Lyell, in his pre-Darwinism years, exercised the silent assumptions of his time and believed that everything in nature was planned, designed, and had a predetermined end. He has stated that Lyell was an essentialist, that he was tied to the limitations of that philosophic position, and that he had a moral obligation under the "perpetual intervention hypothesis" to find additional proofs for the wisdom and constant attention of the Creator. I believe that this connection is tenuous and that it overlooks Lyell's streak of pragmatic sophistry. Lyell had originally assumed that each species had originated (or had been created) and had become extinct separately. Of this, there is no doubt. However, when dealing with issues which were beyond what he felt at that time was part of geology—issues such as his view of how one

---

[7] I use Lyell's exact language here (Lyell, 1872, vol. 2, p. 253) to bring out a point that will later be amplified: that is, species evolve as systems more rapidly than do the organs as subsystems.

examines the origin of inanimate things—he often deferred to a theoretical authority. There was a difference between Lyell the operational essentialist whom (like Cuvier) believed that every species has an everlasting, intrinsic plan that distinguished it from all others, and Lyell the natural philosopher, who conceded (albeit temporarily) that the origins of living phenomena could be explained in a traditional body of religious theory based on first cause.

In the last paragraphs of the *Principles of Geology*, Lyell, assuming the license of Hutton's premise of world history without "a vestige of a beginning," argued the paradox of explaining the origin of the earth, the origin of all species, and the origin of man as a moral and intelligent being. He simply stated that these are problems which he believed to be beyond the arguments from analogy. Next, in a religious context, he affirmed the inescapable presence of the evidence of order in nature, the limits of philosophical inquiries, the finite powers of man, and the clear dominance of a Creative Intelligence. And then, Lyell, the wise old barrister, in an attitude of traditional supplication before judgment, rested his case.

## HISTORY AND CONTINUITY BY ANALOGY

In a recent book by a British historian on medieval warfare, the author states:

> History as it occurs in the field is not ordered or structured. It is only in the minds of historians who introduce order and discipline into the disorder, using such 'measured' means as chronological divisions and periodization. . . . The history of medieval warfare reflects, as all history, continuous flux (Koch, 1978, p. 9).

His discussion begins in the seventh century with the Islamic invasions and ends in the seventeenth century with the Thirty Years War. I mention this only in amazement. If our ancestors actually did fight wars everywhere, all of the time, for a thousand years, then it is no wonder that we fear the catastrophic mentality. Or, perhaps there were places where, and times when, peace prevailed, and this explanatory strategy simply reflects a Heraclitean point of view of the subject.

The confrontation between the thinking of Cuvier and that of Lyell is one that is basic to historical explanation. It touches on our concepts of time, of events, and of where we stand as observers. It results from a basic inadequacy of the logical process in dealing with discrete observations of the historical past. In this section, I want to explore some of the operational aspects of historical reconstruction that may define sudden events. I must warn the reader, however, that only a partial truth will result from this effort. Those who want to find discontinuity in the historical record will find it; those who want to see continuity will see it—albeit with differing clarity; and those who want to avoid the problem will assume that the connection is outside of the data base, if no one knows what is there.

As Diogenes and calculus have both demonstrated, it is possible to know that continuity exists only by first believing in it. Stated more rigorously: it is possible to conceive of continuity from discontinuous observations only by inductive inference based on an *a priori* assumption of continuity. Cuvier rejected this assumption and inferred regional discontinuity caused by systemic disruption. Lyell insisted on this assumption as a conservative measure, anticipated expansion of his data base, and inferred a progressively[8] altered steady state. All arguments about rates, especially about rate changes in geologic or evolutionary processes, are extensions of these positions.

In short, one cannot use deductive logic to construct a continuous history out of discontinuous facts. One must have logically connected facts: that is, they must have relationships of a higher order of generality that identifies them as part of a series. This means that "facts," if they are to be ordered, must be statements about observations that relate to a common process. One can order oranges and apples, if one is ordering fruit, by assuming that they have some common function: for example, both grow on trees, are eaten, or can be used for juggling. In order to see continuity, one must have a concept of structure (see the definitions in the Introduction to this chapter). Those who cannot see structure and function cannot see continuity.

Discontinuity may be real, or it may be an artifact of the qualifying descriptors of the facts in a series. Assuming that not all of the elements of a succession are unique in their properties and that those having properties in common are functionally related, discontinuity is self-evident, if partial continuity is inferred. The ease with which discontinuity is demonstrated in many similar fossil or rock sequences is the foundation of stratigraphy, but this begs the question of the validity of the descriptors. Without structural inference among the descriptors, discontinuity can seem more evident than it is. An example of this would be making a comparison of the correspondences in lists of species in a succession of

---

[8] Perhaps "progressively" should be qualified to mean for historic and biologic systems. Gould's assertion (see Chapter 1) of the non-directional aspect of Lyell's thinking is difficult to accept as a generality. For instance, Lyell's classification of the Tertiary was directional and progressive from ancient to modern, and it was based on statistical replacement.

strata without referring to the developmental relationship among those species. The identity among the lists (presence/absence compilations) can be useful, but the real information about near relationship breaks down. This is a fundamental weakness in stratigraphy where novelty is the basis of most of the taxonomy.

To suggest that the presence of connective evidence exists outside of the examined data set requires a stronger inference than that required to suggest continuity within. It is obvious that, without a firm concept of structure, this inference becomes difficult, if not impossible. To accept a general theory which holds that the real connection is a discontinuous succession usually exists outside of a potentially obtainable data set discourges further inquiry, especially about the yet-undiscovered structural relationships of the data in hand. To assert that there are gaps in the sampling due to a mechanism that always makes such gaps likely requires a very strong systemic theory.

That motion exists is a matter of common sense. That transformation in nature is seldom uniform is also common sense and can be observed. Therefore, it should not be a problem to accept the principle that sudden events of a large proportion could possibly occur in nature. For the historian, the questions are: 1) what are the theoretical constraints, and 2) what kinds of evidence are acceptable? For the war historian mentioned earlier, the assumption that all history is flux reduces the logical possibility of a nuclear holocaust. An alternate theory, that history is pulsatile, admits the possibility of stress. Popper (1957) contended that history was unpredictable except among laterally contemporaneous events. Drooger (1973) likened the geologic record to a movie film clip, with stratigraphy comparable to discreet static frames. The events of history shift from scene to scene, of course, and one has to admit that many of the frames are missing.

This film clip analogy deserves further examination. In the film clip, one senses motion in the passage of the frames, because the rate of passage exceeds the rate at which one can perceive the individual frames. The motion depicted is either continuous or inferred to be continuous. Our sense of inferred motion is very strong; creators of animated films are keenly aware of this. The number of frames used to make Bugs Bunny move are relatively few. In fact, the artist simply creates a series of images in which the subject changes attitude and position, and the viewer makes these images move. The preconceptions of the observed are paramount.

Film clips of real events, especially of events that are not staged for the camera, more closely approach the Drooger analogy. The geologic historian wants to demonstrate continuity and causal connection in sudden, as well as in slowly transforming, change. Take, for example, the events recorded in the Magruder film of the Kennedy assassination. In this film, the subjects moved at various speeds relative to the passage of the frames: before the shooting, the motion of the car was slow and deliberate, and the movement of the passengers was more rapid; when the shots were fired, however, the rate of recording was exceeded by the rate of events, with the result being that, even now, it is not possible to ascertain the number of shots that were fired by noting the reactions that were recorded on the film.

From the historian's point of view, this recorded event implies a structural disturbance of a magnitude sufficient to change murder to assassination. This more general inference tells us something about the importance of understanding systemic structure in reconstructing events. It is said that the Romans feared the armies of the Germanic tribes, not because they were disciplined or well-armed, but because they were natural fighting units. That is, the Goths, the Allemans, or whoever were orgnized into groups composed of one hundred warriors, all of whom were related by birth. When the fighting got rough, they would not desert. As the Roman armies began to consist more and more of mercenaries with various backgrounds, they began to run more and more often. Without this kind of knowledge of structure, the historian's task of understanding why empires collapse, sometimes suddenly, would be significantly hampered.

Once again, by analogy, let us examine the role of inference of continuity and that of sudden alteration of structure in a single organism: a racing horse. Assume that you are an engineer who has never seen a horse, and, suddenly, one is standing there before you. Having heard that animals—especially animals bred to be "perfect"—have no unused structures, would you conclude that the horse was made to run? Considering that three legs are sufficient for a stool to stand upright and motionless, would you conclude that one leg of the horse is redundant? Would you ever guess that, when the horse does run, it actually proceeds from a walk to a trot to a canter[9] to a gallup: a disjunct sequence of stable states of increased motion? When you eventually did see the horse run, could you swear that it always had one foot on the ground? What kind of historical inference would an engineer make if he were told that the horse's "foot" was really a single toe?

We have not strayed so far from historical geology as it must, at first, seem. In history, in mathematics, and in all of science, much of the power in inductive inference comes from analogy. A model is the ultimate anal-

[9] Further, according to the Oxford University Dictionary, in order to judge whether or not the horse did "canter," one would be required to make an unlikely comparison by analogy with the movements of pilgrims on their way to Thomas Becket's shrine at Canterbury.

ogy. Similitude is the condition of showing the relationship between the model and reality. Hans Stille may have had the first glimpse of plate tectonics when he watched the ice beginning to flow in an Alpine stream. The artists of the early eighteenth century thought they saw the mechanics of God's actions in the violence and passivity of landscapes. Now, we infer flexural rigidity in tectonic plates by observing the way in which sea level changes around erupting volcanic islands. We have progressed from Lyell's analogies of common-sense narrative description to more abstract analogies that include larger and larger events.

## MODELS AND DISRUPTION

The record of geologic history has been shown to be full of discontinuities. The better our methods of correlation become, the more geographically widespread some of the hiatuses become. Some are due to the withdrawal of the sea causing worldwide interruptions in the process of entombment of evidence, as well as erosion of the previous record. But even in the deep sea, where sedimentation is potentially continuous, there are both universal breaks in the record and intervals of rapid transition. Lyell's model of universal balance in local geological forces, wherein the negative reactions of one place would be countered by a positive reaction someplace else, assumed also that the geologic record was always somewhere formed or forming. To concede otherwise would be to admit that global catastrophes were possible. Lyell knew very little, however, about lateral geographic continuity in the geologic record or about the forces that extend laterally. Remember that he would not admit to the likelihood of significant eustatic changes in sea level. Any theory of catastrophic events of geological proportions must include the element of widespread lateral change. The mechanics of plate tectonics, ocean water-mass dynamics, climate and species invasion and predation of epidemiological proportions provide this element.

In the previous section, I attempted to deal with the role of inductive inference in reconstructing continuity in discontinuous successions of data. Yet, it should be obvious that too much emphasis on the continuity of structures can allow far more inclusive, but sudden, events to fall through the cracks. An example of this might be the case of the Messinian salinity crisis: for many years (beginning with Bonelli and Lyell), the inference of species continuity or gradual replacement in the Neogene record of the Mediterranean has tended to obscure the nearly cataclysmic event that took place there.

Briefly, one point of view (Bizon and Bizon, 1972; Montenat et al.,

1977; Brolsma, 1975a, 1975b) holds that the planktonic Foraminiferal zonal system, which is oceanic and is used the world over to demonstrate the chronologic sequence, is normal—or, almost normal—in the Italian and Spanish Miocene and Pliocene sections and that forams can be found in the correct order even in the upper Messinian (the uppermost Miocene), where the crisis event is supposed to have occurred. The fact that these sections are often used as standard reference sections somewhat confuses the issue, but it is not determinant. When microfaunas (Benson, 1976b, 1976c, 1978) that could never have developed in the ocean were found all over the Mediterranean and especially in the floor in clay partings within beds of thick evaporites of about the same age (Hsü et al., 1973; Ryan, Cita, et al., 1978), it became apparent, not only that the chemistry of the waters of the last stages of Tethys had changed to concentrate its salts, but also that the geographic continuity of these waters had been severed. The presence of the forams, which are rare, could be explained by reworking; but the ostracodes, which had just evolved in the Paratethyan lakes to the east, are unique. The scale of this event involved the formation of about 1.2 million cubic kilometers of evaporites. A worldwide eustatic change in sea level of 70 meters (Adams et al., 1977) has also been found. It is conceivable that the reflooding by Pliocene marine waters that took place after the Messinian dessication, which lasted about ½ million years, may have occurred in a few thousand years. By geological standards, the Miocene event was catastrophic in its timing; by paleontological standards, since most of the marine life failed to survive, the event was—at least, locally—cataclysmic. The Pliocene marine deluge was even more dramatic, and a satisfactory explanation has not yet been found.

The destruction of the Tethys Ocean took place over most of the Tertiary, while the African plate was colliding with, and in part being subducted under, the complex of European plate fragments. Any model proposed to explain this very complicated history will require a mechanism that includes the accidental interaction of parts with long histories and well-established structures. The diastrophic model proposed in 1910 by T. C. Chamberlin to explain sudden changes on a continental scale was basically isostatic with mobile invading faunas. A modern model can utilize much faster lateral systems.

Modeling—or, at least, the term—is just now becoming popular among geologic historians. It must be admitted that there are many who consider it to be undigested, preanalytical nonsense. Diogenes did not need a model to prove that motion was real, although Galileo used one to prove that motion was not real in the way that Diogenes had imagined. T. H. Huxley, had he known about models as such, would probably have said that modeling was the mechanical engineering of refining common sense. If well done, the model can fail, and the modeler survive.

Catastrophe Theory (Thom, 1972; Zeeman, 1976) attempts to fill a need to describe sudden changes of state within systems under stress by opposing forces. Whether this particular model lives up to its name in the geological sense remains to be seen. It has received mathematical criticism (see Kolata, 1977, for a heated journalistic review), which is itself open to criticism. Regardless, if Catastrophe Theory has a chance to resolve systems of a geologic scale, it is a model worth reviewing.

Thom's model is topological and involves some terms and usages that are unimportant here. It is an extension of elementary calculus that considers in a steady state system the disruptions of smooth and continuous reactions (depicted as tending to proceed along smooth surfaces) caused by differential absorption of stress, or energy, at lower levels of organization. It assumes that systems tend to be internally homeostatic, which is true of most natural systems. Subsystems are reordered to absorb some of the stress coming into the system. This reordering follows the structures that are implicit in the subsystems according to their particular history. Implicit in the geometry of these structures may be alternate reaction pathways which have remained underutilized or redundant. As long as the principle function and general structural configuration of the system is maintained, some slight deformation may be the only result. This can be described mathematically as the first derivative of differential calculus, the slope of the equilibrium surface at any point in the progress of deformation.

The reordering of the subsystems allows the system to be maintained past a point where failure would have occurred if the redundancy[10] had not been present. Instead, a sudden change of state follows the new alignment of old structures. Although the behavior of the system may seem to change suddenly, the alteration in structure had been implicit— or, nearly implicit—in its prior design.

To make this concept of inherent sublevels of structural stability work, two more conditions are necessary. First, in order to produce underutilized or redundant structures, the subsystems must have had a history independent of the present function of the system. In geological or life systems this is not difficult to provide. Even number systems have properties that are artifacts of their invention and underutilized or redundant for any given function. In nature, redundancy is common. Second, the exter-

---

[10] This use of the concept of redundancy recognizes that natural systems are not deterministic, single function systems. Rather, they are complexes of reaction pathways with differing historical derivations and have differing, intrinsic, "fail-safe" capabilities.

nal forces causing the systemic deformation are likely to change their states more suddenly, the closer their resultant approaches alignment with the redundant reaction pathways. This means that they must be capable of producing a resultant. Forces of unlike nature would not work. This limitation is a model constraint that forces the modeler to define the system in similar structural terms.

Once the sudden transformation has occurred, a new state is defined—perhaps, with some initial deformation involved in recovery—along a new surface of equilibrium. The "catastrophe" is indicated by the discordance in this surface from that of the previous state. Because the external forces are usually conceived of as being unequal and, often, unopposed, both the hysteresis effect and the sudden change increase: that is, the new and the old surfaces overlap, and, as the magnitude of the forces increases, the difference in state becomes greater.[11]

This characterization of Thom's model takes some liberty with its expression in two ways: it emphasizes the role of the redundancy in the subsystem structures from an engineering point of view, and it has strong "preadaptive" overtones that are typical of an evolutionist's point of view. Only further consideration will determine whether or not the analogies are strained; but, meanwhile, it provides the heuristic function that models are intended to provide. The engineer is just as interested in providing multiple pathways of reaction in an economical design for future construction as the biologist is in understanding those erected in the past. The problems of defining similitude are identical.

Before returning to a brief discussion of the application of the catastrophe model to the history of the Messinian Salinity Crisis, we should consider how well this model can explain other smaller and larger geologic events. In every instance, the defining structure should be suddenly altered as a consequence of converging external forces being absorbed by history-laden substructures, thereby bringing the main system into a new equilibrium. Lyell was impressed by isostatic and geomorphological processes along the edges of continents. This is a good place to begin this model.

Prévost (see the Charles Lyell section of this chapter) concluded that changes in off-shore bars caused the sudden change in the vertical succession of fresh-water to marine sedimentation: that is, the break between terrigenous clays and marls in the Tertiary section of the Paris Basin was not caused by a sudden change in sea level. This may not be cataclysmic

---

[11] Depending on the rigor of definition, the degree of opposition that is necessary can become a moot point. The Marxist historian can claim that synthesis comes only from contradiction; most natural historians, however, would argue that the majority of living systems avoid direct confrontation.

or catastrophic in the British sense; but is it not, in fact, catastrophic in the catastrophic model sense?

One assumes that in the coastline's lateral profile of equilibrium there are mounting, and ultimately overriding, stresses that are represented by the stream of sand which we call an "off-shore bar." The source of supply that insures its strength (that is, its resistance to change in form) may be diminishing, or its relationship to energy levels in the sea may be changing. We are especially interested, as was Lyell, in the sedimentary succession in the lagoon as the condition of the barrier changes.

Note that in this system the active forces that are capable of developing sudden changes in stress are lateral and stratified (laminar). The passive forces are tidal (wind tides under short-term conditions) and epeirogenetic (a long-term condition, possibly subsidence), both of which are vertical. Lyell and Prévost were correct in assuming that sudden breaks in the sedimentation of a lagoon could be controlled by the barrier bar, but they were wrong to dissociate this from sudden flooding. We now know that, under stable vertical conditions, wave-train systems (Hutton's "cannonade of the sea") cannot by themselves destroy the separation between terrestrial (brackish, limnitic, and so on) and marine sedimentation and that flooding, even temporary flooding, is required.[12]

As a model of a sudden change in stratigraphy on a local scale, three elements become apparent. The first is that the controlling mechanism, the barrier, tends to be homeostatic: that is, it is capable of absorbing energy under most of the extreme lateral stress systems that are of the type which have been responsible for its formation (a kind of geological hysteresity). The general structure of the lagoon is preserved. Secondly, with the addition of a vertical component—the wind tides from the sea or, conceivably, sudden subsidence—the barrier can fail if its form includes no redundancy to absorb the energy from the resultant of forces coming from unexpected directions.

When failure does occur, it is sudden, and the effects in the lagoon are usually catastrophic; these effects are usually due to the loss of biological controls: the killing of oyster reefs, stabilizing grasses, dominate carbonate producers, and the like. Flooding—in this case from the sea—can be equally devastating to marine faunas if the flooding is from the land; or, conversely, if there is a complete cut-off of discharge that starves the sand supply, lowers the vertical potential of the lagoon, and thus allows an even faster deterioration of lagoonal conditions. The third characteristic

[12] Barrier bars are broken, not by high waves, but by the sudden return of flood tides caused by a release of water previously held shoreward by high winds. Long-period waves of duration are more likely to be constructive than destructive. Short-period local storm waves seldom cause long-term damage.

of this model is the sudden dominance of lateral forces once they have been vertically displaced even slightly. To a stratigrapher, this is stating the obvious; but it was not so apparent to Lyell, who thought that such lateral translations would always be on a local scale and, in general, gradual.

By analogy, one can proceed from the lagoonal system to both greater and lesser stratified systems of geologic interest: from the crust itself to the encrusting exoskeleton of the crustaceans that inhabit the crust. Sudden transformation of form is characteristic of both these natural systems; this may not be entirely accidental, but I would not like here to defend the completeness of this analogy. One might simply state that, in both systems, the vertical displacement of form, or morphology, was noted first and thought to be most important; then, it was discovered that lateral forces of potential movement were more important. Growth by accretion, the changing patterns of conservative parts, and the sudden transition of these parts are all changes in form that a paleontologist understands to be normal. In alluding to the potential of catastrophe theory to examine—if not to explain—sudden change in such apparently divergent systems, I prefer here only to suggest that catastrophe is a possible logical extension of the analogy.

The discovery of the Mediterranean Messinian "crisis" event derived its conceptual potential from sources that Lyell, given his concept of uniformity, would have been reluctant to acknowledge. He thought that the crust was rigid to a depth of several thousand miles. Thin pockets of igneous, lavalike masses allowed some vertical displacement at the surface. Lateral movement of significant magnitude was not possible. He did admit lateral movement of water masses into continent-sized lagoons; but, by not allowing for much change in their level, he thereby negated the suddenness of the event.

The general historical model of the Messinian event and the subsequent flooding of the basins is far from complete. Most importantly, it lacks a proven tectonic causal mechanism that triggered the crisis reaction. In theory, the tectonic part of this mechanism must include the raising of a threshold in the Iberian Portal that sealed off the Mediterranean and yet allowed for a rapid reopening. The concept of a Pliocene waterfall that eroded away the Gibraltar isthmus may be necessary (albeit a dramatic) consequence of the desiccation model, but it must have been a secondary effect of the normal tectonic evolution of the Rif and Betic sections of the Gibraltar block complex (Hsü, Montadert, et al., 1978, chap. 58). The flexural rigidity along the Betic-Meseta contact of the leading edge of subduction is thought to have caused an orogenic (Alpine-type) phase of development in the Betic (southern) margin, followed by a taphrogenetic (Germanotype) phase, with uplift culminating in the late Miocene. Internally in the Mediterranean, the Italian-Sicilian blocks were in large part depressed during the Miocene and Early Pliocene; through the Early Pleistocene, they were raised rapidly, forming thresholds into the Eastern Mediterranean. The stratigraphic records of events after the "crisis" contain abundant evidence of rapid vertical, and some lateral, displacement; this suggests flexural, rather than isostatic, rebound. In the areas of critical uplift near the thresholds, there are no long sequences of shallow marine deposits of an antiquity that predates the deep Miocene and Pliocene phases. The key is in a generalized subduction model complicated by transcurrent movements in the areas of the thresholds.

If the crustal configurations of the Mediterranean and the consequent distribution of strata are less passive than Lyell thought them to be, so too are the waters of this sea less fluid in structure than he might have imagined.

The water masses of the world ocean systems are stratified and maintained as steady-state inertial structures not unlike those of the crust, but with different origins and histories. If these subsystems seem relatively more dynamic to us than the crustal systems, it is only because of a different time reference. Their very fluidity is an expression of their ability to absorb stress at a more immediate and fundamental level. Energy is conducted laterally through these subsystems at an almost uniform rate. Yet, at the interface between them, the upset transformation phenomena are the same as those in the crust. This change is most easily noticed in the biologic reactions that historically record and synthesize these energy transfers.

The potential for understanding sudden change is increased if we have a model that explains the failure of different systems at different rates according to concepts of tectonic, hydrologic, and biologic subsystems with differing inertial histories and alternative reaction pathways. Rather than being an inhibitor of further inquiry, as Lyell feared, the idea of catastrophe may stimulate further inquiry.

## EVOLUTION AND SUDDEN EVENTS

In the one hundred years since both Lyell and Darwin died, the discovery and description of many many thousands of past species preserved in the fossil record have failed to show the continuous and gradual transition of form between species that was predicted by the original theories of evolution—irrespective of whether the cause was internal (as argued by Lamarck) or external (as posited by Darwin's theory of natural selection).

For a while, the fossil record itself was thought to be an inadequate reflection of the events of the origins of the species, either because there were gaps in what it preserved or because the isolating mechanism of selection always placed the linkages between species outside of our view. But now, with the tremendous amount of data that is available, the seemingly ordinary sudden appearance of new forms must be considered a natural consequence of the inherent mechanism of inheritance. A rising interest in the controls of regulatory genes, which affect both the proportions of organ structures and the structural genes that make possible the presence of these subsystems, may shed new light on this old problem of the sudden appearance of new species.

The search for paleontologic novelty for classificatory or biostratigraphic purposes has tended to put into a digital mode information about changes in the histories of the various preserved forms of life. If the evidence of evolution is, in fact, analogue, then one must admit that these purposes would have been more directly served by simply identifying levels of character state change. The difficulty in finding common occurrences of such a desired linkage system cannot be ascribed entirely to the arbitrariness of species designations or to the lack of a functional understanding of morphology. We must consider the possibility that most of the morphological transformations between ancestral and descendant species are real and sudden transitions between stable intermediate stages.

The study of allometry is an attempt to describe the quantitative transformations in the changing proportions of organ structures that take place during either the growth or the evolution of a species or a group of related species. Beckner (1959) has characterized the morphologist as being like an engineer who has been commissioned to draw a schematic diagram of device circuits, or—as we might here express it—of the reaction pathways of the principal structures of component organ systems. By analogy, the morphologist conceives of this integrated system as a structural frame—a diagram of forces in equilibrium—being deformed by selective causes that are usually unknown, but may be discovered. When regulator genes change the proportions of some parts of the systemic framework in relation to others, the roles and structural balances of these other parts may be altered, sometimes with unpredictable functional consequences. As mechanics and geometry—such as in a skeletal frame—approach identity, the efficiency of function may pass into a stable phase of shape transformation. The fact that variation in organic form often departs from the most efficient design—that is, the design with the simplest possible geometry conceivable—is testimony, not only to the constraints of inheritance and past history, but also to the indeterminant nature of organ structure. Sometimes confusing to the morphologist, this freedom and inertia in design provides the redundancy necessary for survival in a changing selection process.[13]

The change of one species to another through natural selection is an important event in geologic history. The passage of geologic time is marked through the recognition of the successive production of novel morphological features. If this event is likely to be a rapid transition because of the nature of the mechanics of the evolutionary process, confidence in biostratigraphic correlation is increased. We can then assume that the stratigraphic record—which, under the best of conditions, is marked by faunal discontinuities—is more complete than we had assumed in the past; we can also assume that its levels are more isochronous than we had thought. We can begin to concentrate for environmental reconstruction on the stability of structural arrangements of morphology with a sense that some designs, from an engineering standpoint, are better than others. We already know that only a limited number of all possible designs are likely to be stable. If one assumes that either the design or the selection process or both are not gradual—even if they are continuous— then it may be easier to determine why the successful forms persist.

Reconstruction of a phyletic series—that is, a succession of morphologic states that are usually identified as separate species—requires the kinds of inductive inference and assumptions of continuity that are analogous to those used in the construction of value relationships in a mathematical series. One searches for structural continuity where it may not at first be apparent, and then one guesses the rest by studying comparative behavior (Polya, 1954). The mathematical series is intrinsically less ambiguous than a chain of ancestor-descendant relationships, but if similitude can be demonstrated among the conservative homologous structures,[14] then the analogy will not be strained.

A chain can be constructed among possible candidates by measurements or estimates of the magnitude and direction of morphological transformation. If the network of measurements—or observations of form—approaches an estimate of the strain in the actual reaction pathways of functioning structure, the allometric deformation will be the same as the phyletic transformation. This operation also assumes that the data base expresses the natural variation in the populations. The most difficult part of the operation is trying to guess what the fundamental en-

[13] The question of why a complete rotary motion is rare—if not absent—in animal organs (although described, perhaps, in some flagellate organisms) could be explained either as perpetual redundancy or as the complete absence of redundancy. It would be difficult to anticipate an accidental wheel design or to depart gradually from it.

[14] Homology is a term common to both paleontology and mathematics; but this term has unfortunately come to mean the phylogenetic relationship that it is often used to prove. Here it means, as Richard Owen (Owen, 1847) meant, the same part in different systems.

gineering of the fossilized system was (Benson, 1975, 1975a). Here it is helpful to recognize different levels of inertial subsystems, which are (to some degree) what some paleontologists call primitive, derived, and phenotypic character-state variations.[15]

Although allometric deformation is often measured using the grossest estimates, usable first approximations of relationship are still derived. An example would be using the length-height ratios of parts such as equine teeth, gastropod shells, and arthropod carapaces as estimates of the sizes of the same parts (of the same sex, and so on). Compared to measuring the structures, this method is tantamount to measuring the boxes in which the parts came. Gingerich (1979) extrapolates such estimates to the increase in the size of the total animal through time and concludes that species development was continuous. In my opinion, this is straining—but, perhaps, not breaking—the principle of similitude. Even if the transition in size allometry of these structures seems to be continuous, one wonders how it can be justified to infer that the transition between species was continuous.

Most phyletic judgments of continuity in evolution are inferences drawn from stages of structural development and not from a statistical continuum between populations. This either suggests that the record is incomplete (as Darwin concluded in 1859) and that the speciation event almost always occurs among a few individuals on the periphery of—or, isolated from—the main gene pool (the "punctuated equilibrium" of Eldredge and Gould, 1972, 1977), or it suggests that our understanding of adaptive structure is still incomplete and that paleontologists do not understand, especially at the species level, the difference between structural transformation and change in shape. Another possibility, of course, is that the rush to describe novelty in the fossil record has produced a multitude of unreal species, especially among those used for biostratigraphy.

Therefore, any claim that the fossil record testifies to the presence of microcatastrophism as a general means of species evolution requires that structural similitude be well established within comparable populations.

---

[15] This may approach Abel's (1928) attempt to describe Biological Inertia as a descriptive synthesis and convenience. This was an ill-fated attempt to compete with alternative explanations of orthogenesis. However, with increased interest in "control genes"—as distinguished from "character genes"—to explain positive or negative hypertelic allometry (polygenetic "hypertrophy," Rensch, 1960) related to growth-size-correlated requirements the idea that some organs are out of scale with others because of tolerance in natural selection becomes increasingly attractive. It is absurd to believe that the size of every character is in absolute balance with its current functional requirements, although most, perhaps, are. This hypertrophy—which seems so pronounced in mammals, where the imbalance in function is obvious—may be ultimately justified as useful redundancy, or it could as easily cause extinction. In either case, it would be a cause of a sudden event.

Otherwise, someone else can always infer structural continuity at another level of organization. The point of catastrophe theory, which was described in the previous section, is that, when the system as a whole undergoes sudden transformation, subsystems can be slightly deformed and survive.

The fact that so many groups of organisms and their homologous organ subsystem designs have continued repeatedly to produce new replacement species, even at a rate higher than that of the increase of overall diversity, may be testimony to the efficiency of natural selection, but it is also evidence of an inherent redundancy and a fundamental plasticity in morphological structural change. The genetic mechanism may control the production of variation and also provide the inherent stability of proven structural designs, but it is the multiplicity of functional options within these designs that allows the greatest potential for survival.

Organisms are self-perpetuating, self-regulating, memory-storage systems. The information in the memory takes the form of many integrated subsets of reaction pathways that we recognize as structure. The active physiological systems are seldom fossilized, but the passive skeletal ones leave an inertial record of experiments of support, protection, and movement for survival. Each sample of this record—measured against a selected, more distant past—is a reflection of the organism's more recent history of success as a kind of design. A similitude exists between the principle of information of the design and the overall tendency of evolution in its major structural elements. A similitude also exists between the noise and uncertainty, or unpredictability, in the progress of the external world and a structure that may have lost much of its immediate functional need. The combination of these kinds of information—one realized and the other historical—diminishing and potential, provides redundancy and a possible "fail-safe" design for the future.

## CONCLUSION

Lyell believed in a deterministic physical world of nearly uniform motion—a "steady-state" construction—perhaps analogous to Da Vinci's giant astronomical clock in which the hour hand does not move. Lyell agreed with Cuvier that life was incidental to the functioning of this machine. As if Da Vinci's clock were to be decorated by a ceaseless parade of animated figures, the various kinds of life were introduced and replaced on its stage from somewhere off-stage: new actors, with new costumes, but in essentially the same play. Both Cuvier and Lyell thought that these actors existed somewhere else in the machine, but Cuvier's followers in England insisted that the clockmaker often stopped the clock and replaced the actors. Whereas Lyell was fascinated by the continuous motion of the machine, Cuvier pointed out the way in which it occasionally malfunctions, bends its parts, and "does in" some of the prominent actors. Darwin, after studying the machine from more vantage points, noticed the struggle of the actors with the machine and how some became cognizant of its dangers and survived by adjusting their roles. He concluded that the separate roles which the actors appeared to play from time to time were part of the illusion. In fact, the roles were part of the machine and had gradually improved as time progressed.

The analogy of Da Vinci's wonderful astronomical machine is now out of date, but its replacement, "spaceship earth," still has clocklike properties. We know now that the mechanism, as well as the actors, evolves to redesign itself. We also have in natural history—thanks to the historical sense of Darwin, who extrapolated Lyell's faith in continuity—a better sense of history with which to replace a possibly capricious natural theology. Yet, from Cuvier we have learned that the machine is not in perfect transition and that the steady analogue pace of one of its part assemblies can be engaged with the processes of another assembly having a different history and function, to create a digital motion. This is not unnatural; nor is it, if the whole mechanism is known, a violation of trust in knowledge of its design.

If there is a lesson to be learned from all of this, it is that most of the passengers and parts of this evolving and sometimes machinelike spaceship have survived when some part of the mechanism convulsed, although trying to trace the successful among the alternate pathways requires the conception of a scenario that stresses common sense and may bring forth prejudice. When this happens, remember Lyell, who survived to change his mind by exercising humility before judgment; also remember Cuvier, who admonishes us that although there are always reflecting men who try to reduce knowledge to general principles, when this knowledge is confined to a privileged few, nearly all (before our own generation, of course) will endeavor "to convert their intellectual superiority into a means of domination, exaggerating their merit in the eyes of others and disguising the poverty of their knowledge." (Cuvier, English, ed., 1840, p. 49). History says as much about historians as it does about events.


## ACKNOWLEDGMENTS

I would like to thank the people who helped me prepare this essay, discussed its contents, and reviewed the almost final product; those individuals include: Laurie Brennan, my assistant; Carlita Sanford, my secretary;

John C. Greene, visiting historian at the Smithsonian; and my colleagues: William Oliver, Martin Buzas, Richard Cifelli, Richard Grant, Antonio Russo of Modena (Italy), W. A. Berggren of Woods Hole Oceanographic Institution, S. D. Schafersman of Rice University, Frank Whitmore, and John Van Couvering, who invited me to submit this paper for inclusion here.

I am especially indebted to the late Peter C. Sylvester-Bradley, my longtime friend, sometimes coauthor, mentor, professor, and critic. I would have my style, enthusiasm, and motivation be in the future as good as his was. Over the years, our lively discussions of many of the subjects included in this essay taught me about the value of the dialectic, refutation, and the role of the hypothesis. His encouragement was invaluable, his criticism and disagreement even more so. I shall miss his remarks on this paper.

## LITERATURE CITED

Abel, O., 1928. Das biologische tragheitsgesetz. *Biol. Gen.*, 4:1–102.

Adams, C. G., et al., 1977. The Messinian salinity crisis and evidence of Late Miocene eustatic changes in the world ocean. *Nature*, 269:383–386.

Albritton, C. C., Jr. (ed.), 1963. *The Fabric of Geology* (San Francisco, Calif.: Freeman, Cooper & Co.), 372, pp.

Barlow, N., 1958. *The Autobiography of Charles Darwin* (New York: Harcourt, Brace & Co.), 253 pp.

Benson, R. H., 1975. Morphologic stability in ostracoda. *Amer. Pal. Bull.*, 65 (282):13–45.

⸻, 1976a. The evolution of the ostracode costa analysed by "theta-rho difference." In: G. Hartmann (ed.), Evolution of post-paleozioc ostracoda. *Naturwiss. Ver. Hamburg Abh. Verh.*, 18/19:127–139.

⸻, 1976b. Biodynamics of the Messinian salinity crisis. *Palaeogeogr. Palaeoclimatol. Palaeoecol.*, 20:1–70.

⸻. 1976c. Changes in the ostracodes of the Mediterranean with the Messinian salinity crisis. *Palaeogeogr. Palaeoclimatol. Palaeoecol.*, 20:147–170.

⸻, 1978. The paleoecology of the ostracodes in DSDP Leg 42A. *Initial Reports of the Deep Sea Drilling Project* (Washington, D.C.: U. S. Govt. Printing Office), vol. 42, pp. 777–787.

Bibby, C., 1967. *The Essence of T. H. Huxley, Selected Writings* (New York: Macmillan), 246 pp.

Bizon, J. J., and G. Bizon, 1972. Atlas des principaux foraminifères planctoniques du bassin méditerranean, Oligocène à Quaternaire. *Evol. Technique*, pp. 1–316.

Bourdier, F., 1971. Georges Cuvier. In: C. C. Gillispie (ed.), *Dictionary of Scientific Biography*, vol. 3 (New York: Charles Scribner's Sons), pp. 521–528.

Brolsma, M. J., 1975a. Stratigraphic problems concerning the Miocene-Pliocene boundary in Sicily (Italy). *K. Ned. Akad. Wet., Proc. Ser. B*, 78 (2):94–107.

———, 1975b. Lithostratigraphy and foraminiferal assemblages of the Miocene-Pliocene transitional strate of Capo Rossello and Eraclea Minoa (Sicily, Italy). *K. Ned. Akad. Wet., Proc. Ser. B*, 78:1–40.

Buffon, G. L. L., 1802. *Théorie de la terre, histoire naturelle, genéral et particulière, Part 4* (Paris: S. C. Sonnini), 5 pp.

———, 1939. Epochs of the history of the earth. In: Mather, K. F. and S. L. Mason, *Source Book in Geology, 1400–1900* (Cambridge: Harvard University Press), pp. 65–73.

Chamberlin, T. C., 1910. Diastrophism as the ultimate basis of correlation. In: B. Willis and R. D. Salisbury, *Outlines of Geological History* (Chicago,, Ill., Univ. Chicago Press).

Cuvier, G., 1825. *Recherches sur les ossements fossiles* (Paris: A. Belin), 340 pp.

———, 1840. *Cuvier's Animal Kingdom* (London: Wm. S. Orr & Co.), 670 pp.

———, 1939. Theory of the earth. In: Mather, K. F. and S. L. Mason, *Source Book in Geology, 1400–1900* (Cambridge: Harvard University Press), pp. 194–200.

Cuvier, G., and A. Brongniart, 1939. Essai sur la geographie mineralogique des environs de Paris. In: ibid., pp. 194–200.

Darwin, C., 1859. *On the Origin of Species by Means of Natural Selection; or, the Preservation of Favoured Races in the Struggle for Life* (London: John Murray), 511 pp., facsimile reprint, 1964. (Cambridge, Mass.: Harvard Univ. Press).

Drooger, C. W., 1973. The term Messinian; general framework of the colloquium. In: C. W. Drooger (ed.), *Messinian Events in the Mediterranean* (Amsterdam: North-Holland Publishing Co.), pp. 7–9.

Durant, W., 1954. *Our Oriental Heritage* (New York: Simon and Schuster), 1049 pp.

Durant, W., and A. Durant, 1967. *Rousseau and Revolution* (New York: Simon and Schuster), 1091 pp.

Eldridge, N., 1979. Cladism and common sense. In: J. Cracraft and N. Eldridge (eds.), *Phylogenetic Analysis and Paleontology* (New York: Columbia Univ. Press), 233 pp.

Eldridge N., and S. J. Gould, 1972. Punctuated equilibria: an alternative to phyletic gradualism. In: T. J. M. Schopf (ed.), *Models in Paleobiology* (San Francisco, Calif.: Freeman, Cooper & Co.), pp. 82–115.

———, 1977. Evolutionary models and biostratigraphic strategies. In: E. G. Kauffman and J. E. Hazel (eds.), *Concepts and Methods of Biostratigraphy* (Stroudsburg, Penn.: Dowden, Hutchinson & Ross, Inc.), pp. 25–40.

Gingerich, P. D., 1979. The stratophenetic approach to phylogeny reconstuction in vertebrate paleontology. In: J. Cracraft and N. Eldridge (eds.), *Phylogenetic Analysis and Paleontology* (New York: Columbia Univ. Press), pp. 41–77.

Greene, J. C., 1959. *The Death of Adam: Evolution and its Impact on Western Thought* (Ames, Iowa: Iowa State Univ. Press), 388 pp.

Hsü, K. J., M. B. Cita and W. B. F. Ryan. 1973. The origin of the Mediterranean evaporite. *Initial Reports of the Deep Sea Drilling Project* (Washington, D.C.: U.S. Govt. Printing Office), vol. 13, pp. 1203–1231.

Hsü, K. J., et al., 1978. *Initial Reports of the Deep Sea Drilling Project* (Washington, D.C.: U. S. Govt. Printing Office), vol. 42, pt. 1, 1249 pp.

Hubbert, M. K., 1967. Critique of the principle of uniformity. In: C. C. Albritton (ed.), Uniformity and Simplicity. *Geol. Soc. Amer. Spec. Paper,* 89, pp. 33–50.

Hutton, J., 1788. Theory of the earth, or an investigation of the laws observable in the composition, dissolution and restoration of land upon the globe. *Roy. Soc. Edinburgh, Trans.,* 1:209–304.

———, 1795. *Theory of the Earth with Proofs and Illustrations* (Edinburgh: William Creech), vol. 1, 620 pp.; vol. 2, 567 pp.

Huxley, J., 1960. Evolution of life; a panel discussion. In: S. Tax and C. Callender (eds.), *Evolution After Darwin* (Chicago, Ill.: Univ. Chicago Press), vol. 3, pp. 107–143.

King, C., 1893. The age of the earth. *Amer. Jour. Sci.,* 3rd series, 45:1–20.

Koch, H. W., 1978. *Medieval Warfare* (London: Prentice-Hall, Inc.), 256 pp.

Kolata, G. B. 1977. Catastrophe theory: the emperor has no clothes. *Science,* 196:287, 350–351.

Lamarck, J. de M., 1809. *Philosophie Zoologique* (Paris: Dentu), 2 vol.

Lyell, C., 1872. *Principles of Geology* (New York: I. D. Appleton & Co.), 671 pp.

Montenat, C., et al., 1977. Exemple de continuité marine Mio-Pilocene en Mediterranée accidentale: le Bassin de Vera (Cordilleres Betiques-Espagne Meridionale). *Inst. Français Petrole Rev.* 31 (4):613–663.

Moore, R. C., 1970. Stability of the earth's crust. *Geol. Soc. Amer. Bull.,* 81 (5):1285–1323.

Pendered, M. L., 1924. *John Martin, Painter, His Life and Times* (New York: Dutton & Co.), 318 pp.

Playfair, J., 1802. *Illustrations of the Huttonian Theory of the Earth* (Edinburgh: William Creech), 528 pp.

Polya, G., 1954. *Mathematics and Plausible Reasoning. Volume I. Induction and Analogy in Mathematics* (Princeton, N. J.: Princeton Univ. Press), 250 pp.

Popper, K. R., 1957. *The Poverty of Historicism,* 1966 ed. (London: Rutledge & Kegan Paul, Ltd.), 166 pp.

Prévost, C., 1823. De l'importance de l'étude des corps organisés vivants pour la géologie positive. *Soc. Hist. Nat. Paris Mém.,* 1:259–268.

Rensch, B., 1960. *Evolution Above the Species Level* (New York: Columbia Univ. Press), 419 pp.

Rudwick, M. J. S., 1970. Alexandre Brongniart. In: C. C. Gillispie (ed.), *Dictionary of Scientific Biography* (New York: Charles Scribner's Sons), vol. 2, pp. 493–497.

Russell, B., 1946. *History of Western Philosophy,* 1961 ed. (London: George Allen & Unwin, Ltd.), 842 pp.

Ryan, W. B. F., et al., 1973. *Initial Reports of the Deep Sea Drilling Project* (Washington, D. C.: U. S. Govt. Printing Office), vol. 13, 144 pp.

Smith, S., 1960. The origin of the origin. *Adv. Sci.*, 64:391–401.

Thom, R., 1972. *Structural Stability and Morphogenesis* (Reading, Mass.: W. A. Benjamin), 348 pp.

Werner, A. G., 1939. The aqueous origin of basalt. In: Mather and Mason, *Source Book in Geology* (Cambridge: Harvard University Press), pp. 138–139.

Wilson, L. G., 1973. Charles Lyell. In: C. C. Gillispie (ed.), *Dictionary of Scientific Biography* (New York: Charles Scribner's Sons), vol. 7, pp. 563–576.

Whitehead, A. N., 1970. The origins of modern science. In: P. Cloud (ed.), *Adventures in Earth History* (San Francisco, Calif.: Freeman, Cooper & Co.), pp. 13–20.

Zeeman, E. C., 1976. Catastrophe theory. *Sci. Amer.*, 234 (4)65–83.

Chapter 3

# REFLECTIONS ON THE "RARE EVENT"
# AND RELATED CONCEPTS IN GEOLOGY

PETER E. GRETENER

Department of Geology, University of Calgary

## INTRODUCTION

The human lifespan is a poor yardstick by which to assess geological time spans. In particular, episodic events with a low frequency of occurrence tend to be neglected by this approach. "The present is *the* key to the past" (italics mine) is a statement that reflects the arrogant attitude of the newcomer, *Homo sapiens*. Nobody will dispute the fact that the study of the present can reveal many important aspects of the past; but the above proverbial statement implies that *everything* about the past can be learned from looking at the world around us. This is certainly not the case for several reasons, two of which are:

1) Many factors have been changing gradually. Thus, there was no significant land life before the Silurian, or for about 90% of the earth's total history; terrestrial heat flow has decreased by a sizable factor over the life of the earth; the composition of the atmosphere has undergone drastic changes since the dawn of earth history; and so on.

2) The present provides no—or, only vague—indications of the infrequent events that must have happened during the past. The fossil record gives clear evidence of the fact that "unusual" conditions must have prevailed at certain times during the last 600 Ma,[1] particularly at the time boundaries separating the Precambrian from the Cambrian, the Permian from the Triassic, and the Cretaceous from the Tertiary. Events of such low frequency are not amenable to rigid scientific study. They remain in the realm of speculation, since only their effects can be detected in the fossil record.

[1] We may as well adopt the system: 1 ka = $10^3$ years; 1 Ma = $10^6$ years; and 1 Ga = $10^9$ years.

It is this latter aspect of the spasmodic nature of earth history with which this paper is primarily concerned.

In view of the stated objections to the previously cited proverbial statement, it seems more realistic to say: "The study of the present offers a keyhole view of the past." The present discussion, then, centers largely around the *angle of vision* that this keyhole provides.

From Ager's (1973) compilation and from many other papers (see H. Tazieff, 1976, for a good recent reference), the following conclusion is inescapable: episodic processes play an important role in geology. The question that remains is whether or not this role is an overwhelming one.

## THE CONCEPT OF THE "RARE EVENT"

Although the concept of the rare event in geology is trivial and not new, its importance is largely still unappreciated. The concept can be best demonstrated with the 8-dice game (the number is arbitrary). Probability theory predicts (Gretener, 1967) that the chance of throwing 8 sixes in any one trial is about one in two million. Clearly, in a dice game that permits only a few throws, it is a safe bet that 8 sixes will not appear. As the number of throws gets very large, however, 8 sixes are bound to appear. In fact, probability calculus predicts that for six million throws

| # of sixes | Basic Probability | # of throws required for 95% probability of at least one occurence |
|:---:|:---:|:---:|
| 8 | 1: $1.7 \times 10^6$ | $5 \times 10^6$ |
| 7 | 1: 42,000 | $1.2 \times 10^5$ |
| 6 | 1: 2,400 | $7 \times 10^3$ |
| 5 | 1: 200 | 700 |
| 4 | 1: 30 | 100 |
| 3 | 1: 10 | 30 |
| 2 | 1: 4 | 12 |
| 1 | 1: 3 | 8 |
| 0 | 1: 5 | 13 |

Table 3-1. Probabilities Associated with the 8-dice Game*

* Values supplied by M. Nosal, Department of Mathematics, University of Calgary, Calgary, Alberta, Canada.

there is a 95% probability of 8 sixes showing up at least once (see Table 3–1).

In day-to-day life, probabilities of such a low order tend to be called "impossibilities," not because they are physically impossible—there is no physical law that prohibits 8 sixes from turning up—but rather because the probability for one occurrence is so low as to be safely neglected. Yet, in terms of the earth's history, the "game" has been going on for a long time. What is proper—or, at least, permissible—for one evening's dice game no longer holds when the 8-dice club plays night after night.

In a science such as geology, the failure to make a clear distinction between the terms "impossible" and "improbable" is a fatal mistake. For the modern scientist to relegate the improbable into the realm of the impossible constitutes the coward's approach, which is caused by realizing that to accept the rare event as a viable geological agent forever removes a part of geology from what has become the fetish of the twentieth century: rigid scientific analysis.

# THE "RARE EVENT" OR THE 8-DICE GAME

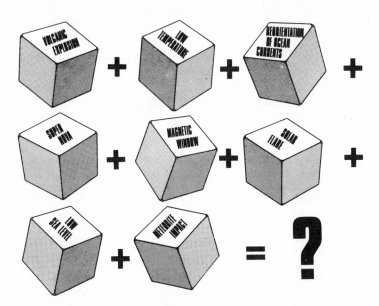

Figure 3–1. This is an illustration and not a model. Listed on the dice are the various factors which have been advanced as being responsible for abrupt faunal changes. The coincidence of several, or all, such factors will be rare and will produce one of these elusive faunal breaks. (See the text for details.)

Many geological papers (see Gretener, 1967) have casually referred to the idea of the rare event in geology, but this idea has never been analyzed in depth. Since the concept is difficult, or impossible, to grasp in a rigid scientific manner, the current tendency is to ignore it. I devoted a short paper (Gretener, 1967) specifically to this much neglected phenomenon and called proper attention to it. The fact that we cannot rationally deal with the idea of the rare event in geology does not mean that it does not exist. In fact, probabilistic expectations force us to anticipate the existence of such phenomena.

One of the better examples of a rare event in geology is that of meteorite impact with the earth (Gretener, 1977), the idea of which has gained strongly in respectability since 1967. There is no longer any doubt that cosmic bodies of various sizes have hit the earth numerous times throughout its history. It also has become evident that the impact of a large meteorite could have global consequences through such secondary features as tsunamis, turbid ocean waters, and/or dust clouds (McLaren, 1970).

In terms of the Phanerozoic faunal breaks, the 8-dice game may take the form suggested in Figure 3-1. Figure 3-1 should not be taken at face value, but rather it should be viewed as an illustration which does not imply that all the factors listed have to coincide in order for a major faunal break to be produced. It is far more likely that the number of decisive factors is smaller and that their probability of occurrence is much less than the one-in-six probability in a simple game of dice. Countless papers have been written on the topic of the faunal breaks; it is unlikely, however, that any explanation will ever be other than reasonable speculation, particularly since the various breaks may have entirely different causes.[2]

---

[2] In Figure 3-1, the labelling on the dice is self-explanatory except for the term "magnetic window." Attention should be called to the fact that it is erroneous to attempt to link magnetic reversals and faunal extinctions—even if the basic cause, increased radiation, is valid—for the following reason: current theory assumes that a reversal is caused by the earth's magnetic field collapsing and then rebuilding in the opposite direction; since there are equal chances during rebuilding for either a reversal or a non-reversal, one must conclude that the occurrences of a zero magnetic field—a magnetic window—are, on the average, twice those of the observed reversals. It is, however, the low, or zero, magnetic field and not the reversal *per se* that has the postulated effect. We cannot, therefore, expect faunal breaks to coincide with magnetic reversals, even if we accept the concept as originally proposed by Uffen (1963). Furthermore, physicists have argued that a current collapse of the earth's magnetic field would lead to only an insignificant increase of radiation at sea level. This same argument, however, may not hold at those times when cosmic radiation is greatly enhanced by either a supernova or an extraordinary solar flare. Increased radiation as a reason for rapid evolution may still be a viable model, but the scenario obviously must be more sophisticated than the ones that have so far been used.

## THE CONCEPT OF PUNCTUATION

This discussion quite obviously centers around the question of the relative importance in geology of episodic processes versus continuous ones. We will later show that it is not realistic to talk about catastrophism in the case of spasmodic processes. The term "punctuation" as proposed by Gould (1977) seems to fit the discontinuous nature of the episodic process without evoking visions of destruction and collapse. Thus, the term appeals as a neutral description of earth history proceeding in fits and spurts.

In my 1967 paper, I included as episodic agents, not only truly rare events, but also such phenomena as major storms, flashfloods, landslides, and others. Clearly, these latter phenomena are rare only by human standards; from a geological point of view, they are regular episodic agents.

In terms of the 8-dice game, one can speculate that the occurrence of 7 sixes will also produce an event, although a less dramatic one than that produced by the occurrence of 8 sixes. Likewise, 6 sixes will not go unnoticed, and so on. As the effects become less shattering, the occurrences will be more frequent. Table 3–1 provides the basic probabilities involved.

From Table 3–1 follows another triviality: Big events are rare, small events are common.

Depending on the frequency of the punctuations, such events on a geological time scale actually simulate continuous processes, because high-frequency punctuation in the geological record is generally not resolvable as such. There is no physical law that in any way restricts frequency of occurrence, and there is no reason to believe that it does not range from hours to gigayears. An attempt to classify discontinuous processes according to their rates of occurrence is shown in Table 3–2 (from Gretener, 1977).

| Types of Discontinuous Processes | # of years required for a 95% probability of at least one occurrence |
|---|---|
| Regular Events | $10^2$ |
| Common Events | $10^3$ |
| Recurrent Events | $10^6$ |
| Occasional Events | $10^8$ |
| Rare Events | $10^9$ |

In Table 3–2, the rationale behind the arbitrary selection of discontinuous processes is purely pragmatic. "Regular Events" figure in the human life (for example, the one hundred year flood of the engineer). "Common Events" must appear in recorded human history, even though usually are no longer perceived as having actually happened but are rather ascribed to the vivid imaginations of the ancient recorders. "Recurrent Events" are important in terms of the fossil stage. "Occasional Events" are the types of events that are responsible for the major faunal breaks; and "Rare Events" are truly just that: events which have occurred, at most, a very few times throughout earth history.

The concept of punctuationism includes all such events. Acceptance of punctuationism assigns a decisive role to the discontinuous process in the earth's history. Because some of the high-frequency processes proceed in front of our eyes and even meteorite impacts—which represent either recurrent or occasional events—take place for all to see, the idea of punctuations as such seems difficult to refute. What remains for the future is the determination of the overall importance of this idea. I suggest that many geological enigmas will vanish once we think in terms of discontinuity, rather than of continuity. Evidence for discontinuity is not restricted to the stratigraphical record (Ager, 1973), but also exists in many fields of geology (Gretener, 1977).

## THE CONCEPT OF THE "HAZE OF THE PAST" AND THE RELATIVITY OF GRADUALISM

The loss of resolution with distance in time and space is, to say the least, a well-known phenomenon; yet, it is also one that is often forgotten. Every human historian must reckon with the "haze of the past": the fact that things get more hazy and fuzzy as one moves toward "the dawn of civilization." This is, of course, even more true for the geologist, who deals with time spans that are almost a million times longer. When comparing the present with the past, to forget about the concept of the haze of the past inevitably leads to false conclusions.

It has been suggested, for example, that magnetic field reversals are more frequent in the recent past than they were in the more distant past, which suggestion implies that the magnetic field is getting more "nervous." One might well ask: "Is this true, or is it simply an apparent effect that results from our sharper perception of the recent past?"

Contemplation of the haze of the past also leads us to accept the concept of the relativity of gradualism. As an example, we can cite the case of isostatic rebound. The last such movement occurred during the last

10,000 years and is still in progress in such places as northern Canada and Scandinavia. Isostatic rebound covers all of man's conscious history and is definitely perceived as a gradual phenomenon. However, if we consider that the earth's skin can completely recover from the unloading of from 1 to 2 kilometers of ice within a period of 15,000 to 20,000 years (Walcott, 1970), we realize that such a response qualifies as instantaneous in any geological period other than the most recent one. With the compression of the time scale that invariably happens with increasing time, gradual phenomena tend to become events. This leads to the question: "What constitutes an event?"

## AN ATTEMPT TO DEFINE THE TERM "EVENT" IN GEOLOGY

The following definition is purely pragmatic and is devoid of any scientific sophistication.

On a time-change plot such as the one in Figure 3–2a (from Gretener, 1977), an event appears as a "jump." Mathematically speaking, we would say that the first derivative of the plot goes to infinity, which means, in turn, that the change is instantaneous or that it proceeds with infinite velocity. This is, of course, nonsense; even meteorite impacts or fault movements have a finite velocity. The jump is, therefore, more apparent than real and is due simply to the compressed nature of the time-axis. When we suitably expand the time, we find that the jump resolves itself into a ramp as shown in Figure 3–2b indicating that the rate of change is high—very much higher than it was before or after—but not infinite.

Thus, one is left with the question: "What qualifies as an event?" A simple answer seems to be the following: the duration of an event occupies no more than 1/100 of the total time span being considered. The rationale behind this profound (sic!) definition is this: on any graph that we might normally draw, 1/100 of the time scale (abscissa) will not be more than the width of a pencil line and will be, therefore, no longer resolvable and will appear as a jump. This definition is also reasonably compatible with the actual time resolution that one can normally achieve. Within the Phanerozoic, this is—at the very best—about 1 *Ma*.

Total time spans considered in geology range from about 1 *Ma* to 1 *Ga*. Consequently, geological processes may have durations ranging from 10 *ka* to 10 *Ma* and still qualify as events. It is immediately apparent that, in using the totally unsuitable yardstick of a human lifespan, only the most extreme punctuations rate as events. Clearly, both the concept of the

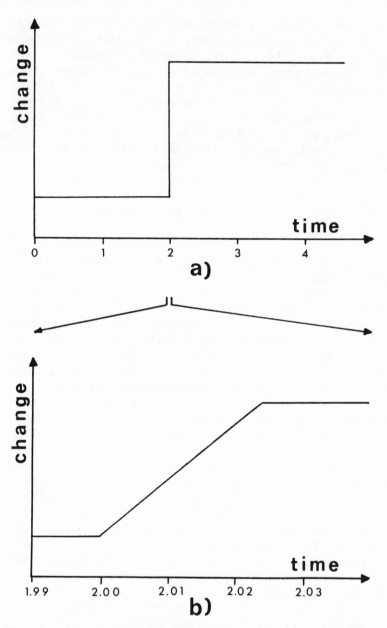

Figure 3–2.  a) An event as it appears on a time-change plot. Due to the compressed nature of the time scale, the instantaneous nature of the event is apparent.

b) Suitable expansion of the time scale demonstrates that events also proceed at finite velocity. (Figure 3–2 from Gretener, 1967.)

event and that of gradualism are relative, since resolution diminishes sharply with increasing time.

The above definition of an event also relates to the rate of occurrence. When a geological event that takes 1 *Ma* is of either the occasional or the rare type (as defined in Table 3–2), it can appear only as shown in Figure 3–2a. For events of a higher frequency, overlapping occurs, and the process becomes continuous. Our definition includes one other qualifier: the years for one occurrence (as shown in Table 3–2) must be at least about 2 orders of magnitude higher than the length of the event.

## "UNIFORMITARIANISM" AND "CATASTROPHISM": EQUALLY MISLEADING TERMS

Endless qualifiers have been attached to the term "uniformitarianism" (for example, Simpson, 1970). There have also been arguments about whether or not the term includes gradualism and actualism. It is quite clear that the term contains the word "uniform," and because geologists are human (at least, we hope that they are), it means precisely that: uniform. Uniformitarianism certainly implies both gradualism and actualism, despite all the arguments to the contrary. Why use a term that is so evidently ill-suited and does not at all portray what geologists think is happening? Punctuation—often, of low frequency—can be demonstrated; merely this fact alone demands that the term "uniformitarianism" be abolished.

And what about the term "catastrophism"? This, too, is an ill-chosen term. The use of this term reveals our anthropomorphic thinking, and the very term itself evokes images of destruction and rubble. It is true that human history does record an unbroken sequence of agony, but human catastrophes are largely self-inflicted. *Homo sapiens'* worst enemy is his own nature. For a reference, one need only consult Toynbee's (1976) *Mankind and Mother Earth.* All of human history rolls past in the 600 pages of this book. One cannot turn many pages without encountering the story of a major population center that got "sacked," not by tsunamis or hurricanes or earthquakes, but rather by human competitors. It is unfair, therefore, to burden mother nature's fast changes with "catastrophism," a term that connotes total disaster. For some species, the fast change may, indeed, spell doom; but for other species, the same change might present a long-awaited opportunity. The fossil record depicts exactly this duplex situation. Ager (1973, p. 100) wrote: "The history of life contains long periods of boredom and short periods of terror." Ager's intent is correct, but his form is overly dramatic, since boredom and terror are largely—if not exclusively—the prerogatives of man. A more correct statement—although a less flamboyant one—would be: The earth's history reveals long periods of tranquillity interrupted by moments of action.

## THE RELEVANCE OF THE PREVIOUS CONSIDERATIONS TO "MODERN MAN'S PREDICAMENT"

Many scholars[3] today feel that nature's first attempt at consciousness must be termed a failure and that the "crown of evolution," like all those species that have gone before, is doomed. There can be little doubt that man's existence is endangered. Even the slightest contemplation of the world around us (not a favorite pastime in our television age) leads to the inevitable conclusion that modern *Homo technicus*—be he either capitalist or communist—is on a collision course with mother nature. This knowledge has so far resulted, however, in much talk and little action. Our inertia can be attributed to many factors, two of which are singled out here:

1) Some people feel that man is, in essence, supernatural and that he can, as such, defy the laws of nature, not only temporarily, but *ad infinitum*. Those who feel this way believe that it only takes another technical "fix" to right the capsizing boat. No natural scientist can accept this notion, and it deserves no further attention.

2) The idea of uniformitarianism is not the exclusive privilege of a few geologists. An arrogant smugness prevails amongst modern societies. This smugness is based on the premise that man has been around for so long that his future is assured, regardless of what he does. This premise is a misconception both of time and of uniformitarianism. But just how long is "so long"? Let us not split hairs; call it a megayear. On the human time scale, this would be 40,000 generations, an incomprehensible amount of time even for human consciousness when one considers that our personal experience seldom exceeds four generations. On the geological time scale, this would be how long? An hour? A minute? It would be exactly the time required for a regular or a common or, possibly, a recurrent event (see Table 3-2). As punctuationists, we cannot possibly share this confidence in the assured continuance of human existence, because it is based on the insufficient yardstick of man's direct experience. There is no natural law that guarantees man's survival. To believe that

---

[3] This note is just a reminder that scientists, humanists, and professors are not necessarily scholars. Very few of the former can, indeed, qualify as scholars. There is, in fact, a dire shortage of this particular subspecies.

there is such a law is to reject personal responsibility, which is, as George Gaylord Simpson (1949) so aptly said, nondelegable. To reject personal responsibility is to reject also what to me is the very essence of human nature: "the capacity to predict the outcome of our own actions that makes us responsible for them and that therefore makes ethical judgment of them both possible and necessary" (Simpson, 1966). To opt out means no less, and no more, than to admit that nature's first attempt at consciousness has, indeed, failed. The total dominance of man has already had drastic effects (Newell, 1963). If these effects continue unchecked for another few decades, life on earth may well be back to square one, with nothing remaining except a marvellous index horizon, as isochronous as a worldwide ash layer. Such a scenario would not be surprising to a punctuationist.

Man's consciousness has reached a level that is evolved enough for him to be able to understand at least the basic concepts of evolution, the most important of which demonstrates to us that there is no uniformity in the process. Only a humble recognition of this fact and an accordingly wise guidance of our destiny can prevent a premature punctuation. Our eventual disappearance, as anticipated by evolutionary theory, is beyond our worry.

## CONCLUSIONS

1) The "rare event" in geology is defined as a spasm, an episode, or a punctuation with such a low rate of occurrence that it has taken place, at most, a few times through all of earth history. No known physical law permits us to relegate such highly improbable events into the realm of the impossible.

2) Acceptance of the rare event as a viable geological agent does not reintroduce the preternatural into geology. It does, however, remove certain—and, possibly, important—happenings from the realm of rigid scientific analysis. A shocking realization for any modern scientist.

3) The concept of the event and that of gradualism are both relative in view of the enormous range of time spans that are considered in geology. What qualifies as an event in early geological history, due to the lack of time resolution, may resolve itself into a gradual process in the Tertiary or the Quaternary.

4) Despite many rescue attempts, the term "uniformitarianism" remains an unfortunate choice. Overwhelming evidence demonstrates that the course of the earth's history is anything but uniform. The term "uniformitarianism" should be abolished because it is misleading.

5) The term "catastrophism" is burdened with the connotation of destruction. It should not be used to designate geological events that simply represent fast changes, which are destructive for some and beneficial for others. The term "spasm," "episode," "event," and "punctuation" are all, in this respect, neutral. "Punctuationism" (Gould, 1977) is the preferred term, because it is both neutral and precise.

6) A belief in uniformitarianism *sensu stricto* is dangerous, insofar as it provides a false sense of security for modern man and encourages him to belittle his current predicament. (This situation is nothing a minor punctuation could not resolve, but hardly to our satisfaction!)

## ACKNOWLEDGMENTS

My thanks to Art Meyerhoff and Tom Oliver, whose criticisms of the manuscript led to significant improvements and to clearer definitions; to Milo Nosal, who provided the statistical computations for the 8-dice game and whose prompt and efficient cooperation was much appreciated; to the organizers of the Woods Hole Symposium, who provided exemplary hospitality and a stimulating environment in which to develop and mature the ideas expressed in this paper; to Mrs. Irish, who typed the manuscript; and, last but not least, to the National Research Council for its courage in continuously supporting such a nonconformist as myself.[4]

## LITERATURE CITED

Ager, D. V., 1973. *The Nature of the Stratigraphical Record* (New York: Wiley & Son), 114 pp.

Gould, S. J., 1977. Evolution's erratic face. *Nat. Hist. Mag.* (May), pp. 12–16.

Gretener, P. E., 1967. Significance of the rare event in geology. *Amer. Assoc. Petrol. Geol. Bull.*, 51 (11):2197–2206.

———, 1977. Continuous versus discontinuous and self-perpetuating versus self-terminating processes. *Catastrophist Geol.*, vol. 2, no. 1, pp. 24–34.

McLaren, D. J., 1970. Presidential address: time, life, and boundaries. *Jour. Pal.*, 44 (5):801–815.

Newell, N. D., 1963. Crises in the history of life. *Sci. Amer.*, 208 (2):76–92.

Simpson, G. G., 1949. *The Meaning of Evolution* (New Haven, Conn.: Yale Univ. Press), 364 pp.

———, 1966. The biological nature of man. *Science*, 152:472–478.

———, 1970. Uniformitarianism. An inquiry into principle, theory, and method in geohistory and biohistory. In: Hecht, M., and W. C. Steere (eds.), *Essays*

---

[4] This courage has, unfortunately, since been lost.

*in Evolution and Genetics in Honor of Theodosius Dobzhansky* (New York: Appleton-Century-Crofts), pp. 43–96.

Tazieff, H., 1976. Horizontal landslides during the 1960 Chile earthquake. *Catastrophist Geol.*, 1 (2):27–32.

Toynbee, A., 1976. *Mankind and Mother Earth* (London: Oxford Univ. Press), 641 pp.

Uffen, R. J., 1963. Influence of the earth's core on the origin and evolution of life. *Nature*, 198:143–144.

Walcott, R. I., 1970. Isostatic response to loading of the crust in Canada. *Canadian Jour. Earth Sci.*, 7 (2):716–734.

Chapter 4

# THE STRATIGRAPHIC CODE AND
# WHAT IT IMPLIES

DEREK V. AGER

University College, Department of Geology, Swansea, Wales

When the above title was wished on me, the first thing I noted was that the adjective "stratigraphic" sounded so ugly in comparison to the equivalent in the Queen's English, "stratigraphical." I, therefore, presumed that "stratigraphic code" (sic) must mean the American (that is, the United States of America) code, which has also come to mean the "International Code" (Hedberg, 1972). The reports that are completed in this code cover every shade of opinion, every method, and every approach; but, nevertheless, they accept the presumption that there is, basically, a holy trinity of stratigraphy: lithostratigraphy, biostratigraphy, and chronostratigraphy. The British stratigraphical committee, of which I was a member for many years, separated the second of these from the trinity (George et al., 1967, 1969; Harland et al., 1972). This committee pointed out that, in effect, biostratigraphy was simply one of many *methods* of correlation (albeit one more used than most) and that, as such, this method was no more worthy of special recognition than, for example, radiometric dating, geomagnetic reversals, climatic changes, volcanic dustfalls, or the "event" stratigraphy of which I am particularly fond. Indeed, some of these aforementioned methods provide—by their suddenness—a more "catastrophic" (and, therefore, a more precise) time-marker. Continental Europeans, especially the French and the Germans, quite logically object to lithostratigraphy on the grounds that it is not stratigraphy at all, but simply the purely descriptive sedimentary petrography that goes first. In fact, some of them have suggested that lithostratigraphy should be called "pre-stratigraphy," because it is what you do before you start on the real business of stratigraphy, which is the correlation of events. This view has been expressed most strongly by Erben (1972) and more diplomatically, in three languages, by Lafitte et al. (1972).

So, we have thus disposed of two parts of the American code's trinity. I, personally, would go a step further and assassinate the third and holiest part—chronostratigraphy—which is, in my view, a bastard hybrid of time and rocks that serves only to obfuscate the whole record and the entire discussion. In my simple way (which, perhaps, only demonstrates that I am simpleminded), I can see only two basic concepts in stratigraphy: one is that the rocks accumulated in one way or another, and the other is that this happened during the passage of time. In other words, there are rocks, which still remain, and there is time, which has passed and can never be recovered. All the rest is semantic confusion.

We need time terms and rock terms and nothing more. We can talk about Devonian granites or Permian orogenies or Cretaceous fossils, just as we can talk about Elizabethan plays or Georgian houses or Victorian morals. I see the need for nothing else and can only hope that chrono-stratigraphy, like the Marxist state, will eventually wither away.

But what bearing does all of this have on "uniformitarianism as a scientific concept"? I believe that it is, psychologically, of very great importance. By chronostratigraphy we imply, perhaps unintentionally, that the whole record of geological time is preserved in the form of rocks. One of the first things we learn in stratigraphy is that the Cretaceous system in rocks is the equivalent of the Cretaceous period in time. But this is almost certainly not true. In my view, the rock record is so episodic in its accumulation and so incomplete in its preservation that there have almost certainly been worldwide gaps during which no sediments were deposited; or, if sediments were deposited during these gaps, they were later eroded away. Even in the deep oceans (which supposedly represent the epitome of continuous, layer-cake sedimentation), Tony Ramsay (1977) has drawn attention to immense lacunae in Caenozoic deposits. To expect otherwise is similar to saying that the whole political history of our two countries is preserved in the *Congressional Record* or *Hansard,* while forgetting the vacations, the weekends, the nights, and even those rare occasions when our professional talkers have less than usual to say.

To pursue this analogy, how can we divide up the political record? We could divide it chronologically into presidential terms of four years, as is done in the American system, regardless of how inconvenient and inappropriate the changeover might be. Alternatively, we could divide it organically by using the more "natural" British method of replacing the prime minister either when he retires or when the majority of his colleagues and/or the people think that he has outlived his usefulness. In fact, we can pursue the analogy further: in the British system, a change in prime minister is not necessarily the same as a change in government (it wasn't, for example, in 1978), just as in the American system a change in president is not necessarily a change in the effective government by a majority in Congress. Thus, in the political record, as in the geological record, we have three systems operating side by side.

The succession of American presidents is directly related to time, in four-year cycles; although complicated by periodic catastrophes, the basic time relationship is unchanged. This, then, is straight chronology. The succession of British prime ministers is directly related to organic changes (whether of the body politic or of their minds) and can be compared, therefore, to organic evolution in the geological record. This, then, is biostratigraphy.

The replacement of British governments—or, I suppose, of majorities in Congress—is a comparatively rare event and may be regarded as a periodic catastrophe (at least by the large proportion of the population who voted for the unsuccessful party). Also rare events and periodic catastrophes are the assassinations and the "Watergates" which complicate the regular cyclic replacement of American presidents.

In fact, if we think about it, the real markers in the historical record are the much more important catastrophes: the revolutions, the wars, the economic depressions, and so on. We think of Winston Churchill as a wartime prime minister—the preeminent index fossil of the war years—but, in fact, he was prime minister neither at the beginning of the Second World War nor at its end. So, it is catastrophes and not index fossils (be they prime ministers or presidents or kings) that define our historical record, and I think that the same is true for the stratigraphical record.

To me, the whole record is catastrophic, not in the old-fashioned apocalyptic sense of Baron Cuvier and the others, but in the sense that only the episodic events—the occasional ones—are preserved for us. Thus, in the western High Atlas of Morocco, I found a thick succession in the Upper Jurassic which seemed to consist wholly of records of severe storms hitting a low-lying coastal landscape of lagoonal deposits (Ager, 1974). Hayes (1967) has shown that there is a 95% chance of a hurricane hitting any particular point on the Gulf Coast once every 3000 years. We know how much sediment is likely to have accumulated there during that time, and we know from a number of recent studies the amount of sediment likely to be shifted by a hurricane. The inevitable conclusion is that, along the Gulf Coast, the rare event—that is, the hurricane—is likely to be the phenomenon most commonly recorded in the future stratigraphical record. Since seeing these things in Morocco, I have—as is always the way in science—been seeing similar situations everywhere.

But let us look at the fossil record, which is of chief concern to me. We always seem to be brainwashed by the presumption that the rocks are full of different kinds of fossils, all of which are busily evolving side by side. We all know that this is just not true. I cannot speak for microfossils, which to me are only so much sediment (and behave like it), but how many evolving lineages do any of us know in any one place? I have discussed this before (Ager, 1976), but I think it needs restating. My first predecessor at Swansea was Sir Arthur Trueman, who worked out in the local Jurassic cliffs the classic evolutionary story of *"Ostrea"—Gryphaea*. This became so popular in the paleontological literature that I once called it the *Drosophila* of paleontology. It was proven statistically; and then, later, it was disproven statistically and entirely rejected.

When I followed in the Swansea chair nearly half a century later, I could hardly avoid going out to look at the classic sections; when I did, it seemed obvious to me, as a paleoecologist (with the benefit of hindsight), that they did not record an evolutionary story, but only an ecological one. The *"Ostrea,"* or *Liostrea*, was clearly adapted (like modern oysters) to cementing itself onto hard surfaces, in this case, onto chert clasts in the conglomeratic basal beds of the Jurassic. The catastrophe that had happened to these oysters was the end of the supply of such clasts. So, *Liostrea* came to a nasty muddy end and was replaced by *Gryphaea*, which was clearly adapted to living on a soft seafloor.

Every paleontological succession of which I know is basically similar. In the Middle Jurassic of the Jura Mountains in southeast France, I recorded ten or more different assemblages within 24 meters of strata (Ager, 1963, p. 302). In a recent paper, Walker and Walker (1976) recorded a fascinating succession of brachiopod assemblages in the Ordovician of Tennessee. Each fossiliferous level had gone through a little ecological sere before being smothered in sediment that had probably been generated by a storm, and the area was then recolonized from elsewhere. The vast majority of abundant organic accumulations represent, not unusually prolific life at one particular place, but slow and often mechanical concentration over a long period, though one that is always short in geological terms.

To illustrate my point, let me take just one paleontological section that I know well: the late-Devonian sediments described by Peigi Wallace (1969) in the Carrière du Bois near Ferques in northern France. This section may be summarized as follows in upward succession:

1) Lamellar stromatoporoid assemblage (3 beds)   2.50 meters
2) Massive stromatoporoid assemblage (1 bed)   .50 meters
3) Rugose coral assemblage (2 beds)   .60 meters
4) Atrypid-spiriferid assemblage (1 bed)   .20 meters
5) Stropheodontid assemblage (6 beds)   6.50 meters
6) Productellid-spiriferid assemblage (9 beds)   12.00 meters
7) Large brachiopod-coral assemblage (9 beds)   8.00 meters

These beds are of Frasnian age and represent a time interval of, perhaps, some five million years. "Equivalent" strata elsewhere are vastly thicker (for example, the 10,000 feet of Frasnian strata in Arctic Canada). This simple observation alone would be enough to demonstrate the vast gaps that must exist in the Carrière du Bois succession, even if the extremely fossiliferous nature of all the beds were to be taken as evidence of very slow deposition. Some 30 meters of sediment here in five million years averages out to only two-thirds of a millimeter of sediment per thousand years. Clearly, this succession must have been episodic.

Each bed is reasonably consistent within itself, both in lithology and in fossil content, and I submit that each represents a combination of environmental and geophysical factors that, for a very short time, permitted the accumulation and preservation of the sediments and organisms that happened to be around in that area. Each bed, as well as each assemblage represents, in effect, a little catastrophe that led to preservation and was separated by other catastrophes about which we know nothing. Bed 7 represents an abrupt change in sedimentary environment accompanied by a corresponding change, not particularly in the kind of fossils present, but in their average size. This is a bigger catastrophe.

Higher up again, one comes to another formation—the Schistes de Fiennes—which represents a completely different state of affairs. The sedimentary setting is different, the fossils are completely different, and the rocks at which we look were laid down in a different environment. A deeper, muddier sea swept catastrophically across the shell banks. Higher again, shallow water conditions return with the Gres de Ste. Godeleine, but we are now in a new world—the Famennian—where something fundamental happened, not only in this one inlier, but also in Famennian rocks all over the world. Usually (though not always) there is evidence at this time of a major paleogeographical change: all the atrypid and pentamerid brachiopods disappear, as do nearly all of the corals, stromatoporoids, trilobites, tentaculitids, orthids, and stropheodontids. This change is far more important than the one at the end of the Famennian, which happens to mark the boundary between the Devonian and the Carboniferous.

In other words, we are looking at evidence of a whole series of little catastrophes, followed by a bigger one, then by a still bigger one, and then by a very big one indeed. But how does stratigraphical nomenclature handle these data? Lithostratigraphically, Beds 1 to 7 are lumped together into one formation, the Calcaire de Ferques; biostratigraphically, they are lumped together with the formation above them into the Biozone of *Manticoceras;* and chronostratigraphically, they are lumped together as the Frasnian Stage. All of these approaches ignore the essentially episodic nature of the record and, thereby, give a false solidity to what is nothing more than a shadow.

The fact that the Frasnian-Famennian boundary coincides with a major catastrophe is almost coincidental. You would hardly notice it in the nomenclature. Lithostratigraphically, they are often lumped to-gether—for example, in the classic sections in Windyana Gorge in west-ern Australia, where we now know that the corals and stromatoporoids of the famous reefs are replaced at the critical horizon by algal stromato-lites. The same is true in the splendid escarpment of Chinaman's Leap near Banff, Alberta, Canada. Biostratigraphically, this escarpment is only one goniatite zone following another. Chronostratigraphically, most peo-ple are content simply to call the entire assemblage the Upper Devonian.

If we were really to look for "natural breaks" in the organic succession of earth history, we would certainly put one of the major boundaries be-tween the Frasnian and the Famennian. This boundary is, in many ways, a much better one than that between the Devonian and the Carbo-niferous; it is somewhat better than the one between the Lower Paleozoic and the Upper Paleozoic; and it could even be said that it was better than the one between the Paleozoic and the Mesozoic.

And people, especially in the Soviet Union, are still expecting natural breaks to define their stratigraphy. Recently, for example, Meyen dis-cussed three alternatives for defining "stratons," the stratigraphic units of the international stratigraphic scale:

1. The stratons . . . should be natural units reflecting natural stages in the evolution of the earth and its organic life. . . .
2. The stratons . . . are natural only insofar as they reflect natural stages in the evolution of organic life, but not historical-geologic events. . . .
3. The stratons . . . are natural only insofar as they are based on the natural stages of development of the stratotype regions . . . (Meyen, 1974).

If the history of the earth in physical terms—either global or local—will not serve us, then can we not surely say that the evolution of organic life, as still our best method of correlation, will ultimately provide the synchronous events that will solve all our problems? Now that we know so much more about almost every major group in the organic world throughout the greater part of its history, is this not true? What do we do to piece together those evolutionary lineages that we feel so certain really exist? (I include the last phrase to protect myself from California funda-mentalists who have already clasped to their bosoms a most-unwilling me!) With the vast resources of paleontological research now available to us, we can wander the world—preferably in person but, failing that, also in our libraries and museums—and slowly piece together the complex pattern of whatever fossil group happens to appeal to us. The result will be the same: we do not find tall evolutionary redwoods reaching higher and higher with occasional branches; rather, we find a dense thicket of short, spiky branches that lead nowhere. This was certainly the conclusion I came to after I spent many years studying the Mesozoic Rhynchonellida (Ager, Childs, and Pearson, 1972). Every little evolutionary branch very soon came to an end and was replaced ecologically by another branch of another genus or another family. As far as I can see, work on every other phyla has yielded that same conclusion. Sometimes a main stem does persevere for a long period of time, but it almost invariably does so with very little change, while the real changes are taking place in the outermost branches. But this subject is not my primary concern in this paper. Suffice it to say that I find every stratigraphical succession to be nothing more than a series of sudden spurts and replacements that are equally catastrophic for the organisms involved.

How, then, does "the code" deal with all these problems? The American code, for all of its words, is singularly ambiguous about everything. It recommends type sections for lithostratigraphy (that is easy enough), but for the vital time divisions, it tries to please everyone by recommending stratotypes and markers without either making them firm requirements or indicating how they should be used. On the continent of Europe, from France to the Soviet Union, there is a fixation with the stratotype, which is the absolute criterion for the stage. Further, stratotypes that happen to have been named in a particular motherland arouse all sorts of nationalist sentiments.

I have seen too many of these stratotypes to have much faith in them. Thus, to see the type section of the Aquitanian, which defines the limit between the Paleogene and the Neogene, one dons rubber boots, wades down a stream, and pushes aside the undergrowth to see a tiny exposure of atypical strata. The top stage of the Jurassic is disputed between the Portlandian, the Tithonian, and the Volgian. Many of us were quite happy to abandon the Portlandian (which was founded—like so many of the Jurassic stages—by a Frenchman, Alcide d'Orbigny, on English localities, although he had never visited Britain). The Tithonian made the traditionalists unhappy because it was named after a mythical god (significantly, the husband of Tethys) and thus had no obvious stratotype. I was once at a meeting where someone seriously suggested that one of the streams in the Ardèche should be renamed the "Tithon," so that the stage could have a home. The Volgian stratotype, situated on the Volga not far from Lenin's birthplace, is particularly valued by the Russians; but it contains at least one—and probably several—major breaks, which probably equate with major parts of the succession elsewhere.

What, then, do stratotypes represent? I once asked a Frenchman to define a stratotype. He replied simply: "A fragment of the true cross." This seems to me an excellent, if blasphemous, definition, but not one that really helps us very much. All we know is that some tiny fragment of the strata laid down during a particular period of time is present at a particular place. But we must have an international language with which to communicate.

The British code tried to go beyond this; indeed, some of us would have liked it to have gone further than it did. First of all, it said, in effect, that stratigraphical columns are invented by man and not handed down to us by Moses or Chairman Mao. Although many of us—perhaps, most— would agree that there were major catastrophic events in the history of the earth, I think a few of us would suggest that there are enough of them on a worldwide scale to build a stratigraphy. We must, therefore, be pragmatic and adopt a Jeremy Bentham-type approach—not the greatest good for the greatest number, but rather the greatest convenience for the greatest number. This is where the crunch comes, for it means that geologists must drop fixed ideas and feelings such as those of national pride. The Ruritanians must stop using the term "Zendarian," even though it was named after one of their most famous castles. Geologists must abandon cherished preconceptions; the Russians, by way of example, must relinquish having the beginning of the Callovian as the beginning of the Late Jurassic, despite the fact that this might make the mapping of their vast country more difficult. Paleontologists must accept the notion that the most marked and visible changes in their faunas or floras are not necessarily the best points at which to place major stratigraphical boundaries. (In fact, these points are likely to be some of the worst places for major boundaries.)

The British stratigraphical committee was being, perhaps, unreasonably idealistic when it advised, in effect, that the best level at which to place a boundary is, paradoxically, the level at which it is least obvious. Even the international working part that attempted to mark the base of the Devonian by hammering in an imaginary "golden spike" at Klonk in Czechoslovakia became frightened at the last moment. Instead of driving in the spike for eternity, they defined the base of the Devonian as that point at which *Monograptus ultimus* first appears in a particular, thin limestone band. This allows some future paleontologists, with his own interpretation of the involved species, the liberty to move the spike as he sees fit. Any deviation could only be a matter of centimeters, but the committee's qualification still spoiled a splendid principle. At the last moment, as it were, the committee surrendered to the idea that there is an absolute truth in the beginning of the Devonian. It is as if the Devonian period began with a flash of heavenly lightning at 6:15 in the evening on October 26, 405,000,000 BP. This is nonsense, and we know it; yet, the idea is still there subconsciously, even in those of us who are the most sceptical.

Ultimately, there are not enough big catastrophes to use as our reference points, and there are too many little ones; so, we must be arbitrary and seek only to cause the least possible confusion and inconvenience.

I have long supported the "golden spike" approach in principle, and I will continue to do so; nevertheless, I wonder if it might not—in the long run—be more logical to adopt a more pragmatic and independent arbiter: namely, time itself. It seems to me that uniformitarianism has become what I have called "catastrophic uniformitarianism" (Ager, 1973) and that catastrophes are so commonplace that they have become stepping stones rather than milestones. On that basis, I cannot see any argument against our eventually switching to a purely chronometric scale.

We can look at the example of archaeologists who are now—thanks to new techniques—fitting together the histories of civilizations as far apart as China and Peru. They are now—thanks to the bristlecone pine—revising the early history of Europe. One result is that the Orkneys and Malta now have the distinction, I believe, of containing the oldest buildings on our continent. I suspect that the reason for this was that the old red sandstone of the one area and the Miocene limestone of the other were both so easily worked that the blocks were not worth stealing for later buildings. But this only illustrates my point that, although opinions and techniques change, we will sooner or later have to come back to a simple date.

After all, why do we make such a fuss about the last 15% of the earth's history? And why aren't we concerned with the last few thousand years of that? Our Precambrian colleagues, with their vastly greater time-scale, already think in years rather than in historical divisions. Is it not logical that we should aim at eventually doing just that and adapt our stratigraphical codes accordingly? Our techniques cannot cope at the moment, but it is a purely pessimistic presumption to think that they will never be able to cope, especially if we bear in mind the advances that have been made over the past few years in isotope chemistry.

Let us say, then, that the Devonian period began on January 1, 405,000,000 BP, at one minute past midnight—Greenwich Mean Time, of course!

## LITERATURE CITED

Ager, D. V., 1963. *Principles of Paleoecology* (New York: McGraw-Hill), 371 pp.
————, 1973. *The Nature of the Stratigraphical Record* (London: Macmillans/Halstead Press), 114 pp.
————, 1974. Storm deposits in the Jurassic of the Moroccan High Atlas. *Palaeogeogr. Palaeoclimatol. Palaeoecol.,* 15:83–93.
————, 1976. The nature of the fossil record. *Geol. Assoc. Proc.,* 87:131–160.
Ager, D. V., A. Childs, and D. A. B. Pearson, 1972. The evolution of the Mesozoic Rhynchonellida. *Geobios.,* 5:157–235.
Ebren, H. K., 1972. Replies to opposing statements. *Newslett. Stratigr.,* 2:79–95.
George, T. N., et al., 1967. Report of the stratigraphical code subcommittee. *Geol. Soc. London Proc.,* 1638:75–87.
————, 1969. Recommendations on stratigraphical usage. *Geol. Soc. London Proc.,* 1656:139–166.
Harland, W. B., et al., 1972. A concise guide to stratigraphical procedure. *Geol. Soc. London Jour.,* 128:295–305.
Hayes, M. O., 1967. Hurricanes as geological agents: case studies of Hurricane Carla, 1961, and Cindy, 1963. *Univ. Texas Bur. Econ. Geol. Rept. Invest.,* 61, 56 pp.
Hedberg, H. D. (ed.), 1972. Introduction to an international guide to stratigraphic classification, terminology, and usage. *Lethaia,* 5 (3):283–295.
Laffitte, R., et al., 1972. Some international agreement on essentials of stratigraphy. *Geol. Mag.,* 109:1–81.
Meyen, S. V., 1974. The concepts of "naturalness" and synchroneity in stratigraphy. Akad. Nauk SSSR. Izvest. Ser. Geol., 5:79–90. (English trans., 1976: *Int. Geol. Rev.,* 18:80–88.)
Ramsay, A. T. S., 1977. Sedimentological clues to palaeo-oceanography. In: A. T. S. Ramsay (ed.), *Oceanic Micropalaeontology* (New York: Academic Press), pp. 1371–1453.
Walker, K. R., and W. C. Parker, 1976. Population structure of a pioneer and a later stage species in an Ordovician ecological succession. *Paleobiology,* 2:191–201.
Wallace, P., 1969. The sedimentology and palaeoecology of the Devonian of the Ferques inlier, northern France. *Geol. Soc. London Quart. Jour.,* 125:83–124.

Chapter 5

# STATISTICAL SEDIMENTATION AND MAGNETIC POLARITY STRATIGRAPHY

CHARLES R. DENHAM

Department of Geology and Geophysics,
Woods Hole Oceanographic Institution

## INTRODUCTION

Magnetic polarity zonation is now being widely used in the study of fossil vertebrate chronology within continental sedimentary deposits (for example, Johnson et al., 1975; Butler et al., 1977). Hiatuses and variable rates of sediment accumulation can easily disrupt the expected magnetic polarity pattern and lead to either uncertain or invalid stratigraphic correlations. In the study of continental sediments, statistical sedimentation is a concept that can usefully provide insights into the resolution of the magnetostratigraphic/biostratigraphic method.

We will describe in statistical terms the meaning of a record in which fewer polarity intervals are observed than were expected. We assume the sedimentation to be a simple two-state—ON again, OFF again—model. Technically, it is an alternating renewal process (Cox, 1962), characterized by the ON parameter, $T_1$ (the average duration of an interval of net-accumulation), and the OFF parameter, $T_2$ (the average duration of an hiatus, whether erosional or non-sedimentary). We assume that the geomagnetic polarity intervals have an average duration of $T_0$, due to a Poisson reversal process (Cox, 1969).

## POISSON PROCESSES

Background on the various formulas used in this report can be found in D. R. Cox's *Renewal Theory* (1962). In a simple Poisson process, with parameter T, the probability density function for the length t between renewals (that is, polarity reversals or ON episodes or OFF episodes is:

$$PDF(t) = \frac{1}{T} e^{-t/T} \tag{1}$$

The probability of n renewals within time t is:

$$P(n,t) = \frac{(t/T)^n e^{-t/T}}{n!} \tag{2}$$

The expected number of renewals and its variance are both t/T.

Formulas such as these can be combined to describe the features of a statistical magnetostratigraphical record. Computer simulations of the record are easily accomplished by a Markov chain (for example, Harbaugh and Bonham-Carter, 1970) and may provide helpful illustrations for some, but we do not include any here.

## POISSON MAGNETOSTRATIGRAPHY

*A) Chance of recording part of a given interval*   The probability that a polarity interval of length t will be at least in part recorded, given $T_1$ (ON) and $T_2$ (OFF), is:

$$P_3(t) = 1 - \frac{T_1}{T_1 + T_2} e^{-t/T_2} \tag{3}$$

This formula derives from the chance that sedimentation is ON at the beginning of interval t, or that it is OFF but will turn ON sometime therein.

*B) Chance of recording the entire interval*

$$P_4(t) = \frac{T_1}{T_1 + T_2} e^{-t/T_1} \tag{4}$$

This formula describes the probability that sedimentation is ON at the beginning of interval t and will not run OFF until time t has elapsed.

*C) Chance of recording both polarity transitions*   The probability that sedimentation will be ON at both the beginning and the end of time interval t is:

$$P_5(t) = (\frac{T_1}{T_1 + T_2}) \ (\frac{T_1}{T_1 + T_2} + \frac{T_2}{T_1 + T_2} e^{-/T_1} e^{-/T_2}) \tag{5}$$

For small t, the limit is just $T_1/(T_1 + T_2)$. For large t, it is $(T_1/(T_1 + T_2))^2$, as expected. The first fraction in (5) is the probability of ON at an arbitrary time, given no other information. Its complement, $T_2/(T_1 + T_2)$, is the chance of OFF under the same circumstances.

*D) Chance of proper neighbors*   Because of hiatuses, a particular polarity interval may not be bounded by sediment from its closest neighbors in the magnetic timescale. The chance that interval $t_i$ exists and is properly bounded is:

$$P_6(t_i) = P_3(t_{i-1}) \times P_3(t_i) \times P_3(t_{i+1}) \qquad (6)$$

The right-hand side is calculated by (3).

*E) Completeness of the recorded polarity sequence*   Geomagnetic reversals occur randomly in time. The process is described most accurately by

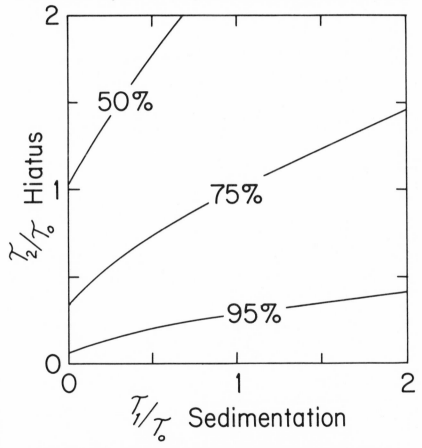

Figure 5–1. A graph of text Equation (7), showing the fraction of geomagnetic polarity intervals that would be recorded after infinite time, for various values of $T_0$ (the average length of a polarity interval), $T_1$ (the average duration of an episode of net-accumulation), and $T_2$ (the average duration of an hiatus).

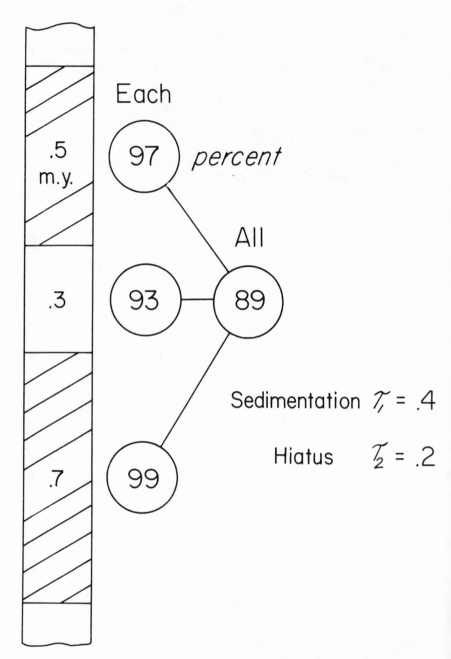

Figure 5-2. An example of the chance of recording a polarity interval and its nearest neighbors in the polarity timescale. Details are described in the text.

a low-order gamma distribution, but the Poisson model can be used for a good approximation (Cox, 1969; Phillips, 1977). Given $T_0$ and $T_1$ and $T_2$, the fraction of geomagnetic polarity intervals that would be recorded in the long term is:

$$F = 1 - ( \frac{T_2}{T_0 + T_2} ) \ ( \frac{T_2}{T_1 + T_2} ) \qquad (7)$$

Contours of this expression are shown in Figure 5-1.

*F) An Example*   A short polarity timescale is shown in Figure 5-2. The three successive polarity intervals are durations of 0.7 myr, 0.3 myr, and 0.5 myr. For the purpose of illustration, we have assumed that $T_0 = 0.5$ myr, $T_1 = 0.4$ myr, and $T_2 = 0.2$ myr.

The chance that the middle polarity interval is at least in the part recorded is 93% (Equation 3); the chance that it is complete is 31% (Equation 4); the chance that both of its bounds are actual polarity transitions is 47% (Equation 5); and the chance that it is properly bounded is 89% (Equation 6). In the long term, about 93% of polarity intervals such as these would be represented in the magnetostratigraphic recording (Equation 7).

## APPLICATION TO THE OLIGOCENE EPOCH

The geomagnetic polarity sequence that is represented in Figure 5-3 by marine magnetic anomalies #6c through #13 is assigned to the Oligocene and lower Miocene epochs. The twenty-five inferred polarity intervals in this sequence that we are studying can be found in mammal-bearing strata of the White River and Arikaree groups in the northern Great Plains. Identification of the magnetic anomaly polarity pattern in these rocks would be extremely helpful for accurately dating the fossils. In addition, if the polarity patterns can be recognized, the presence of radiometrically datable ashes would make the area very attractive as a mid-Tertiary calibration point for the timescale based on marine magnetic lineations.

We hope to find the long reversed-polarity interval that corresponds in Figure 5-3 to the interval between anomalies #12 and #13, as well as the one that corresponds to the interval between anomalies #6c and #7. We do not, however, expect to observe all twenty-three of the other intervening polarity reversals. What can we say about the strata, if the magnetic record is, indeed, not complete? Let us calculate some probabilities to answer questions such as those raised above.

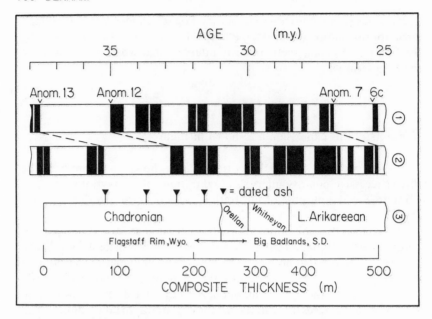

Figure 5-3. A comparison of geomagnetic polarity timescales and the Oligocene-lower-Miocene fossil mammal stages.
1) is the Heirtzler et al. (1968) magnetic timescale.
2) is the LeBrecque et al. (1977) magnetic timescale.
3) is the Tedford (1975) land-mammal chronology.
The triangles (▼) indicate the approximate stratigraphic positions of K/Ar dated ashes at Flagstaff Rim, Wyoming (Everndon et al., 1964).
The composite thickness of sediments of the White River group at Flagstaff Rim and the Arikaree group at the Big Badlands is shown at the bottom of the diagram.

The average number of successes for our set of polarity intervals is simply the sum of their individual probabilities:

$$\mu = \sum_{i=1}^{N} P_i \qquad (8)$$

The variance is:

$$\sigma^2 = \sum_{i=1}^{N} P_i(1 - P_i) \qquad (9)$$

The minimum number of successes at 95% confidence is $\mu - 1.65\,\sigma$. These formulas hold because the number of intervals is large, which assures us, through the central-limit theorem, that both the standard binomial treatment and the normal distribution approximation are accurate.

For various combinations of $T_1$ and $T_2$ up to 1 myr, the minima are contoured in Figures 5–4a through 5–4d as follows:

A) *Number of recorded polarity intervals* Figure 5–4a illustrates the minimum number of polarity intervals which are recorded at 95% confidence. With small $T_2$ (OFF), most of the expected twenty-five intervals appear at least in part. We hope that permanent contributions of sediment were made more frequently than about once every 0.2 myr on the average. Additionally, since the long reversed intervals at the beginning and the end of the sequence are so important as markers, we hope that they did not coincide with rare long episodes of erosion or non-deposition. Although the chance of such an occurrence is statistically unlikely, it is not impossible. The possibility of the rare destructive event always exists to cloud the convenient, average picture beyond recognition.

B) *Number of complete intervals recorded* In Figure 5–4b, it is seen that the number of intervals recorded in their entirety is generally low. Recording part of an interval is far easier than recording all of it.

C) *Number of actual polarity transitions recorded* The number of transition-bounded intervals (see Figure 5–4c) is low, except for small values of $T_2$ (OFF). With hiatuses averaging several tens of thousands of years, most of the transitions appear. As $T_2$ grows, however, most of the observed polarity boundaries represent hiatuses. If $T_2$ is large, then the best way to find an hiatus in the strata is to look for a polarity boundary.

D) *Number of properly bounded intervals* Figure 5–4d shows that for small values of $T_2$ (OFF), almost all of the observed intervals are properly bounded. The number diminishes rapidly as $T_2$ rises. Substantial uncertainties result if $T_2$ is of an order from 0.2 myr to 0.3 myr, because so many of the individual intervals that are seen in the strata represent more than one interval in the magnetic timescale.

DISCUSSION

Small values of $T_2$ (OFF) cause small losses in the number of recorded polarity intervals, but they cause large losses in the number of those intervals that are either complete or bounded by transitions or bounded by their proper neighbors. Figures 4a through 4d show how similar results can be obtained from a wide range of $(T_1, T_2)$ combinations.

If only half of the expected number of polarity intervals were observed, then virtually none of them would be either complete or bounded by

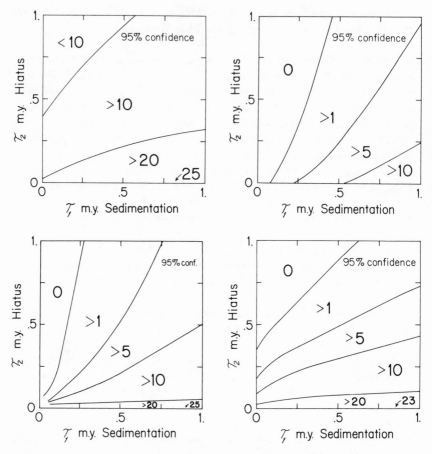

Figure 5-4. The statistical interaction of the marine magnetic anomaly sequence (represented by #6c through #13 in Figure 5–3) with the alternating-renewal sedimentation process described in the text. The time-constants are $T_1$ (ON) and $T_2$ (OFF), the same as in Figure 5–1. The LaBrecque et al. (1977) timescale was used in these calculations.

a) The number of polarity intervals that would be at least in part recorded, at 95% confidence (Equation 3).

b) The number of completely recorded polarity intervals (Equation 4).

c) The number of intervals in which both polarity transitions would be recorded (Equation 5).

d) The number of intervals that would be recorded along with their nearest neighbors in the geomagnetic polarity timescale, each being recorded at least in part (Equation 6).

transitions or bounded by their adjacent neighbors in the geomagnetic polarity sequence. That is to say, the patterns ought not to be recognizable, except in cases where unusually long polarity intervals have occurred (for example, in Figure 5-3, the interval between anomalies #6c and #7, as well as the one between anomalies #12 and #13). This is why long polarity intervals are so important. They isolate the interval in between them, so that inferences can be made about the completeness of the stratigraphic record contained therein. The number of observed polarity intervals is a welcome clue to how well the strata represent the time that they span.

The number of reversals can be used to estimate the time span, as was done by Kono (1973) for lava sequence. For example, the 50 and 60 meters of sediment assigned to the Orellan mammal stage (Figure 5-3) was thought by Clark et al. (1973) to span only about 10,000 years, rather than 1 myr as had been thought by others. Clark based his estimate mainly on the number of distinct sedimentary layers and on the myth of the one-hundred year flood. He also noted the apparent absence of evolutionary change in several fossil mammalian orders, as well as the lack of erosional unconformities in the section.

Our reconnaissance work has revealed about four polarity intervals in the Orellan sediments at three localities, with differing patterns in the western and eastern Badlands (see Figure 5-5). If $T_0$ were 0.5 myr, then at 95% confidence, the unit must have spanned at least 0.77 myr. On the other hand, if the unit had accumulated in less than 26,000 years, then no reversals at all would be expected at 95% confidence. Thus, from a statistical point of view, Tedford's (1975) assignment of 0.75 myr to the Orellan stage is compatible with the paleomagnetism, whereas Clark's estimation of 10,000 years is not. This is a useful—albeit a somewhat extreme—example of how the number of polarity intervals can help to resolve disputes between various geological disciplines, in which estimates of durations and rates are often conjectural.

Many of the actual geomagnetic polarity reversals may not appear in the standard magnetic timescale. In a mortality study, Aldridge and Jacobs (1974) showed that as many as 50% of the actual polarity intervals may be missing, most likely because their short durations prevent their discovery in the marine magnetic lineations. If this is so, then the present estimates of $T_0$ are minima. Hence, estimates on the completeness of the stratigraphic record from the number of observed polarity zones are maxima.

The versatile Poisson statistical process, as the best-guess model for sedimentation when no other information is available, may help to clarify thinking on the possible impact of hiatuses in magnetostratigraphic stud-

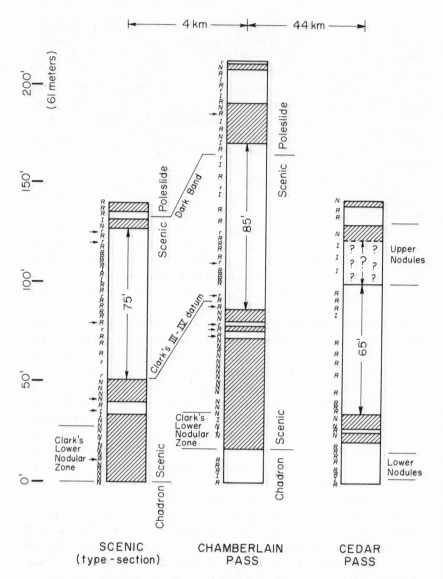

Figure 5-5. Reconnaissance magnetic polarity stratigraphy observed at three localities of the Scenic Member (Orellan mammal stage) of the Brule Formation in the South Dakota Badlands. The polarity of each specimen is annotated at the left of each column. The most reliable determinations are labeled N and R; less certain polarities are labeled n and r; and an arrow denotes the trend during progressive AF and/or thermal demagnetization. The thick normal and reversed polarity zones are antiparallel at 95% confidence. For a few samples, the polarity was indeterminate (I).

ies. Many problems—such as, the likely proximity of a particular polarity boundary to its expected position in the stratigraphic column—can be explored at, say, 95% confidence. Magnetic reversals occurred frequently in the past, and their record can be used advantageously for learning about the duration and continuity of sedimentation in cases where the sedimentological evidence itself may be either lacking or undiagnostic.

## ACKNOWLEDGMENTS

I thank Christopher Tapscott, Alan Chave, Hans Schouten, John Van Couvering, and Bill Berggren for dicussions. The work is supported by National Science Foundation grant #76–13359 (Geology).

## LITERATURE CITED

Aldridge, K. D., and J. A. Jacobs, 1974. Mortality curves for normal and reversed polarity intervals in the Earth's magnetic field. *Jour. Geophys. Res.*, 79:4944–4947.

Butler, R. F., E. H. Lindsay, L. L. Jacobs, and N. M. Johnson, 1977. Magnetostratigraphy of the Cretaceous-Tertiary boundary in the San Juan Basin, New Mexico. *Nature*, 267:318–323.

Clark, J., J. R. Beerbower, and K. K. Kietzke, 1973. Oligocene sedimentation, stratigraphy, paleoecology, and paleoclimatology in the Big Badlands of South Dakota. *Chicago Nat. Hist. Mus. Fieldiana Geol. Mem.*, 5:1–158.

Cox, A., 1969. Geomagnetic reversals. *Science*, 163:237–245.

Cox, D. R., 1962. *Renewal Theory* (London: Methuen), 142 pp.

Everndon, J. F., D. E. Savage, G. H. Curtis, and G. T. James, 1967. Potassium-argon dates and the Cenozoic mammalian chronology of North America. *Amer. Jour. Sci.*, 262:145–192.

Harbaugh, J. W., and G. Bonham-Carter, 1970. *Computer Simulation in Geology* (New York: Wiley-Interscience), 575 pp.

Heirtzler, J. R., G. O. Dickson, E. M. Herron, W. C. Pitman, III, and X. LePichon, 1968. Marine magnetic anomalies, geomagnetic field reversals, and motions of the ocean floor and continents. *Jour. Geophys. Res.*, 73:2119–2136.

Johnson, N. M., N. D. Opdyke, and E. H. Lindsay, 1975. Magnetic polarity stratigraphy of Pliocene-Pleistocene terrestrial deposits and vertebrate fossils, San Pedro Valley, Arizona. *Geol. Soc. Amer. Bull.*, 86:5–12.

Kono, M., 1973. Geomagnetic polarity changes and the duration of volcanism in successive lava flows. *Jour. Geophys. Res.*, 78:5972–5982.

LaBrecque, J. L., D. V. Kent, and S. C. Cande, 1977. Revised magnetic polarity time scale for Late Cretaceous and Cenozoic time. *Geology*, 5:330–335.

Phillips, J. D., 1977. Time variation and asymmetry in the statistics of geomagnetic reversal sequences. *Jour. Geophys. Res.*, 82:835–843.

Tedford, R. H., 1975. Correlation of North American mammal ages: Chadronian to Blancan. Unpublished chart, 4 pp.

# Part II

# THE CRETACEOUS/TERTIARY BOUNDARY:
# A CASE IN POINT

Chapter 6

# MASS EXTINCTION:
# UNIQUE OR RECURRENT CAUSES?

NORMAN D. NEWELL

American Museum of Natural History

Extinction is a normal and continuing aspect of competition and replacement in organism communities; hence, it must be regarded as an essential component of organic evolution.[1] However, the world demise of ecologically diverse members of a biota—such as the extinctions toward the end of the Cretaceous period—naturally calls to mind some overriding external environmental agent, since purely biological factors—such as disease or accumulations of lethal genes—are not likely to affect in concert a whole assemblage of organisms.

An influx of immigrants can decimate a native fauna in a brief span of time (as happened, for example, when placental mammals were introduced into South America during the Pleistocene), but in such cases, the sequence of events leading to mass extinction is usually evident in the geologic record. Van Valen and Sloane (1977) are inclined to the view that something like this may have occurred at the end of the Cretaceous.

The passing of the ammonites and dinosaurs at the close of the Mesozoic calls to mind the similar snuffings out of dominant higher categories at several levels in the Phanerozoic: episodes near the close of the Cambrian, Ordovician, and Devonian periods; ones at the end of the Permian, Triassic, and Miocene intervals; and the many lesser ones that are scattered throughout the column. Characteristically, these extinction events were followed by the loss of overall diversity, a reduction in provinciality, and the elimination of the most specialized groups (Newell, 1967a). Each of these events, of course, involved different geographic and climatic conditions, and each was, in this sense, unique. Nevertheless, some evident common denominators encourage a search for general causes. The

[1] This paper and the six that follow formed a symposium—*Biological Changes Near the Mesozoic-Cenozoic Boundary*—held on August 8, 1977, at the Second North American Paleontological Convention in Lawrence, Kansas.

principle of parsimony suggests that the hypothesis most likely to be correct is the one that accounts for a maximum number of observations with a minimum number of assumptions.

More than a century and a half ago, Cuvier and Brongniart, in their celebrated study of the Paris Basin, recognized remarkable biological changes at the Cretaceous-Tertiary boundary and connected them in some unspecified way with diastrophism. Early in the nineteenth century, virtually all paleontologists shared the Cuvier-d'Orbigny view of the fossil record: a world succession of distinctive faunas and floras that was punctuated by abrupt changes, not only in taxonomic content, but—more importantly—by gross changes in community structure that suggested significant changes in the whole biosphere. Even before Darwin's time, biostratigraphers learned how to find their way through a fossiliferous sequence by associating distinctive morphologies; they paid only limited attention to species and genera.

When the knowledgeable biostratigrapher John Phillips, nephew and disciple of William Smith, named the erathems in 1840–1841, many of the standard geological systems were already in use and recognized by their fossil content (Figure 6–1). Because of the uncritical acceptance of general catastrophism in those days, geologists did not doubt that a natural classification of the fossiliferous rocks could be based on mass extinction. Their view was based on empirical data and was not simply a superstitious adherence to biblical revelation. The Bible did not support multiple catastrophes and recreations.

By the end of the nineteenth century, most geologists strictly interpreted Lyell's uniformitarianism to exclude all unfamiliar processes. This reaction to unfettered mysticism had the unfortunate effect of discrediting unusual or concealed world events. There had always been a few highly respected exponents of diastrophic rhythms (Suess, Chamberlin, Grabau, Stille, and others—all professed uniformitarians), but they were voted down by a majority of their peers, who felt uncomfortable when confronted by what appeared to be a mystical general catastrophism that lacked a solid basis in scientific principles.

Chamberlin rationalized the connection between mass extinctions and diastrophism through the mediation of eustatic changes in sea level, but skeptics expended a considerable effort stressing his lack of data and the probability that paleontological gaps were time-transgressive and probably artifacts of method. Meanwhile, biostratigraphers were busy redefining the limits of biostratigraphic stages to correspond with the datum events provided by biological invasions and extinctions.

The development of the paradigm of plate tectonics has had a salutary effect on geological thought by opening up the way to world episodes of

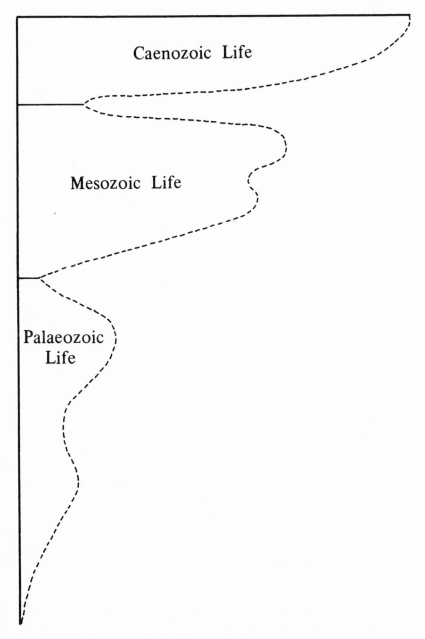

Figure 6-1. John Phillip's mid-nineteenth-century view of catastrophic extinctions of organisms at era boundaries.

diastrophism and revolutionary geographic, as well as climatic, changes. Thus, for the first time, the great oscillations in climate and the sweeping changes in habitat that had already been identified by nineteenth-century geologists as probable causes of mass extinctions are provided with a theoretical mechanism (Valentine and Moores, 1971; Schopf, 1974: Simberloff, 1974). Synchroneity of widespread orogenies and significant in-phase oscillations of sea level no longer seems unreasonable. While persuasive arguments are plentiful, however, there are unanswered ancillary questions.

Assuming that plate movements are sporadic and have, at times, taken place at high rates, where is the evidence that they did, in fact, cause major oscillations in sea level, destroy major habitats, and disrupt climatic patterns? Can mass extinction be precisely correlated in time with the maxima or minima of plate tectonics? It seems that plants were not greatly disturbed by the agencies that removed the dinosaurs. Why not? And, in view of the hypothesis that much of the world's carbon dioxide may be tied up in Cretaceous coals and limestone, what about the postulated greenhouse effect of volcanic dust and a high concentration of carbon dioxide in the Cretaceous atmosphere?

Two promising lines of inquiry stand out. In the first place, the Paleocene was a time of worldwide continentality as compared to the Cretaceous. As indicated by oxygen-isotope paleotemperatures in belemnites and foraminifera (Saito and Van Donk, 1974), there was a mean drop in temperature from late-Cretaceous to early-Paleocene times of some 5 or 6 degrees Celsius (Figure 6–2). The causes of this climatic shift are debatable and could, in theory, be attributed to either extraterrestrial or terrestrial causes.

The decline of reef corals and the virtual cessation of reef-building in the earliest Danian supports the idea of climatic deterioration. Hermatypic corals are probably the best climatic indicators that we have in the fossil record (Figure 6–3). Elsewhere, I have suggested that a high average temperature is not the critical factor that favors hermatypic organisms (Newell, 1971). In the deep tropics, the relative uniformity of temperature and freedom from stress favors $k$ selection for high diversity and specialization. At the present time, most species and genera of reef corals are confined to areas with an annual average temperature range of only 2 or 3 degrees Celsius, and they drop out rapidly wherever the range exceeds 8 degrees Celsius.

The complete draining of the interior seas in the Maastrichtian has long been cited as a likely contributing cause of the Cretaceous-Paleocene biotic revolution (Figures 6–4 and 6–5). A predictable effect of this regression should be a shift from a mild maritime climate to a more rigor-

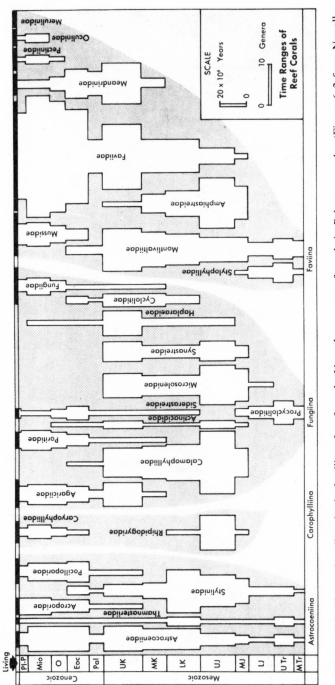

Figure 6–2. Fluctuating diversity in families of reef corals. Note the poverty of corals in Paleocene rocks. (Figure 6–2 from Newell, American Museum of Natural History Novitates 2465, 1971.)

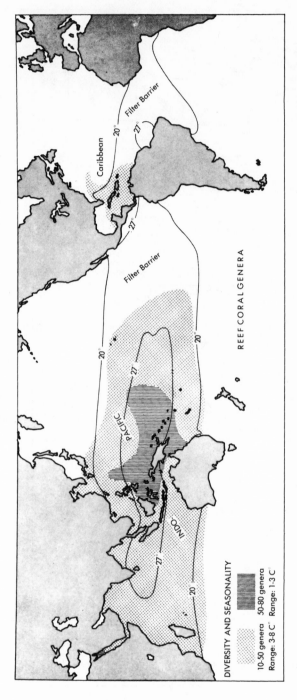

Figure 6–3. The correlation of climate and diversity of reef corals. (Figure 6–3 from Newell, American Museum of Natural History Novitates 2465, 1971.)

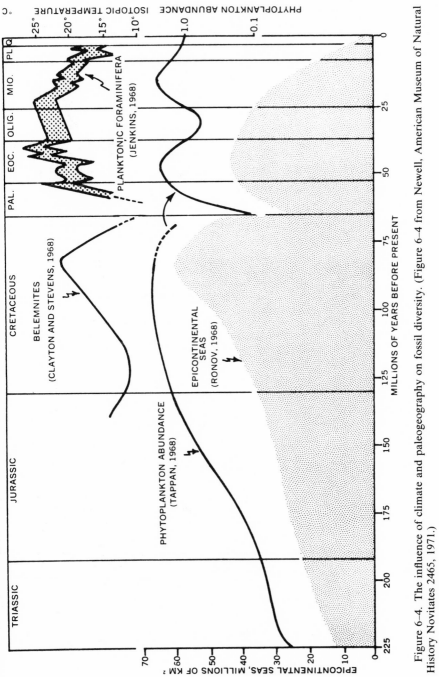

Figure 6–4. The influence of climate and paleogeography on fossil diversity. (Figure 6–4 from Newell, American Museum of Natural History Novitates 2465, 1971.)

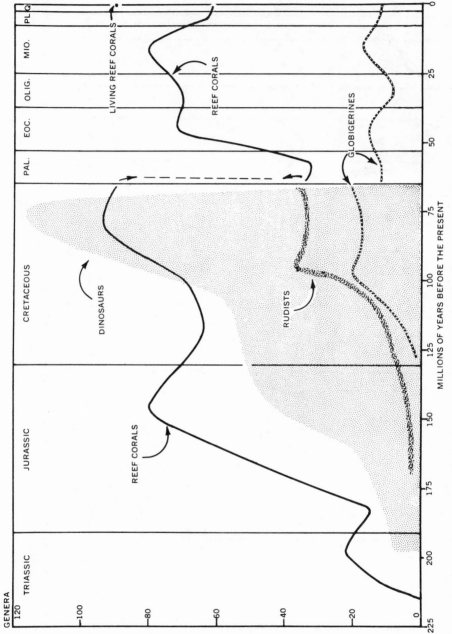

Figure 6–5. The concurrent extinction of some major groups of animals at the close of the Cretaceous. (Figure 6–5 from Newell

ous continental one. It has been suggested that Cretaceous organisms had long enjoyed an equable climate, even at the comparatively high latitudes to which they had become adapted. Thus, they have been vulnerable to climatic deterioration, especially with respect to their reproductive cycles (Axelrod and Bailey, 1968; Pinna and Arduini, 1977).

Since diversity in any group of organisms fluctuates with area of range, it can be argued that the loss of habitat beyond a threshold limit might be disastrous (MacArthur and Wilson, 1967; Schopf, 1974; Simberloff, 1974; Flessa, 1975). MacArthur and Wilson (1967) explained how large areas provide greater environmental diversity and more numerous refuges than do small areas. Hence, extinction rates are, on the average, higher in small populations than in large ones. Insurance against extinction increases with the multiplication of habitats and provinces.

Figure 6-6. As with organisms, word diversity of newsprint is cumulative and increases with area. An arbitrary sample divided into deciles.

Word Diversity

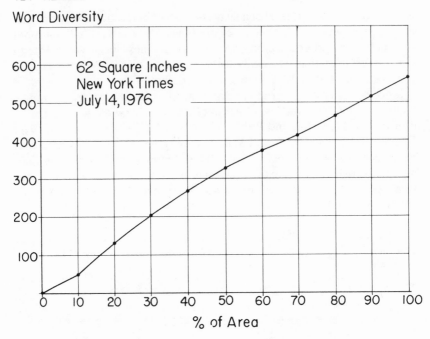

Figure 6–7. A cumulative plot of word diversity based on Figure 6–6.

A note of caution is in order here. Apparent diversity in more-or-less homogeneous habitats is positively correlated with sample size (Figures 6–6 and 6–7). When we uncritically include in a general analysis the heterogeneous habitats and diverse provinces of the whole earth, the situation is rendered complex.

The greatest mass extinction of which we know occurred during a major withdrawl of the seas in the Permian period. According to Schopf (1974), shallow seas shrank in the latest Permian from 43% of the entire marine realm to 13%. This was very likely followed by further losses of epicontinental marine habitats at the very end of the period, the overall regression being matched by accelerating invertebrate extinctions. It might be added that the chemistry of both shallow seas and open oceans was probably also increasingly unfavorable for a period of time extending into the Triassic.

This picture contrasts with the disappearance of the shelf seas during the Pleistocene low levels, which amounted to a reduction of about 6% in the total area of marine waters. As a result of these events, some extinctions of shallow water benthos may have taken place, but these extinctions would have been nothing in comparison to those of the Late Permian. Why not? Perhaps, at a global scale, the hydrographic and climatic changes in offshore waters were really not severe.

A puzzling characteristic of the erathem boundaries and of many other major biostratigraphic boundaries is the general lack of physical evidence of subaerial exposure. Traces of deep leaching, scour, channeling, and residual gravels tend to be lacking, even where the underlying rocks are cherty limestones (Newell, 1967b). These boundaries are paraconformities that are usually identifiable only by paleontological evidence. If we interpret these paraconformities as submarine surfaces of bypassing, what are the implications of episodes of draining of the epicontinental seas? The problems of paraconformities are, in some way, allied to the problems of mass extinction.

Were mass extinctions spread over hundreds, thousands, or millions of years? Dennis V. Kent (1977), in a study of the sedimentation rates and magnetostratigraphy of the relatively complete sequence of pelagic limestones at Gubbio, Italy, concludes that there the interval of crisis for the marine plankton may have been as short as 10,000 years, and Van Valen and Sloan (1977) arrive at an interval of 100,000 years for Montana dinosaurs. This would be a very short time for the kinds of world changes in climate and geography that some writers have envisaged. Butler et al. (1976) have reported dinosaur remains in New Mexico that were dated by magnetostratigraphy at one-half million to one-and-a-half million years younger than the plankton event at Gubbio; these dates seemingly confirm Van Valen and Sloan's (1977) prediction that dinosaur extinction slowly followed the expanding cool temperature belt. This interpretation is appealing. Yet, what if the correlations are inexact? Or, what if the New Mexican dinosaurs have been reworked into Paleocene deposits? Further, how can we explain an apparently earlier extinction event that happened in the open sea rather than in the more rigorous climate of the continental interior?

In many papers (including Chapter 13 in this volume), Dale Russell has long and ably argued for an extraterrestrial catastrophic event— probably a burst of hard radiation from a nearby supernova—at the end of the Maastrichtian. The difficulty with this hypothesis has been the problem of obtaining credible evidence.

Luis and Walter Alvarez and their co-workers (in press), while rejecting the supernova hypothesis, have evidence that points to a different kind of extraterrestrial event. At Gubbio, Italy, and at Stevns Klint, Denmark, they found thin layers of clay at precisely the systemic boundary; the clay from both sites contained the element iridium, in quantities that were, respectively 30 and 160 times the amount that would normally be expected in sedimentary rocks. This element is known to drift in continuously from outer space with meteoric dust and meteorites. They suggest that a small asteroid colliding with the earth could throw enough dust into the atmosphere to darken the sky for several years. Predictably, photosynthesis on land and sea would be strongly inhibited; this inhibition, in turn, would have disastrous effects on the base of the food chain, causing the mass destruction of many kinds of animals. Confirmation of this hypothesis could be provided by finding iridium excesses that mark other mass extinctions.

All of this adds up to the conclusion that the problem of mass extinctions is still unsolved, but we are learning how to ask questions. The causes of mass extinctions possibly lie in the chance coincidence of multiple factors such as the area of habitat and the climate and chemistry of the atmosphere and ocean. Very likely, the Permian and the Cretaceous extinctions were fundamentally unalike. The simplest hypothesis is not necessarily the best, and a single explanation may not cover all mass extinctions.

## LITERATURE CITED

Axelrod, D. I., and H. P. Bailey, 1968. Cretaceous dinosaur extinctions. *Evolution*, 22:595–611.

Butler, R. F., E. H. Lindsay, and L. L. Jacobs, 1977. Magnetostratigraphy of the Cretaceous-Tertiary boundary in the San Juan Basin, New Mexico. *Nature*, 267:318–323.

Flessa, K. W., 1975. Area, continental drift and mammalian diversity. *Paleobiology*, 1:189–194.

Kent, D. V., 1977. An estimate of the duration of the faunal change at the Cretaceous-Tertiary boundary. *Geology*, 5:769–771.

MacArthur, R. H., and E. O. Wilson, 1967. *The Theory of Island Biogeography* (Princeton, N. J.: Princeton Univ. Press), 203 pp.

Newell, N. D., 1967a. Revolutions in the history of life. In: Uniformity and simplicity. C. C. Albritton, Jr. (ed.), *Geol. Soc. Amer. Spec. Paper*, 89:62–92.

———, 1967b. Paraconformities. In: Teicher C., and E. Yochelson (eds.), *Essays in paleontology and stratigraphy* (Lawrence: Univ. of Kansas Press), Kansas Univ. Dept. Geol. Spec. Publ. vol. 2, 349–367.

———, 1971. An outline history of tropical organic reefs. *Amer. Mus. Nat. Hist. Noviates*, 2465:1–37.

Pinna, G, and P. Arduini, 1977. Osservazioni sulla crisi biologica del Cretacico terminale. *Milano, Soc. Italiana Sci. Nat., Atti*, 118:17–48.

Saito, T., and J. Van Donk, 1974. Oxygen and carbon isotope measurements of Late Cretaceous and Early Tertiary foraminifera. *Micropaleontology*, 20:152–177.

Schopf, T. J. M., 1974. Permo-Triassic extinctions. *Jour. Geol.*, 82:129–143.

Simberloff, D. S., 1974. Permo-Triassic extinctions. *Jour. Geol.*, 82:267-274.

Valentine, J. W., and E. M. Moores, 1970. Plate tectonics regulation of faunal diversity and sea level: a model. *Nature*, 228:657-659.

Van Valen, L., and R. E. Sloan, 1977. Ecology and the extinction of the dinosaurs. *Evol. Theory*, 2:37-64.

Chapter 7

# THE TWO PHANEROZOIC
# SUPERCYCLES

ALFRED G. FISCHER

Department of Geological and Geophysical Sciences,
Princeton University

## INTRODUCTION

The philosophy of historical geology, swept 150 years ago by Lyell's actuocentric viewpoint, has since then undergone a progressive change: in
one field of geology after another, an explanation of individual geological
events on an "actualistic" basis adds up to a history in which the outer
earth has deviated markedly in state and behavior from the one in which
we live—a state which cannot be considered a norm. While most of these
changes are of the gradual type, the occurrence of global catastrophes is
now also supported by theory and data.

The proposition advanced in this paper is that these deviations consist
of a mixture of random drift and an hierarchy of oscillations—or
cycles—that are irregularly punctuated by catastrophic events. This
paper is focused on the longest of these cycles, ones that have a period in
the range of 300 myr. The temporal patterns of eustasy, continental dispersion and aggregation, plutonism, and sedimentation suggest two
cycles of mantle convection: one Paleozoic, the other Mesozoic-Cenozoic.
Each of these cycles contains one phase in which the atmosphere is
enriched by carbon dioxide, leading to greenhouse climates, and another
phase in which this gas is depleted, leading to icehouse climates susceptible to the development of ice sheets. Some of the biotic crises of earth
history are viewed as marking the inversion points between these climatic
regimes.

## THE PERCEPTION OF LONG-TERM CHANGE

Whatever meanings may be read into the terms "actualism" and "uniformitarianism" (see Gould, Chapter #1; also Rutten, 1971), Lyell dealt with earth history from an "actuocentric" perspective. The present state and functioning of the earth were taken as the norm. While local change was the very fabric of historical geology, this change was to be wholly accounted for by processes within the range of human experience. Furthermore, the state of the whole was considered to be invariant or, after the discovery of Pleistocene glaciations, to involve only temporary deviations from the present state (Lyell, 1867).

Since Lyell's day, most historical geologists have proclaimed a Lyellian philosophy, but their practices have commonly deviated from it. Biostratigraphy, for example, developed along the lines laid out by the catastrophist d'Orbigny, simply because the paleontological record is neither uniform nor gradualistic in character, but is rather a record of sharp discontinuities. Global synchroneity of paleontological discontinuities is a fundamentally non-Lyellian concept.

Biostratigraphers were not the only ones who discovered a lack of uniformity in earth history. It became clear that the origin and very early history of the earth involved states that were very far removed from the present one (for example, Rutten, 1971). It also became evident that the evolution of organisms implied changes in the composition of the atmosphere (Berkner and Marshall, 1965) and in the nature of geomorphic and sedimentational processes. The discovery of ancient glaciations implied intervals of major climatic deterioration, while the existence of former atmospheric greenhouse states was suggested by Arrhenius (1896) and by Chamberlin (1897, 1898, 1899). Periodicity of orogenic events became widely accepted. Sea-level changes are now generally attributed to global eustatic influences, as well as to local effects (Grabau, 1940; Vail et al., 1977). Major rearrangements of continents and oceans, which certainly had global effects on the state of the outer earth, were proposed by Wegener (1915) and confirmed by modern plate tectonics. Changes in the thermal structure and behavior of the oceans have been deduced (for example: Emiliani, 1954; Douglas and Savin, 1975; Fischer and Arthur, 1977). In short, based on data largely gathered within the frame of Lyellian thought, every aspect of historical geology has come to reveal historical patterns at variance with the actuocentric view that sees the earth as a whole persisting in a largely invariant state, of which the present is satisfactorily representative.

While most of this change has come about slowly—at what might be thought of as a uniformitarian pace—the role of catastrophe cannot be dismissed as it was by Lyell. It now seems not only possible, but also

likely that stellar events (see Russell, Chapter 13), collisions with asteroids and/or meteorites (Alvarez et al., 1980; Smit and Hertogen, 1980;
Hsü, 1980; Emiliani, 1980), and giant volcanic explosions have brought
sudden changes to the outer earth as a whole. In principle, the possibility
of yet other catastrophes—such as the spillout of great freshwater bodies
into the world ocean (Gartner and Keaney, 1978; Thierstein and Berger,
1978)—must also be taken seriously.

Emancipation from the confining bounds of the actuocentric viewpoint
has raised historical geology from a science concerned with local history
to one whose ultimate aims encompass the evolution of the earth as a
whole. This emancipation began in Lyell's lifetime, and though it has had
to fight its way step-by-step against Lyellian conservatism, it has grown
steadily. We may now view earth history as a matter of evolution in
which some changes are unidirectional (at least, in net effect), others are
oscillatory or cyclic, and still others are random fluctuations, while the
whole is punctuated by smaller or greater catastrophes. The prime tasks
of modern historical geology are to separate the local signals from the
global ones, to plot the relationships of global patterns both to time and
to each other, and to search for the forces that drive these varied processes.

The present paper is an attempt in that direction. Whereas in an earlier
paper (Fischer and Arthur, 1977), the focus was on patterns in the range
of 30 myr, here I am concerned with longer-range processes. Cycles in
eustacy, volcanism, and climate seem to exhibit a first-order periodicity in
the range of 300 myr; essentially, the record is of two great tectonic-cli-
matic cycles: beginning with the Late Proterozoic ice ages, followed by
the inferred greenhouse conditions of the Early-Middle Paleozoic, next
by the Late Paleozoic ice ages, then by the Mesozoic greenhouse state,
and finally by the present icehouse state. I propose that these cycles are
driven by two cycles in mantle convection.

The long-range cycles are complicated by shorter oscillations—such as
the cycle of 30 myr that was proposed by Fischer and Arthur (1977)—and
by climatic cycles, with periods in the range of 20,000 to 400,000 years,
that are driven by orbital perturbations (Imbrie and Imbrie, 1981;
Fischer, in press). These cycles may also involve modulation of atmospheric carbon dioxide content.

## THE EUSTATIC CURVE AND ITS IMPLICATIONS

Figure 7-1 shows, among other things, the global first-order eustatic
curves of Vail et al. (1977) and of Hallam (1977), modified by me to present a compromise between the Russian and the American records. The
curves are bimodal, showing exceptionally high sea-levels during two pe-

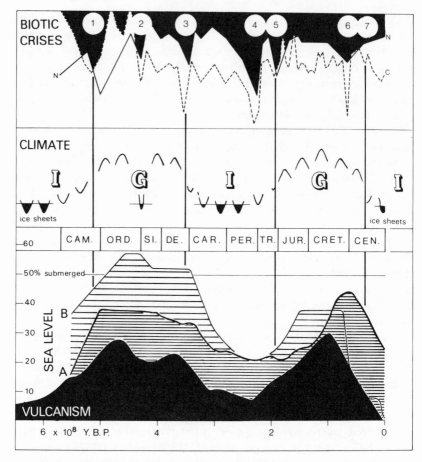

Figure 7-1. The two supercycles of Phanerozoic history. Biotic crises (numbered) as shown by drops in the number of marine animal families, N after Newell (1967), C after Cutbill. Climate inferred as alternating between icehouse states (I) and Greenhouse states (G), in each of which occur minor fluctuations. Time of ice sheets shown by blackened troughs. Sea levels (A) according to Vail et al. (1977), (B) according to Hallam (1977). Vulcanism as judged from American granite emplacement after Engel and Engel (1964). Figure adapted, with modifications, from Fischer (1981).

riods—one mid-Paleozoic, the other Cretaceous-Eocene: times that have long been recognized as "thalassocratic" and that contrast with the widespread "epeirocratic" emergence in the latest-Precambrian, Permo-Triassic, and Oligocene-Neogene times.

The height to which the sea rises on the continental blocks depends on the volume of water and on the size of the ocean basins. If we assume that the total volume of surficial water (including ice) has not changed appreciably during Phanerozoic time and that the development of a large cryosphere is a temporary phenomenon, then this eustatic curve is primarily an index to the volume of the ocean basins.

The theory of plate tectonics offers three main ways by which this volume may be appreciably altered: 1) by changes in the length of the oceanic spreading ridge system, 2) by changes in the width of oceanic ridges (equivalent to changes in spreading rate), and 3) by changes in the mean thickness of continents.

*Changes in the length of spreading ridges*   The ridges are thermal bulges that displace sea water (Sclater and Francheteau, 1970). Russel (1968) suggested that the migration of a ridge from the ocean to a continent—such as occurred in the North American West in Neogene times—increases the volume of the ocean bases and induces a drop in sea level. This argument is readily expanded: the number of lithospheric plates varies in time, and the total length of the ridges increases with the number of plates. It seems likely that the number of plates and the length of the ridge system are maximal during times of continental dispersal and minimal during times of aggregation (Pangea). Thus, sea levels may be expected to be high during times of continental dispersion and low during pangeal episodes.

*Changes in the rate of seafloor spreading*   Since fast spreading produces broad ridges, and since ridges decay when the spreading rates decrease (Sclater and Francheteau, 1970), Hays and Pitman (1973) suggested that universal changes in spreading rates must produce global changes in sea level. They assumed that spreading rates could change sufficiently to move sea level by some hundreds of meters; but Cretaceous spreading rates are still imprecisely known, and those of earlier times are probably lost beyond recall.

Whatever relative roles might be played by change in the ridge length versus change in the spreading rate, it seems clear that "plate activity," a concept that covers both factors, must have a strong influence on long-term eustasy. As Pitman (1978) has shown, such volume effects are too gradual to be the principal cause of eustatic changes of the second or third order on Vail's curve, but they are adequate to explain the first-order effects with which we are here concerned.

*Changes in continental thickness*   A third factor that must contribute to long-term eustasy lies in changes in the mean thickness and area of continents. If we assume that the volume of the continental lithosphere stays essentially constant, then the mean thickness of the continental blocks will be inversely proportional to their area. However, these parameters must change during earth history. We begin with a pangeal aggregation of continents that are grouped into a single emergent block having a comparatively short perimeter of submerged transitional lithosphere. As this supercontinent is broken up by rifting, its mean thickness is decreased. The total perimeter of the resulting fragments will be larger, and the total volume of submerged intermediate crust will be greater. Thus, a given volume of continental matter becomes more widely spread, mean elevations are decreased, the volume of the ocean basins is lessened, and sea level must rise. When dispersal gives way to reaggregation, the thinned margins of meeting continents will be overthickened into mountain welts, the mean thickness of continents will increase as their area shrinks, and sea level will drop. In essence, we may regard this effect as being an indirect response to plate activity.

*The eustatic curve as an index to plate activity*  If we view the eustatic curve in this light, we find that the Late Paleozoic Pangea is, indeed, a time of low sea-levels. This could be attributed either directly to reduced plate activity or indirectly to its expression through continental thickening. It probably represents a combination of both. The great rise in sea level during the later parts of the Mesozoic roughly coincides with continental dispersion. The rise in Vail's curve seems somewhat delayed, while Hallam's curve for Russia appears a better fit to the history of continental breakup and scatter, as well as to the inferred increase in the number of plates and length-of-ridge.

The drop in sea level following the Eocene may mark the beginnings of aggregation: namely, the collapse of Tethys, as well as the shortening of the East Pacific Rise. Considering the still-rather-dispersed distribution of continents, the drop of sea level that is pictured by Vail et al. appears to be rather large. Perhaps, there has been a greater reduction in plate activity than is generally recognized. Alternatively—or, in addition—the buildup of ice in the Antarctic region may have been sufficiently persistent to have contributed markedly to the sea-level drop in Oligocene and Neogene times. Presumably, there will be a further drop in sea level when the already overaged Atlantic closes and produces a Novopangea.

Prior to the Pangaea-associated eustatic low, the mid-Paleozoic sea-level high generally matches the dispersed pattern of continents (Scotese et al., 1979) and the inferred high rate of plate activity. The drop in sea level appears to have been delayed relative to continental aggregation; judging from orogenies in eastern North America, sea level remained high through the Devonian, when aggregation was proceeding. However, the accuracy of paleogeographic reconstruction fades rapidly back through the Paleozoic era.

No reliable geographic reconstruction exists for the Late Precambrian.

However, the Late Precambrian-Cambrian stratigraphy of both the Atlantic and Pacific sides of North America indicates Late Precambrian rifting and the succeeding onlap of geosynclinal sequences onto the continental margins—the typical picture of a young, linked margin (Kinsman, 1975). The American craton thus appears to have been hemmed in by other continental masses, and a Late Precambrian Protopangea appears not unlikely.

In summary, since the mid-Paleozoic, first-order eustatic lows have been associated with continental aggregation, and first-order highs with continental dispersion. During the Late Precambrian low sea-level stand, a Protopangea can also be inferred.

It seems reasonable to think of continental aggregation as being the result of a simplification of plate patterns that involves a reduction in the number of plates and a shortening of total ridge length, as well as—possibly—a decrease in mean spreading rates. Conversely, continental dispersion requires an increase in the number of plates and a lengthening of ridge length. The failure of first-order eustasy to match precisely the aggregation and dispersal history of continents may be taken as evidence that ocean-basin volume change results, not only from continental thinning and thickening, but also from the development and disappearance of submarine spreading ridges and from changes in spreading rates. Specifically, since newly developed rift zones on continents take time to grow into oceanic spreading ridges, sea-level rise is likely to lag behind a quickening of plate activity: thus, rifting in Triassic time preceded sea-level rise by some tens of millions of years. Similarly, sea levels may remain high for some time after plate activity has diminished, because of the slow decay of ridges (Hays and Pitman, 1973) and because the major continental collisions follow diminution of plate activity with some delay and may be stretched out over a long period (Late Paleozoic).

If the aggregation and dispersal of continents is but another expression of varying plate activity, as is here suggested, then the first-order sea-level curve may be viewed as a somewhat delayed index to first-order variations in plate activity.

## VOLCANISM

Independent of such effects, the model of changing plate regimes implies historical fluctuations in volcanism. Emplacement of mantle-derived basalts is necessarily at a maximum during peak plate activity, as is the generation of andesitic volcanism and associated deep granitic plutons that are derived from the partial melting of crust in the subduction zones.

The record of basalt generation throughout earth history has been largely destroyed by subduction, while the record of andesite generation has been widely obliterated by erosion of the mountain belts. More enduring are the large granitoid plutons, exposed in the cores of mountain belts. Although the initial ages of such plutons tend to be masked by subsequent thermal history, they nevertheless afford the best index to volcanicity. Figure 7-1 shows a plot—redrawn from Engel and Engel's (1964) compilation—of the history of Phanerozoic granite emplacement in North America. This plot closely matches the first-order sea-level curve. Maximal volcanism is not associated with the times of continental aggregation, as has commonly been believed; instead, it matches the high sea-levels, which we have previously shown to correspond to continental dispersion and maximal plate activity. This correspondence confirms the general hypothesis: both sea level and volcanism are functions of the same process—namely, plate activity—which underwent two major cycles in Phanerozoic time.

## CARBON DIOXIDE IN THE ATMOSPHERE

Though present only in the amount of 0.03%, carbon dioxide is one of the most important constituents of the atmosphere. In the first place, the thermal insulating properties of carbon dioxide control the climates of the biosphere. Secondly, carbon dioxide in combination with water serves as the most effective agent in weathering and is, therefore, vital to geomorphic and sedimentological cycles. Thirdly, carbon dioxide serves as the ultimate source of carbon for organisms and, thus, for the generation of organic byproducts such as carbonaceous rocks and atmospheric oxygen.

Holland (1978) comprehensively dealt with the carbon cycle. The relatively small reservoir of atmospheric carbon dioxide is buffered by exchange with the much larger reservoir of oceanic carbon. Yet, even if we consider the total amount of carbon in the earth's fluid envelope (40,690 $\times 10^{15}$g), the combined terrestrial and marine cycling by organisms (83 $\times 10^{15}$g/year) yields a residence time of only 490 years. Changes in organic productivity are, therefore, likely to have effects on atmospheric carbon dioxide and—to a lesser extent—on oxygen concentrations.

Tappen (1968) has invoked changes in phytoplankton productivity as a cause of biotic crises. Ancient changes in productivity, however, are exceedingly hard to establish, inasmuch as they need not bear a relationship either to biotic diversity or to the carbon content of sediments (Fischer and Arthur, 1977). In addition, the biotic carbon system is filled with negative feedbacks: an increase in carbon fixation by producers is likely to be followed immediately by an equivalent increase in carbon dioxide production by consumers and decomposers. Furthermore, the efficiency of terrestrial photosynthesis is positively coupled to carbon dioxide availability: a reduction of atmospheric carbon dioxide by rampant plant growth is limited by an attendent drop in the efficiency of photosynthesis, and vice versa (for literature, see Holland, 1978). I find the matter of paleoproductivities to be quite intractable at the present level of knowledge. I am also inclined to the view that internal feedbacks within the organic system tend to keep it in balance, as well as to the view that the role of organisms is more stabilizing than destabilizing. To find the cause of larger fluctuations in carbon dioxide pressure, we must look to its ultimate sources and sinks: volcanism and weathering.

*Volcanic supply and weathering demand* The carbon dioxide cycle and its many ramifications have been comprehensively treated by Holland (1978). The organic cycle, due to its rapidity, has attracted the attention of most investigators. This cycle is, however, only a sub-cycle within the larger geological one. The ultimate source of atmospheric and hydrospheric carbon dioxide is volcanism, which brings carbon dioxide out of both the mantle and subducted parts of the lithosphere. The ultimate sink of atmospheric and hydrospheric carbon dioxide is return to the lithosphere. This cycle is vastly slower than the organic one. Holland calculates loss to the lithosphere at $0.8 \times 10^{14}$g/year; assuming a steady state, he posits an equivalent input by volcanism. In these terms, carbon dioxide in the oceanic/atmospheric reservoir has a residence time of about 500,000 years. Yet, this is not long in terms of the time scales here considered: an imbalance in input versus loss that continued over tens of millions of years could effect considerable changes. In principle, this was clear to Chamberlin (1897, 1898, 1899), to Ronov (1964), and to Budyko (1977).

*Fluctuations—qualitative* The foregoing discussion leads to the conclusion that plate activity and, therefore, volcanism have undergone notable fluctuations during earth history. This implies that the rate at which carbon dioxide has been supplied to the atmosphere has varied. The sea-level curve implies that during earth history the exposed land area of the globe has expanded and shrunk by what appears to be a factor of 2. This must have had a notable effect on rates of carbon dioxide withdrawal by weathering. Coupled in time, volcanism and sea-level change reinforce each other in perturbing the carbon dioxide level of the atmospheric-hydrospheric fluid reservoir: during times of intense plate activity and carbon dioxide supply, the lands are shrinking, and withdrawal by weathering is reduced. The level of carbon dioxide in the reservoir must accordingly rise, until increasing *intensity* of weathering once more balances the higher rate of supply. As plate activity, volcanism, and the carbon dioxide supply decrease, sea level falls, the land area expands, and weathering withdraws carbon dioxide at increased rates, until a lower level in the reservoir establishes a new balance.

Putting this into the historical framework, it would appear that carbon dioxide pressure was relatively low during the latest Precambrian; that it rose during the Early Paleozoic and into the Middle Paleozoic; that it dropped in the Late Paleozoic-Triassic Pangea regime; that it rose again in the course of Jurassic, Cretaceous, and Paleogene times; and that it dropped thereafter.

I do not wish to imply that the carbon dioxide level remained constant within each of these intervals. Prior reference has been made to shorter cycles in sea level, organic community complexity, volcanism (possibly), and other factors that are likely to have produced shorter—and, perhaps, notable—fluctuations in carbon dioxide pressures: fluctuations riding on the broader baseline that corresponds to the first-order curves in sea level and in volcanicity.

Another approach to the history of atmospheric carbon dioxide content through Phanerozoic time is that of Ronov (1964; also see Budyko, 1977). Ronov has suggested a general decrease of carbon dioxide pressure through Phanerozoic time, but he also has focused on the addition of carbon dioxide by volcanism and on the withdrawal of carbon dioxide by sedimentation. However, he has attempted to reconstruct the balance between these by reference to the record of layered rocks exposed on the continents. This approach—without considering submarine volcanism, subduction, and the latitudinal displacements of plate tectonics—is, in my view, unreliable and, in part, misleading. He envisions no rise in carbon dioxide pressure in the early Paleozoic. We agree, however, on the general drop through Carboniferous-Permian time and on the rise to a high in late-Cretaceous time, followed by a decline in the Cenozoic. Some of Ronov's minor peaks may be the result of the shorter cycles to which we have previously referred.

*Fluctuations—quantitative*   So far, our discussion has been cast in strictly qualitative terms. Ronov has assumed an overall decrease: from about ten times present values at the beginning of Cambrian time to an all-time low in recent times; in between there were higher spikes. The arguments for or against such an interpretation are largely climatic and will be taken up below.

The best available control is the one that is afforded by the physiology of living plants (for details and literature, see Holland, 1978). Most higher plants can function over a wide range of carbon dioxide concentrations. For most of these plants, the minimum concentration lies at one-half to one-tenth of the present atmospheric value, and the optimum concentration is reached at double or triple present atmospheric value. Further, most terrestrial plants continue to function well up to pressures that are about sixty times the present value. Plant physiology, thus, does not bound the field very tightly. The optimal adaptation of photosynthesis to concentrations more than twice as great as the present ones, together with the present proximity of carbon dioxide pressure to the lower threshold of plant tolerance, suggests that carbon dioxide pressures have been appreciably higher through most of Phanerozoic time. Thus, the range of values suggested by Ronov (0.03% to 0.4%) seems reasonable.

## CLIMATES AND CARBON DIOXIDE REGIMES

The realization that some of the climates of the past differed markedly from those of today goes back to two discoveries: one being the discovery of the Pleistocene and earlier ice ages; and the other, that of ancient temperate—or, even, subtropical-appearing—floras in regions which are, or were, at high altitudes. Other climatic anomalies were subsequently inferred from such evidence as the distribution of evaporites, while geochemical interpretations of the temperatures of ancient seas have added quantitative data. Some of the supposed anomalies have been resolved by the recognition of latitudinal displacements of lithospheric plates (Smith et al., 1973; Smith and Bryden, 1977; Scotese et al., 1979), but other anomalies still remain.

The many sorts of explanations that have been advanced to explain these apparent climatic changes have been concisely reviewed by Budyko (1977). Many of the known factors that control climate are variable and must contribute to the observed changes. The chief problems are sorting out the relative contributions of each of these known factors and evaluating the existence of those factors which remain hypothetical or unknown. Factors may be divided into three main groups: 1) astronomical variations, 2) changes in the nature of the earth's surface, and 3) changes in atmospheric composition.

*Astronomical factors*   Possible astronomical factors include: long-range changes in the frequency and/or the intensity of radiation due to solar evolution, hypothetical short-term variations in the solar radiation flux, variations associated with galactic rotation, and changes in the solar energy received that result from the earth's orbital perturbations. In addition, it appears possible that transitory climatic shocks may be caused by cosmic accidents such as supernovas or collisions with comets or asteroids.

Studies of stellar evolution (Schwarzschild, 1958) suggest that solar radiation has increased at a rate of perhaps 1% per 80 myr. At first glance, this suggestion seems geologically absurd, for it implies a thoroughly frozen earth throughout earlier Precambrian history. However, a lesser supply of solar heat could have been offset by differences in the volume and the composition of the primitive atmosphere, which was richer in heat-trapping components such as carbon dioxide and water vapor.

The existence in solar radiation of oscillations that are longer and stronger than the sunspot cycle remains conjectural.

The repetition of major glaciations at intervals in the range of 200 to 300 myr has suggested to various investigators that the solar energy budget may be linked in some way with the rotational period of the galaxy (see review by Pearson, 1978). In my view, cyclic mantle conviction on this time scale offers a more plausible explanation.

The effect on climate of astronomical cycles in the earth's orbit and rotation was first suggested by Adhemar (1842) as being a cause of the Pleistocene glacial-interglacial oscillations. Subsequent work has more precisely established the nature and effects of these cycles and has essentially confirmed their roles in driving climatic changes with periodicities ranging from 18,000 to 500,000 years (Hays, Imbrie, and Shackleton, 1977; Imbrie and Imbrie, 1979, 1980). These oscillations, however, have neither the amplitude nor the frequency of the larger variations in Phanerozoic climates.

*Changes in the earth's surface*   Stronger perturbations in climate and sea level with longer periods are more readily attributed to changes in the earth's surface; such changes include: changes in the total area of land (probably by a factor of two), changes in land distribution, changes in the elevation and the disposition of mountain ranges that affect atmospheric circulation, and changes in the configuration of the oceans that affect patterns of marine circulation. Even at first glance, times of first-order sea-level lows—such as the latest Precambrian, the Late Paleozoic, and the Neogene—appear to be favorable for glaciation, and the intervening times would be those in which climatic gradients were lower. Many paleoclimatologists (for example, Brooks, 1949) have viewed such changes in paleogeography as the main driving forces of paleoclimatic change.

*The amplitude of climatic variation*   Are such topographic changes adequate to explain the magnitude of paleoclimatic variations? Paleomagnetic work indicates that the regions into which large ice sheets extended during the late Precambrian glaciations were in low latitudes (Frakes, 1979), far beyond the equivalent extent of Pleistocene ice. Recent work has shown that Cretaceous dinosaurs reached paleolatitudes of about 60° (Charig, 1973) and that Jurassic, Cretaceous, and Paleogene plants in such latitudes were of temperate to subtropical character (Barnard, 1973; Koch, 1963; Schweitzer, 1974). The Eocene biotas of Axel Heiberg and the Ellesmere islands, formed in similar paleolatitudes, contain very large broadleaf evergreen trees and animal assemblages that include salamanders, lizards, snakes, turtles, crocodilians, primitive equids, colugos, and primates: an assemblage of warm-temperate type, in which winter frosts were either light or absent (West and Dawson, 1978). Kollman (1979) and colleagues are presently studying a huge warm-temperate molluscan fauna found in Paleocene strata on the west coast of Greenland. Numerous isotopic temperature determinations on benthonic foraminifera from the deep sea floor (for example, Douglas and Savin, 1975) show that the mean temperature of the Cretaceous ocean masses as a whole was some ten degrees higher than today; these determinations imply either much warmer polar regions or a completely different oceanic circulation, in which the deep waters were formed principally in lower latitudes (Fischer and Arthur, 1977). One possible reason for such differences lies in a different geographic disposition of continents and oceans, but the modelling experiments that have been carried on to date (K. Bryan, pers. comm.; R. Barron, pers. comm.) suggest that this reason alone may be insufficient. A change in atmospheric composition is one of the alternative reasons.

*Changes in atmospheric composition and paleoclimates* The possible effects on climate of changes in atmospheric composition have been reviewed in detail by Budyko (1974, 1976). While dustiness of the atmosphere may cause temporary deviations of climate, the main agents for long-term changes are carbon dioxide and water vapor, both of which trap heat in what has been generally referred to as a "greenhouse effect." A rise in carbon dioxide content would raise the mean temperature of the atmosphere as a whole; such a temperature rise, in turn, would lead to an increase in water vapor content, with positive feedback. Temperatures would rise far less in the tropics, where energy would be expended on increased evaporation, than in the higher latitudes, where heat would be released by increased rainfall (Manabe and Wetherald, 1975).

The greenhouse effect was first postulated by Tyndall (1861). Arrhenius (1896) calculated that a halving of the present carbon dioxide pressure would drop the mean annual temperature by some 4° Celsius, and that a doubling of present pressure would increase mean temperatures by 8° to 9° Celsius. Manabe and Wetherald (1975) calculated that doubling the carbon dioxide pressure would raise mean annual temperatures by some 2° in the tropics, by about 3° at 50 degrees latitude, and by more than 10° at 80 degrees latitude. Further increases in carbon dioxide concentration would have little effect (Budyko, 1976).

*The fit of the carbon dioxide model to the stratigraphic record*   A geological model for variations in atmospheric carbon dioxide content has been constructed above. This model suggests that carbon dioxide pressures should have been relatively low in the latest Precambrian, in the Late Paleozoic, and in the Oligocene-Neogene: all of these being times during which widespread glaciation episodically occurred (Frakes, 1979). The only widely accepted Phanerozoic glaciation which fails to coincide with these times is the one at the Ordovician-Silurian boundary that is known chiefly from high paleolatitudes in Africa.

The model also suggests higher carbon dioxide pressure from Late Cambrian to Devonian times and from the Jurassic into the Eocene. We have previously outlined the overwhelming evidence for warm climates in high latitudes during the later Mesozoic and Paleogene. Our knowledge of Paleozoic climates remains too deficient for us either to corroborate or to disprove the efficacy of the model relative to that interval, but two aspects of the interval—the Ordovician-Silurian glaciation and the incidence of "black shales"—call for comment.

*The Ordovician-Silurian glaciation*   The glacial deposits which occur in Africa near the Ordovician-Silurian boundary (Beuf et al., 1971; Frakes, 1979) argue against a general greenhouse state having existed during that time. Since glacial sediments of that age have been conclusively identified only at high paleolatitudes in northern and southern Africa, this glaciation seems to have been restricted to the polar region. Additionally, there is no evidence for a long-continued, multiple glaciation as in the other glacial episodes. Three possibilities come to mind: 1) that an essentially non-glacial world was briefly plunged into glacial conditions by one of the above-mentioned minor cycles that ride on the first-order oscillations; 2) that this glaciation could have been the result of some "catastrophic" event of external origin—some support for this possibility may be found in the fact that it is the only glaciation known to be associated with a major faunal crisis (see Figure 7–1); and 3) that this glaciation resulted from an "overshoot" of the greenhouse that led to a state in which heavy cloud cover raised the earth's albedo to a level at which much solar energy was reflected, and the earth cooled (Fischer, 1981).

*The incidence of black shales*   A characteristic feature of the Jurassic and Cretaceous periods is the large extent of black shales, as well as concurrent evidence of widespread anoxia in the oceans and seas of those periods. Fischer and Arthur (1977) have interpreted these phenomena as being results of climatic greenhouse conditions, during which the oxygen-carrying capacity of the oceans was diminished by higher temperatures, and oceanic circulation was slowed due to a lessened density-drive. A similar tendency toward stagnation is displayed in the widespread black shales of the Middle and Late Cambrian, in the graptolite shales of the Ordovician and Silurian, and in the oil shales of the Devonian. Strøm (1939) was the first to point out that these deposits are too widespread to be the products of barred basins; he suggested that the deposits were caused by lowered climatic gradients. Berry and Wilde (1979) offered another explanation: that these deposits were the products of a transition from the anoxic ocean of early earth history to the well-oxygenated ocean of later times. This explanation, however, runs afoul of evidence that the seas of both the latest Precambrian and early-Cambrian times were particularly well oxygenated, as shown by extensive red marine sediments in the Lie de Vin series in Morocco and in the Tommotian and Atdabanian deposits of Siberia, northern Europe and the Appalachians. I, therefore, cannot accept Berry and Wilde's interpretation; rather, I agree with Strøm that these deposits, like their Mesozoic equivalents, resulted from a greenhouse climate.

*The alternation of greenhouse and icehouse states*   The "greenhouse theory" is the most satisfactory explanation that I can suggest for long-range changes in paleoclimatology. In this theory, the earth's climates are postulated to have oscillated between two basic states: one, the "greenhouse state"—characterized by elevated atmospheric carbon dioxide levels that are, at least, twice their present value—and the other, the "icehouse state" (Fischer, 1981), in which we presently live. The effects of carbon dioxide variation are reinforced by associated changes in geography that are induced by the same basic cause: plate activity.

The greenhouse state is characterized by low gradients in latitudinal temperature, relatively warm polar regions, high mean ocean temperatures, sluggish oceanic circulation, and a tendency toward marine anoxia. Large lowland ice sheets and sea ice are generally absent, although plateau ice caps and mountain glaciers may actively enlarge due to increased precipitation from an atmosphere containing appreciably more water. Since we do not know the greenhouse state from personal experience, the actualistic method is not strictly applicable, and we can only reconstruct this climate by inference.

The icehouse state is distinguished by strong gradients in latitudinal temperature, low mean ocean temperature, active oceanic convection, highly oxygenated oceans, and a tendency for continental ice sheets and sea ice to be developed.

The general time-distribution of these states is shown in Figure 7-1. Superimposed on this long-range, first-order, climatic oscillation are smaller oscillations, which may involve significant short-term deviations from the base-line states that are plotted.

In Ronov's (1964) view, atmospheric carbon dioxide levels lay above 0.5% during all of pre-Oligocene times. According to Budyko (1977), this implies continued greenhouse conditions through all of that time. I do not follow Ronov's basic assumption and find it easier to believe that carbon dioxide levels have fluctuated through the Cenozoic in reference to a level rather than a declining base line, and that they have modulated climates through the Paleozoic as suggested above.

## CLIMATIC OSCILLATIONS AND BIOTIC CRISES

In the course of earth history, the general tendency to develop more complex faunas and floras was interrupted time and again by "biotic crises" of varying magnitudes. The smaller ones have been the daily concern of biostratigraphers from Cuvier and d'Orbigny onward; the big ones have posed special geological problems. Newell's (1967) compilation of major biotic crises, based on extinctions of animal families, is replotted in Figure 7-1. Most of us generally accept the reality of major crises occurring at the Cambrian-Ordovician boundary, at the Ordovician-Silurian boundary, in the Late Devonian, at the Permo-Triassic boundary, at the Triassic-Jurassic boundary, and at the Cretaceous-Tertiary boundary.

Previously (Fischer, 1981), I related all of these crises to changes in climatic state. Neither the Permo-Triassic crisis nor the Cretaceous one, however, match long-range inversions in climate. Furthermore, there is now a plausible argument that the crisis at the Cretaceous-Tertiary boundary was the result of the earth's colliding with some extraterrestrial object (Alvarez et al., 1980). I still believe that a climatic inversion from either basic climatic state to the other, involving step-function changes in atmospheric and oceanic dynamics, would be likely to prove so upsetting to the world's organic communities that it would cause biotic crises, and that such crises might be the best available signposts of the crossover points in the climatic cycle. We turn, therefore, to examine the crises in sequence.

*The Cambrian-Ordovician crisis* fell into the time interval during which the latest-Precambrian–Early Cambrian icehouse state might have given way to the mid-Paleozoic greenhouse state.

*The Ordovician-Silurian crisis* was coincident with a glaciation and might have been due to a short—but temporary—excursion into another climatic state or to some external factor that brought about both the glaciation and the biotic crisis.

*The Late Devonian crisis* seems to have coincided with the crossover from the mid-Paleozoic greenhouse state to the Late Paleozoic icehouse state.

*The Permo-Triassic crisis* occurred in the latter part of the Late Paleozoic-Triassic icehouse interval, after the glaciations, but well before the breakup of Pangea. Among its suggested causes are an extreme drop in sea level for unknown reasons (Newell, 1967) and the development of brackish conditions in the upper waters by various mechanisms (Fischer, 1965; Thierstein and Berger, 1978), but the matter remains very uncertain.

*The Triassic-Jurassic crisis* is a good candidate for having been a crossover in climatic state, occurring shortly after rifting had begun and preceding the development of greenhouse conditions.

*The Cretaceous-Tertiary crisis* differed from any of the others in its abruptness: even in areas of essentially constant marine sedimentation, it can be defined to a bedding plane or a thin "boundary clay." The occurrence of exceptionally high concentrations of iridium in this clay (Alvarez et al., 1980; Smit and Hertogen, 1980; Hsü, 1980) makes a strong case for this crisis having been precipitated by a collision with some planetary object (an asteroid? a comet?). Whether or not the dinosaurs died out at the time of the marine crisis remains uncertain, as does the manner in which the collision stressed the biosphere (Emiliani, 1981). This crisis assuredly was not associated with a long-term inversion of climatic state, inasmuch as the Mesozoic greenhouse state persisted into Eocene times.

*The Late Eocene-Oligocene biotic crisis* is a second-order crisis in Newell's plot (see Figure 7-1), but one that was strong in the pelagic realm (Fischer and Arthur, 1977). It greatly reduced the diversity of planktonic foraminifera and of nannoplankton and brought about sharks replacing whales as the dominant marine super-predators. Like the

Cretaceous-Tertiary crisis, it resulted in oceanic blooms of the opportunistic problematicum *Braarudosphaera.* It marks the crossover from the Mesozoic-Eocene greenhouse state to the subsequent icehouse one. The reason for its comparatively low intensity might lie in its having been anticipated by the Late Cretaceous catastrophe, from which the communities had not yet fully recovered.

In summary, I take the crisis in Late Cambrian time (at about 500 myr), the one in the Late Devonian time (at about 355 myr), the one at the end of the Triassic (192 myr), and that in the late Eocene (40 myr) to mark the crossover points between the major climatic episodes in earth history. The intervals between these crossover times are: 145 myr, 163 myr, and 152 myr. The regularity of these intervals appears programmatic and suggest the possibility that the inciting cause—plate activity and, ultimately, mantle convection—is a cyclical process with a period of about 300 myr: roughly the periodicity that is commonly attributed to the period of galactic rotation (see the summary in Pearson, 1978).

## SUMMARY AND CONCLUSIONS

Since the Late Precambrian, the earth has passed through two major cycles, each about 300 myr long. These are held to be fundamentally dependent on cycles in mantle convection.

Each of these cycles had an initial phase of rapid convection with numerous plumes or cells that found expression at the surface in the breakup of larger lithospheric plates and the rifting and dispersal of pangea-type supercontinents. This phase resulted in raised sea levels and broad flooding of the continental blocks. Intense volcanism (both basaltic volcanism in rift zones and granitic-andesitic volcanism at convergent plate margins) led to maximal release of carbon dioxide from the mantle and crust. At the same time, atmospheric carbon dioxide losses to the lithosphere via weathering were reduced because of the decrease in land area. As a result, atmospheric carbon dioxide content rose until the increased intensity of weathering once again balanced the inflow from volcanism.

The result was a greenhouse climate, characterized by low latitudinal gradients, warmish and humid poles, and oceans that were relatively warm throughout and that convected sluggishly. Such climates appear to have characterized two segments of Phanerozoic history: one, the period that extended from the end of the Cambrian to the end of Devonian time; and the other, the interval from the beginning of the Jurassic period through most—or all—of the Eocene.

The second phase of the convection cycle was characterized by a lessening of convection and a simplification of convective patterns. The number of independently active lithospheric plates was reduced and continents underwent accretion. Sea level dropped. A decrease in volcanic activity reduced the supply of carbon dioxide to the atmosphere, while a growing land area increased the demands made by weathering. As a result, the carbon dioxide in the atmosphere-hydrosphere reservoir sought a new equilibrium at a lower level. This broke the climatic greenhouse and brought about the icehouse state, characterized by high latitudinal climatic gradients, cold and dry polar regions, a susceptibility to the development of sea ice and continental ice sheets, and oceans that were cold, active, and highly oxygenated. Such states seem to have existed in the latest-Precambrian–Early Cambrian; in the Late Paleozoic-Triassic; and the latter half of the Tertiary period, in which we now live.

Within each of these climatic states, other cyclic and random events produced deviations in climate that induced incidents such as the high-latitude glaciation in the midst of the greenhouse episode at the Ordovician-Silurian boundary.

In historical sequence, at the end of Cambrian time, following rifting, the icehouse state of the latest Precambrian gave way to greenhouse conditions that were coincident with a major sea-level rise. Except for a brief excursion at the Ordovician-Silurian boundary, this mid-Paleozoic greenhouse state persisted up to the Late Devonian, when climates inverted to the Late Paleozoic-Triassic icehouse. Greenhouse conditions were reestablished at the end of Triassic times, and persisted through most—or all—of Eocene times, after which the present icehouse commenced. Greenhouse and icehouse states thus alternated at intervals of about 150 myr, in response to a cycle in mantle convection of about 300 myr. The next major greenhouse episode lies more than 100 myr in the future, though man's return of fossil carbon into the atmosphere might bring about an artificially induced—and, hopefully, temporary—condition in the next century.

## ACKNOWLEDGMENTS

In various lectures over the last several years, I have presented forms of the hypothesis that I elaborated in this paper, a version of which has been published (Fischer, 1981). In the course of this time, the hypothesis has undergone considerable modification. Discussions with colleagues have played a large role in this continuing education. I owe particular thanks to the Geological Society of London, as well as to M. A. Arthur, W. A.

Berggren, Kirk Bryan, R. M. Garrels, R. A. Hallam, L. Hickey, H. D. Holland, Isabella Premoli-Silva, F. T. McKenzie, S. Manabe, N. D. Newell, F. B. Van Houten, and John Van Couvering. Their aid does not imply their endorsement of the product, either in whole or in part. The considerations that have been here presented are also, in part, outgrowths of a Guggenheim Fellowship (1969–1970) and of sediment research supported by a series of grants from the National Science Foundation Geological Program.

## LITERATURE CITED

Adhémar, J. A., 1860. *Les Révolutions de la mer,* déluges périodiques. (Paris: Lacroix-Comon), 2 vols. in 1.

Alvarez, L. W., W. Alvarez, F., Asaro, and W. V. Michel, 1980. Extraterrestrial cause of the Cretaceous-Tertiary extinctions. *Science,* 208:1095–1108.

Arrhenius, S., 1896. On the influence of the carbonic acid in the air upon the temperature of the ground. *Philosophical Magazine,* 41:237–275.

Barnard, P. D. W., 1973. Mesozoic floras. In: N. F.. Hughes (ed.), *Organisms through time* (London: Pal. Assoc. of London), pp. 175–188.

Berkner, L. V., and L. C. Marshall, 1965. On the origin and rise of oxygen concentration in the Earth's atmosphere. *Jour. Atmos. Sci.,* 22:225–261.

Berry, W. B. N., and P. Wilde, 1979. Progressive ventilation of the oceans—an explanation for the distribution of the Lower Paleozoic black shales. *Amer. Jour. Sci.,* 278:257–275.

Beuf, S., B. Biju-Duval, O. de Charpal, P. Rognon, O. Gaviel, and A. Bennacef, 1971. *Les Gres du Paleozoique Inferieure au Sahara* (Paris: Editions Technip), 464 pp.

Budyko, M. K., 1977. *Climatic Changes* (Washington, D.C.: American Geophysical Union), 261 pp.

Chamberlin, T. C., 1897. A group of hypotheses bearing on climatic change. *Jour. Geol.,* 5:653–683.

———, 1898. The influence of great epochs of limestone formation upon the constitution of the atmosphere. *Jour. Geol.,* 6:609–621.

———, 1899. An attempt to frame a working hypothesis of the cause of glacial periods on an atmospheric basis. *Jour. Geol.,* 7:545–584.

Charig, A. J., 1973. Jurassic and Cretaceous dinosaurs. In: A. Hallam (ed.), *Atlas of Paleobiogeography* (London: Elsevier), pp. 339–352.

Crowell, J. C., and L. A. Frakes, 1975. The late Paleozoic glaciation. In: K. S. W. Campbell (ed.), *Gondwana Geology* (Canberra: Australian Natl. Univ.), pp. 313–331.

Douglas, R. G., and S. M. Savin, 1975. Oxygen and carbon isotope analyses of Cretaceous and Tertiary microfossils from the Shatsky Rise and other sites in the North Pacific Ocean. *Initial Reports of the Deep Sea Drilling Project* (Washington, D.C.: U.S. Govt. Printing Office), vol. 32, pp. 509–521.

Emiliani, Cesare, 1980. Death and renovation at the end of the Mesozoic. *Eos*, 61:505–506.

Engel, A. E. J., and C. G. Engel, 1964. Continental accretion and the evolution of North America. In: A. P. Subramaniam and S. Balakrishna (eds.), *Advancing Frontiers in Geology and Geophysics* (Hyderabad: Indian Geophysical Union), pp. 17–37e.

Fischer, A. G., 1965. Brackish oceans as the cause of the Permo-Triassic marine faunal crisis. In: A. E. M. Nairn (ed.), *Problems in Plaeoclimatology* (London: Interscience). n.p.

———, 1981. Climatic oscillations in the biosphere. In: M. H. Nitecki (ed.) *Biotic Crises in Ecological and Evolutionary Time* (New York: Academic Press), pp. 103–131.

Fischer, A. G., and M. A. Arthur. 1977. Secular variations in the pelagic realm. In: H. E. Cook and P. Enos (eds.), *Deep Water Carbonate Environments. Soc. Econ. Pal. Min. Spec. Pub.*, 25:18–50.

Frakes, L. A., 1979. *Climates throughout Geologic Time.* (London: Elsevier), 310 pp.

Gartner, S., and J. Keany, 1978. The terminal cretaceous event: a geological problem, with an oceanographic solution. *Geology*, 6 (12):708–712.

Grabau, A. W., 1940. *The Rhythm of the Ages* (Peking: Henry Vetch), 561 pp.

Hallam, A., 1977. Secular changes in marine inundation of USSR and North America through the Phanerozoic. *Nature*, 269:769–772.

Hays, J. D., J. Imbrie, and N. J. Shackleton, 1976. Variations in the Earth's orbit: pacemaker of the ice ages. *Science*, 194:1121–1132.

Hays, J. D., and W. C. Pitman, III, 1973. Lithospheric motion, sea level changes and climatic and ecological consequences. *Nature*, 246:18–22.

Holland, H. D., 1978. *The Chemistry of Oceans and Atmospheres* (New York: Wiley & Sons), 351 pp.

Holser, W. T., 1977. Catastrophic chemical events in the history of the ocean. *Nature*, 267:403–408.

Hsü, K. H., 1980. Terrestrial catastrophy caused by cometary impact at end of Cretaceous. *Nature*, 285:201–203.

Imbrie, J., and K. P. Imbrie, 1979. *Ice Ages, Solving the Mystery* (Short Hills, N. J.: Enslow Pub.), 224 pp.

Imbrie, J., and J. Z. Imbrie, 1980. Modeling the climatic response to orbital variations. *Science*, 207:943–952.

Kinsman, D. J. J., 1975. Rift valley basins and sedimentary history of trailing continental margins. In: A. G. Fischer and S. Judson (eds.), *Petroleum and Global Tectonics* (Princeton, N. J.: Princeton Univ. Press), pp. 83–126.

Koch, D. E., 1963. Fossil plants from the lower Paleocene of Agatdalen (Angmartusut) area, central Nuqssuaq Peninsula, northwest Greenland. *Meddelelser om Gronland*, 172:1–120.

Kollman, H., 1979. Distribution patterns and evolution of gastropods around the Cretaceous-Tertiary boundary. In: W. K. Christensen and T. Birkelund (eds.), *Copenhagen Univ., Proc. Sympos. "Cretaceous-Tertiary Boundary Events"*, vol. 2, pp. 83–87.

Lyell, C., 1867. *Principles of Geology*, 10th ed. (London: John Murray), 2 vols.

Manabe, S., and R. T. Wetherald, 1975. The effect of doubling the $CO_2$ concentration on the climate of a general circulation model. *Jour. Atmos. Sci.*, 32:3–15.

Newell, N. D., 1967. Revolutions in the history of life. *Geol. Soc. Amer. Spec. Paper*, 89:63–71.

Pearson, R., 1978. *Climate and Evolution* (New York: Academic Press), 265 pp.

Pitman, W. C. III, 1978. "Relationship between Eustacy and Stratigraphic Sequences of Passive Margins," *Geol. Soc. Am. Bull.*, 89:1389–1403.

Ronov, A. B., 1964. General tendencies in the evolution of the composition of the Earth's crust, ocean and atmosphere. *Geokhimika*, 8:715–743.

Russell, K. L., 1968. Oceanic ridges and eustatic changes in sea level. *Nature*, 218:861–862.

Rutten, M. G., 1971. *The Origin of Life by Natural Causes* (Amsterdam: Elsevier), 420 pp.

Schweitzer, H. J., 1974. Die Tertiaeren Koniferen Spizbergens. *Palaeontographica, Abt. B*, 149:1–89.

Sclater, J. G., and J. Francheteau, 1970. The implications of terrestrial heatflow observations on current tectonic and geochemical models of the crust and upper mantle of the earth. *Roy. Astr. Soc., Geophys. Jour.*, 20:509–542.

Scotese, C., R. K. Bambach, C. Barton, R. Van der Voo and A. Ziegler, 1979. Paleozoic base maps. *Jour. Geol.*, 87:217–277.

Smit, J., and J. Hertogen, 1980. An extraterrestrial event at the Cretaceous-Tertiary boundary. *Nature*, 285:198–200.

Smith, A. G., J. C. Briden and G. E. Drewry, 1973. Phanerozoic world maps. In: N. E. Hughes (ed.), Organisms and continents through time. *Pal. Assoc. London Spec. Papers Pal.*, 12:1–42.

Smith, A. G., and J. C. Briden, 1977. Mesozoic and Cenozoic palaeocontinental maps. (Cambridge: Cambridge Univ. Press), 63 pp.

Strom, K. M., 1939. Land-locked waters and the deposition of black muds. In: P. D. Trask (ed.), *Recent Marine Sediments* (Tulsa, Okla.: Amer. Assoc. Petrol Geol.), pp. 356–372.

Tappan, H., 1969. Primary production, isotopes, extinctions and the atmosphere. *Palaeogeogr. Palaeoclimatol. Palaeocol.*, 4:187–210.

Thierstein, H. R., and W. H. Berger, 1978. Injection events in ocean history. *Nature*, 276:461–466.

Vail, P. R., R. M. Mitchum, Jr., and S. Thompson, 1977. Seismic stratigraphy and global changes of sea level, part 4. In: C. E. Peyton (ed.), Seismic Stratigraphy *Amer. Assoc. Petrol. Geol. Mem.*, 26:83–97.

Wegener, A., 1915. *Die Entstehung der Kontinente und Ozeane*. Sammlung Vieweg, 23. Branunschweig.

West, R. M., and M. R. Dawson, 1978. Vertebrate paleontology and Cenozoic history of the North Atlantic region. *Polarforschung*, 48:103–119.

Chapter 8

# THE FABRIC OF CRETACEOUS
# MARINE EXTINCTIONS

ERLE G. KAUFFMAN

Department of Geological Sciences, University of Colorado

## INTRODUCTION

The "mass extinction" event at the end of the Cretaceous period is widely regarded as having been a biological catastrophe that involved the near-synchronous global annihilation of the structurally and ecologically diverse taxa which characterized Mesozoic biotas (see, for example, Newell, 1967). This image of the event has been enhanced by a widespread Cretaceous-Tertiary boundary disconformity, as well as by representations of this extinction in simple graphic plots that make use of unweighted lines, representing major groups, which extend to—but not across—the depicted boundary (for example, see Schindewolf, 1962; also Figure 8–1A). The reptiles, ammonites, rudist and inoceramid bivalves, scleractinian corals, planktonic and larger benthic foraminifera, and calcareous nannoplankton are symbolic "victims" of this crisis, which extensively shocked marine and continental ecosystems. The Mesozoic-Cenozoic contact has consequently served as a model for an era boundary, as an outstanding example of a biological crisis, and as a focus for many hypotheses on the causes of widespread extinction (see the papers in Christensen and Birkelund, 1979).

Yet, the "fabric" of this great extinction event—that is, the patterns of biological and ecological change at and around the Cretaceous-Paleocene boundary—has not been documented in detail except among the calcareous microbiatas of a few exposed marine sequences (Gubbio, Italy; Zumaya, Spain; Tampico, Mexico; and so on), in numerous cores that have been recovered during deep sea drilling, and among principal elements of the Danish biota. These studies have clearly depicted a global catastrophe among pelagic warm-water calcareous microbiotas; broader interpretations of the Cretaceous extinction event, however, have

151

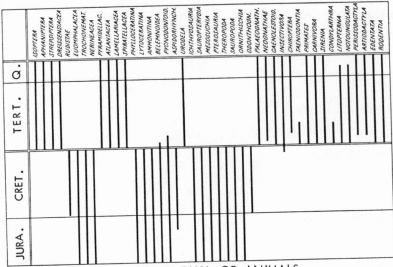

A. HIGHER TAXA OF ANIMALS

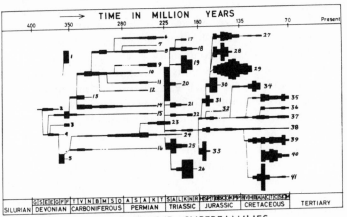

TIME IN MILLION YEARS

B. AMMONOID SUPERFAMILIES

Europe (Type Areas and SW-France)

North America (Alabama)

C. CALCAREOUS NANNOPLANKTON SPECIES

overemphasized this particular catastrophe. In point of fact, available data on other marine groups, especially the macrofauna, do *not* widely support a "catastrophic" interpretation of the extinction event. Most groups of organisms that are classically depicted as having become extinct at the Cretaceous-Paleocene boundary actually died out one million to several million years before the end of the Maastrichtian; or, they had been on the evolutionary wane, and at the time of the boundary event, they were represented only by a few generalists. The total evidence suggests a complex extinction pattern that was spread over millions of years and not a wholesale "catastrophe."

Furthermore, the concept of a global catastrophe that involved taxonomically and ecologically diverse organisms runs counter to the biological theory which predicts that organisms with widely differing adaptive strategies and levels of ecological tolerance should respond differently to changing environmental factors and, thus, should show distinct extinction patterns. In fact, detailed analyses of evolutionary histories at narrow stratigraphic intervals do demonstrate such distinct patterns; but these important data are lost in the simple graphic plots that so many catastrophists employ to depict the Cretaceous extinction (compare the depiction of ammonite history in Figure 8-1A with that in Figure 8-1B). When subject to earthbound forces of natural selection, "mass extinction" is unnatural in both evolutionary and ecological terms.

The assumption of a terminal Cretaceous biotic catastrophe also fails to recognize the limited quantity and quality of evidence available from the marine realm. Many workers have misinterpreted existing data by not placing the data into the detailed temporal, stratigraphic, environmental, and geographic frameworks that are currently available for the Maas-

---

Figure 8-1. Typical graphic representations of the "catastrophic" terminal Cretaceous extinction event that show the coincident annihilation at the Mesozoic-Cenozoic boundary of numerous, ecologically diverse taxa.

A) Diverse higher taxa of animals (from Schindewolf, 1962).

B) Superfamilies of ammonoids; the varied widths of the bars depict the various generic diversities—for example, #26 = 48 genera (after House, 1963).

C) Species of European and North American Gulf Coast coccolithophorids (from Bramlette and Martini, 1964).

Note the implications in these graphic presentations of data: all taxa range to the very top of the Cretaceous before their "synchronous" extinctions; there is a lack of designation of disconformities; and (except for the ammonoids in Figure 81-B) all taxa are represented by uniform dark lines, implying that there had been no loss in diversity before the terminal Cretaceous extinction.

In Figure 8-1B, the plotting of diversity along the chart shows a marked decline in ammonites throughout the Upper Cretaceous; this rendering presents a much more realistic picture of their graded extinction than the one used in Figure 8-1A (Phylocera-tina, Lytoceratina, and Ammonitina).

trichtian-Danian succession. For example, about ninety percent of the world's exposed boundary sequences lack uppermost Cretaceous and/or lowermost Paleocene strata (biotas)—an hiatus of 2 to 5 myr—owing to a widespread regressive disconformity. This hiatus was of a duration sufficient for the development of a variety of extinction fabrics. In the few boundary sequences that are relatively complete, only the microbiotas have been studied in detail; macrofaunas are commonly either sparse or absent.

Thus, most fossil marine groups that have been classically depicted as becoming extinct at the end of the Maastrichtian have no extensive fossil record in the Upper Maastrichtian. The North Temperate boundary sequences of northwestern Europe—for example, that of Denmark—are obvious exceptions. In these sequences, however, the levels of macrofaunal and benthic microfaunal extinction are moderate to low, and they mainly involve changes at the species level within lineages, as well as the terminal extinction of those groups—such as the ammonites and inoceramid bivalves—that were, by the Late Maastrichtian, already in the late stages of evolutionary decline. Furthermore, remarkably similar ecosystems bound the Cretaceous-Tertiary contact in northwest Europe, in the northeastern United States, and, probably, in Greenland: the Atlantic part of the North Temperate realm. The same observation has been made in Antarctica by Zinsmeister (pers. comm. in 1979). In these areas, the marine catastrophe is recorded only in the calcareous pelagic microbiota; non-calcareous microplankton show low to moderate extinction levels in the same strata (Hansen, 1979, p. 141). By contrast, the benthic tropical shelf biota of Tethys—especially reef-associated organisms—was decimated during the Middle to Late Maastrichtian; many tropical groups underwent widespread and rapid Maastrichtian extinction at, or near, a peak in their evolutionary development. This tropical versus temperate differentiation of the terminal Cretaceous extinction event is significant, and it has not been previously interpreted. The key to the mystery of the terminal Cretaceous extinction event may lie in explaining what happened in Tethys and its warm-water margins.

Despite its classic status as a major catastrophe, therefore, the biological fabric of the terminal Cretaceous extinction is largely unknown. Many questions arise from the apparent contradictions in existing data: What is the detailed sequence of events that led up to the terminal extinction? How can this terminal extinction be interpreted from such an imperfect record? Do similar events occur elsewhere in the Cretaceous? If they do, can these other events shed light on this terminal event? How can the great magnitude of the tropical Cretaceous extinction be explained in view of the more subtle changes across the North and South Temperate boundaries? How can the abrupt extinctions of many elements in the calcareous pelagic microbiota be related to the poor record of macrofossils and to their apparently more gradual demise in the same sections prior to the end of the Cretaceous? Which of the many theoretical causes for this mass extinction—namely: rapid temperature changes, salinity decline, widespread regression, diminished ecospace, a massive extraterrestrial radiation event, a meteor impact, changing water chemistry, changing atmospheric chemistry, and so on—can be called upon, either singly or in combination, to best account for this extinction event?

This study is an attempt to explore some of these questions in light of current evolutionary and ecological theory. The actual record of the marine extinction event is analyzed in detail for relevant sections, and its imperfections are documented. Since so few complete marine boundary sequences are known, especially ones that contain macrofossils, answers to the terminal Cretaceous extinction are sought in the well-preserved records of major intra-Cretaceous extinctions. The most dramatic of these extinctions occurred at the Albian-Cenomanian boundary, the Cenomanian-Turonian boundary, and the Turonian-Coniacian boundary. The biological fabrics of these somewhat less severe—but still important—extinctions reveal generally consistent patterns that allow a new interpretation of the terminal Cretaceous extinction event.

## THE CRETACEOUS-TERTIARY BOUNDARY EVENT: AN HISTORICAL PERSPECTIVE

Schindewolf (1955, 1962) depicted the terminal Cretaceous extinction event as having been an essentially synchronous annihilation of numerous, ecologically diverse, major groups of Mesozoic organisms (Figure 8-1A); from this depiction, he derived his concept of "Neokatastrophism." Newell (1967) referred to the Cretaceous-Tertiary boundary as marking both a "mass extinction" and a "revolution in the history of life." Percival and Fischer (1977) termed this event the "Cretaceous-Tertiary biotic crisis." Apparently, the theories of catastrophism leading to mass extinction that were originally espoused by Cuvier (1769-1832) and d'Orbigny (1802-1857) are still with us, although they have been tempered by careful scientific observation. The concept of an abrupt mass extinction having occurred at the Cretaceous-Tertiary boundary is thoroughly entrenched in contemporary science.

This concept is not without evidence to support it. In most parts of the world, a major biologic discontinuity that involves numerous and ecologically diverse groups of marine and terrestrial organisms does occur at the

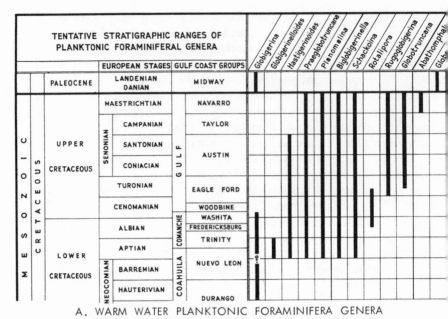

A. WARM WATER PLANKTONIC FORAMINIFERA GENERA

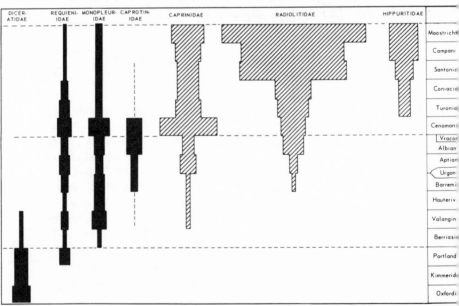

B. TROPICAL RUDIST BIVALVE FAMILIES

Figure 8-2. Graphic representations of extinction patterns among tropical-subtropical marine orgnisms at the end of the Cretaceous.
A) shows the complete and abrupt extinction of numerous important genera of planktonic foraminifera (after Loeblich et al., 1957).

Cretaceous-Tertiary boundary; in fact, many typical Cretaceous stocks disappear completely in the Maastrichtian. The biological break has been magnified, however, by widespread regressive disconformities that produced an hiatus of 1 to 5 myr at the boundary in most exposed sequences (Figure 8–4). In continuously deposited pelagic boundary sequences—such as those at Zumaya in Spain and at Gubbio in Italy—the abrupt extinction of marine pelagic calcareous microbiota within a meter or so of the boundary has been well documented (for example: Percival and Fischer, 1977; Premoli Silva, 1977; papers in Christensen and Birkelund, 1979; Birkelund and Bromley, 1979). The clustering within the Maastrichtian stage (65.6 to 70.5 Ma) of extinctions among terrestrial and marine animals, benthic and pelagic biotas, and very simple to very complex organisms clearly suggests that there was a broad biotic crisis at the end of the Cretaceous. I do not argue about the existence of this extinction event; rather, based on analyses of the extant marine paleobiological and paleoenvironmental record, I question its abruptness, examine its "fabric" (that is, the nature of its biological and ecological patterns), and postulate its causes.

Most of the imagery of a catastrophic terminal Cretaceous extinction has been based on a lack of evidence, has not taken into account major stratigraphic omissions, and has been formulated without detailed study of the macrobiota at the boundary. Misconception of the Cretaceous extinction event has also been generated by the way in which the subject has historically been presented. Variations of such distorted representations are shown in Figures 8–1, 8–2, and 8–3.

Generalized earlier summaries of the terminal Cretaceous event (for example: Schindewolf, 1955, 1962; Newell, 1967) have depicted the extinction of the typical Mesozoic biota at the family, superfamily, and ordinal levels without differentiating between those taxa that disappeared at an evolutionary peak and those that were represented only by their last surviving lineage(s). There can be no comparison between the signifi-

---

B) displays the similar extinction of families of rudist bivalves at a peak in their diversification (after Coogan, in Moore, 1969).

In both representations, the extinction event has been plotted near, or at, the Maastrichtian-Paleocene boundary. Although field data do support such a plotting for planktonic foraminifer, they do not for the rudists, which apparently had their major extinction near the Middle-Upper Maastrichtian boundary.

In Figure 8–1B, the widths of the plots for rudist families reflect generic diversity; the smallest width equals one genus. The solid black patterns represent older, more conservative families; the oblique patterns designate the more plastic, rapidly evolving Late Cretaceous groups. This extinction pattern is a typical one for the majority of important warm-water groups that were affected by the terminal Cretaceous biotic crisis (see Figure 8–1C for coccolithophorids).

| GENERA OF CHEILOSTOMATA | CRETACEOUS | | | | | | | TERTIARY | |
|---|---|---|---|---|---|---|---|---|---|
| | LOWER CRET. | CENOM. | TURON. | CONIAC. | SANT. | CAMP. | MAAS. | DANIAN | MONT. |
| Pyripora.......... | | | | | | | | | |
| Hapsidopora...... | | | | | | | | | |
| Anaptopora....... | | | | | | | | | |
| Dacryopora........ | | | | | | | | | |
| Onychocella....... | | | | | | | | | |
| Aechmella........ | | | | | | | | | |
| Tylopora.......... | | | | | | | | | |
| Rhabdopora....... | | | | | | | | | |
| Stamenocella...... | | | | | | | | | |
| Pnictopora........ | | | | | | | | | |
| Ichnopora......... | | | | | | | | | |
| Hexacanthopora.... | | | | | | | | | |
| Pliophlœa......... | | | | | | | | | |
| Tricephalopora..... | | | | | | | | ? | |
| Pelmatopora....... | | | | | | | | | |
| Discoflustrellaria... | | | | | | | | | |
| Floridina......... | | | | | | | | | |
| Lunulites......... | | | | | | | | | |
| Pavolunulites...... | | | | | | | | | |
| Acoscinopleura..... | | | | | | | | | |
| Escharicellaria..... | | | | | | | | | |
| Coscinopleura...... | | | | | | | | | |
| Decurtaria........ | | | | | | | | | |
| Platyglena........ | | | | | | | | | |
| Hoplitæchmella.... | | | | | | | | | |
| Tænioporina...... | | | | | | | | | |
| Systenostoma...... | | | | | | | | | |
| Cryptostomella..... | | | | | | | | | |
| Exochella......... | | | | | | | | | |
| Puncturiella....... | | | | | | | | | |
| Porina........... | | | | | | | | | |
| Pachythecella...... | | | | | | | | | |
| Beisselina......... | | | | | | | | | |
| Beisselinopsis...... | | | | | | | | | |
| Escharifora........ | | | | | | | | | |
| Bactrellaria....... | | | | | | | | | |
| Nephropora....... | | | | | | | | | |
| Lecythoglena....... | | | | | | | | | |
| Bathystomella...... | | | | | | | | | |
| Allantopora....... | | | | | | | | | |
| Marssonopora..... | | | | | | | | | |
| Pachydera........ | | | | | | | | | |
| Stichocados........ | | | | | | | | | |
| Frurionella........ | | | | | | | | | |
| Columnotheca...... | | | | | | | | | |
| Diacanthopora..... | | | | | | | | | |
| Haplocephalopora.. | | | | | | | | | |
| Psilosecos......... | | | | | | | | | |

cance of the relatively rapid Maastrichtian extinction of the rudist families Radiolitidae and Hippuritidae at the peak of their radiation (Figure 8–2B) and the significance of the final extinction of the equally important Cretaceous bivalve family Inoceramidae, which was represented worldwide near the end of the Maastrichtian by only two genera, a few species, and sparse individuals. These two extinctions have entirely different implications to the theory of extinction and its causes; yet, many authors cite them as having been equally important components of the terminal Cretaceous "crisis" (see, for example, Percival and Fischer, 1977, p. 1). To be meaningful, interpretations of extinction events must involve analyses of lower order taxa and the details of their time-stratigraphic occurrences both below and above the extinction boundary (for example, Figures 8–1C and 8–2A; also, Percival and Fischer, 1977).

The way in which taxa, especially higher taxa, are depicted graphically—as unvarying straight lines extending from their points of origin to their points of extinction (Figures 8–1A, 8–1C, and 8–2B)—produces misconceptions about the nature of the extinction event. Such graphic techniques mask the patterns of radiation, diversity, evolutionary decline, and preterminal extinction events that are the data upon which the interpretation of any extinction event must depend. The previously cited example of rudist and inoceramid bivalves applies equally here. So does the ammonite story: Schindewolf's straight line representation of the ammonite superfamilies (Figure 8–1A) implies that their wholesale extinction occurred at the end of the Cretaceous. Yet, generic diversity plots (Figures 8–1B and 8–5) demonstrate that, by the middle of the Cretaceous, the ammonites had already reached their evolutionary peak and undergone their greatest extinction; very few genera and species remained in the Late Maastrichtian (see Birkelund, 1979), and their final extinction was not dramatic.

The failure to present detailed biostratigraphic data for organisms that became extinct within the Maastrichtian also prevents the comprehensive analysis of the terminal Cretaceous event. With the notable exceptions of

---

Figure 8–3. A typical latest Cretaceous extinction pattern for benthic organisms of the North Temperate realm. Data has been taken mainly from North European cheilostomate bryozoan genera (after Voight, 1959).

Note that 73% of the genera that were present in the Maastrichtian also passed into the Danian (Paleocene), and that 53% also passed into the Montian. At the Cretaceous-Tertiary boundary, therefore, in the majority of lineages, the only breaks that occurred were at the species level, and rates of extinction were only moderately increased. A similar pattern is apparent within benthic macrofossil groups, especially mollusks and certain groups of echinoids (Rosenkrantz, 1960, 1970, and pers. comm. in 1971; Brotzen, 1959, pp. 35–36).

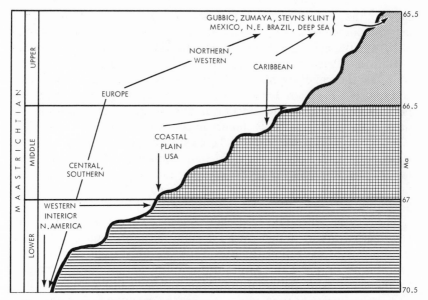

Figure 8–4. A diagrammatic representation of the Cretaceous-Tertiary boundary and the major disconformity—or paraconformity (hiatus)—that over most of the world separates Maastrichtian beds from Paleocene ones; indicated are some examples (highly generalized) from the Euroamerican region and the Caribbean realm.

The disconformity is rendered as white; the three different patterns designate Lower, Middle, and Upper Maastrichtian strata; the radiometric scale to the right (after Obradovich and Cobban, 1975; Kauffman, 1977a) has been modified to reflect new decay constants (Obradovich, pers. comm. in 1978).

In most of the areas that are shown, from 0.5 to 5.0 myr of Maastrichtian evolutionary and extinction history are missing. Considering the rapid evolutionary turnover among Cretaceous macro-organisms (Kauffman, 1977c, 1978a), these intervals of missing history were of sufficient duration for major evolutionary and extinction events to have taken place *before* the end of the Maastrichtian. In many areas, part— or all—of the Early Paleocene record is also missing, which adds to the duration of these gaps. These missing intervals greatly accentuate the abruptness of the terminal Cretaceous extinction event and mask the true fabric of the extinction. Only in a few areas of the world (upper right corner of Figure) is the boundary exposed without significant stratigraphic omission. In most of these areas, as well as in some deep-sea cores, macrofossil groups either drop out in a graded fashion well below the boundary or are absent altogether, and only the plankton shows an abrupt change at the Cretaceous-Paleocene contact. (See the text for references and discussion.)

the analyses of warm-water calcareous plankton (see Herm, 1965; also Percival and Fischer, 1977), as well as those of northwest European macrofaunas (papers and references in Christensen and Birkelund, 1979; Birkelund and Bromley, 1979), most published discussions of the Maastrichtian marine biotas have been, at best, generalized. Most graphic data that has been utilized in the discussion of the extinction event (for example, Figures 8–1, 8–2, and 8–3) only plot the stratigraphic ranges of taxa at levels no finer than the substage; in some cases, these data plot only at the stage level. Thus, regardless of a taxon's evolutionary history, if it is present at the base of the Maastrichtian, its extinction point will be automatically plotted at the Cretaceous-Tertiary boundary. Yet, the Maastrichtian stage had a duration of 5 myr (Kauffman, 1977a; Van Hinte, 1976). In view of the rapid rates of evolution and extinction that have been calculated for Cretaceous marine organisms that represent diverse ecological and structural (genetic?) levels (Kauffman, 1970, 1972, 1977c, 1978a), there was sufficient time within the Maastrichtian for major evolutionary changes to have taken place. Furthermore, during an interval of 5 myr, extinction episodes could just as well be randomly drawn out as clustered so as to indicate a catastrophe. Such plots as Figures 8–1 to 8–3 tell us little about this extinction event and create an artificial image of a biological catastrophe, for which no proof is offered. These kinds of plots stand in contrast to the excellent data that has been presented for the mass extinction of the calcareous planktonic microbiota at the end of the Maastrichtian (for example: Percival and Fischer, 1977; papers in Christensen and Birkelund, 1979; Birkelund and Bromley, 1979). We lack these kinds of data for most of the macrobiota lineages that are also envisaged as having become abruptly extinct at the same time.

In treating the marine extinction event, many authors have failed to note that, in exposed sequences throughout the world, a major disconformity or paraconformity has eliminated millions of years of marine biological history of the Cretaceous-Tertiary boundary zone (Newell, 1967). This disconformity reflects erosion during and following a major Maastrichtian eustatic fall in sea level and the regression of the strand to perhaps 100 meters below the level of the present (Kauffman, 1979a, 1979b; Hancock and Kauffman, 1979). On world cratons, the gap is commonly exaggerated by nonmarine epicontinental sequences near the contact. In epicontinental sequences that still preserve marine Maastrichtian and Paleocene strata, however, it is common for *at least* the Upper Maastrichtian and the Lower Paleocene marine record (a span of 1 to 2 myr or more) to be missing (Figure 8–4). This gap would have been of sufficient duration for the development of considerable evolutionary changes and variable extinction histories; thus, the greater the gap, the more accentuated the differences between preserved Campanian-Maastrichtian and Paleogene biotas. Such plots as those shown in Figures 8–1, 8–2, and 8–3 do not show a diastem at the boundary; for this reason, they misrepresent the extinction event.

Figure 8–4 attempts to generalize the extent of the omissions of Upper

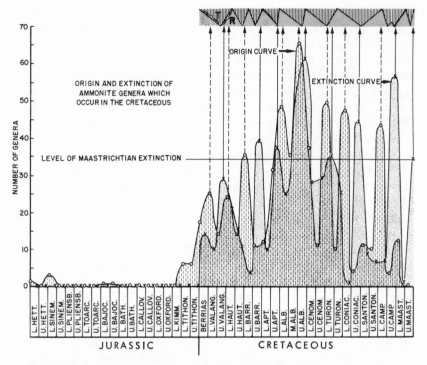

Figure 8-5. An evolutionary summary of Cretaceous ammonite genera that shows levels of origin and of extinction plotted against the history of Cretaceous eustatic fluctuations (after Kauffman, 1977a).

The origin and the extinction curves trace the number of genera (scale on left) that appeared or disappeared during the chronostratigraphic divisions (ranged along the bottom of the Figure).

Eustatic fluctuations produced alternating transgressions (T) and regressions (R) where indicated at the top of the Figure.

The vertical dashed lines show the relationships between the major radiation events and eustatic history; note that six of these events, including the four largest, occurred either during transgressions or at transgressive peaks (times of maximum eustatic rise). The solid vertical lines indicate the relationships between the major extinction events and eustatic history; note that six of these eight events occurred during regressions (mainly during their middle or later parts) and that the other two occurred near peak transgressions—times of temperature stress (rapid changes) in epicontinental areas (see Figures 8-10 through 8-13).

The proliferation of new genera correlates with expanding ecological opportunities during times of eustatic rise and transgression. Extinction correlates mainly with times of increasing stress brought about by regression, diminishing ecospace, and significant changes of temperature and/or salinity occurring at, and following, peak transgression.

Note that the major extinction events for Cretaceous ammonite genera occurred *within* the Cretaceous; also note that nearly all of these events were of greater magni-

Cretaceous marine strata in the Euroamerican region and other selected sites. Cratonic interiors—such as central and south-central Europe and the main part of the western interior of North America—generally have preserved no Maastrichtian and Paleocene marine rocks; or, they have preserved—at best—only the Lower to early-Middle Maastrichtian rocks, which are overlain by nonmarine Cretaceous-Paleocene strata that have been cut by one or more disconformites and, in turn, by locally restricted Lower to Middle Paleocene marine/brackish water facies (for example, the North American sequence of marine Fox Hills, nonmarine Hell Creek/lower Lance, marine Canonball). In the best-preserved American intracratonic record, there is a gap of 2 to 3 myr between marine faunas.

Similar—but, predictably, smaller—gaps exist in continental shelf sites. Throughout much of the United States Gulf and Atlantic coastal plain, rich Middle Danian marine biotas lie in disconformable, or paraconformable, contact on rich Middle Maastrichtian marine biotas. Although, in some cases, the contact lies within glauconitic sand facies with little evidence of an erosional break (in Maryland, for example, the Danian Brightseat Formation resting on the Maastrichtian "Monmouth" Formation), the coastal plain microbiotas indicate that the youngest Cretaceous rocks are of early-Middle Maastrichtian age (*Globorotruncana gansseri* zone) and that the overlying Paleocene rocks are Middle Danian in age (*Globigerina daubjergensis–Globorotalia pseudobulloides* zone). A minimal gap of 2.5 to 3 myr is indicated; the gap narrows offshore. European shelf sites are similar, except in regions of active tectonism (the Mediterranean and North Sea basin margins); in these regions, the subsiding continental margins and tectonic basins have preserved nearly complete pelagic and outer-shelf sequences. In South America and Africa, predominantly incomplete Cretaceous-Tertiary boundary sequences seem—despite a paucity of data—to be similar to those of the North Temperate realm. Core Tethyan areas—such as the Antilles, Central America, and the Caribbean margins—were tectonically active sites with island arcs; yet, the carbonate platform environments of these areas lack Late Maastrichtian strata and, in most cases, Paleocene ones. On the other hand, pelagic deep-water facies in the Tampico area of Mexico, in northeastern Brazil, and in Hispañola have preserved relatively complete boundary sequences, but ones without a significant number of preserved macrofaunas.

---

tude than the one of the latest Maastrichtian, which has previously been envisioned as the point of the "catastrophic" extinction of the ammonites.

Data for Cretaceous ammonite genera have been taken primarily from Arkell, et al. (1957), but are modified by information from numerous other sources.

Thus, almost everywhere, large stratigraphic gaps characterize the exposed epicontinental Cretaceous-Tertiary boundary sequence. A *few* exposed pelagic sequences and deep-sea cores have preserved the boundary zone; but, in most cases, these sequences and cores allow biological analysis only of the microbiota, because macrofossils—other than burrows—have rarely, if ever, been preserved in these facies. The Scandinavian chalk sequences are notable exceptions. For the widely accepted mass extinction of larger marine organisms at the very end of the Cretaceous, therefore, little actual evidence has been preserved. By contrast, excellent documentation supports the sudden catastrophic extinction of the warm-water calcareous microplankton during the terminal "event."

Our best hope for documenting the terminal Cretaceous marine extinction among the entire biota will result from our identifying and studying in detail *all* the fossil groups in exposed sections from those deep epicontinental basins, continental shelves, and continental margin sites that were deposited above the ACD (Aragonite Compensation Depth) and have essentially preserved an uninterrupted sequence of marine sediments across the Late Maastrichtian-Danian interval. A few such sections do exist; most have been extensively studied for their microbiota; but, with the exception of the northwest European sequences—the shallowest and most northerly of the better boundary zone sections—the macrobiota are still poorly known. Yet, even a broad survey of these sequences will yield new information that is relevant to the biological fabric of the latest-Cretaceous extinction.

## THE PATTERN OF LATE CRETACEOUS MARINE EXTINCTION: AN OVERVIEW

The concept of a catastrophic terminal Cretaceous extinction that involved morphologically and ecologically diverse organisms is neither supported by many detailed phylogenetic studies nor predictable by current ecological and evolutionary theory, which suggests that patterns of even major extinction events should be graded. At one extreme, small isolated populations of stenotopic organisms with extreme niche-specificity and limited dispersal may evolve rapidly and, when subjected to even low levels of stress, abruptly become extinct. At the other extreme, in the same environments, ecological generalists with large and widely dispersed populations and low niche-specificity may show low rates of evolution and prolonged, graded extinctions. All intermediate stages of these patterns should exist, and these relationships should hold for any taxonomic level. Theoretically, a catastrophic extinction is possible only as a result of a geographically widespread, rapidly developed environmental crisis that exceeds the tolerance levels of some eurytopic groups (for example, the last inoceramid bivalves) *and*, by implication, also exceeds the tolerance levels of most coexisting stenotypes. Such a catastrophe would either require the interjection of extraterrestrial forces—such as supernovas and radiation levels that are higher than those predictable on earth—or, it would demand that there be rapid and nearly simultaneous large-scale changes in various environmental parameters which, taken collectively, exceed the survival limits of ecologically diverse taxa. Both of these scenarios have been postulated to explain the catastrophic extinction of diverse groups at the end of the Cretaceous. However, detailed analyses of environmental decline and extinction patterns for most taxa during the latest Cretaceous indicate that these changes were graded over millions of years and thus do not support the concept of a single, abrupt, global biotic crisis.

It is true that many major environmental changes did converge at the end of the Cretaceous: the eustatic lowering of sea level to 100 meters below its present stand; global epicontinental regression that resulted in a great diminution of prime marine ecospace and in the elimination of important niches; rapid fluctuations in the marine temperatures of some regions; and the restructuring of oceanic circulation, which possibly involved widespread chemical changes (oxygen, salinity, trace elements, and so on). But these events were also time-graded over, at least, the last few million years of the Cretaceous. Final eustatic fall and regression began in the Late Campanian and, after a last Middle Maastrichtian transgressive reversal ($T_{10}$, Kauffman, 1979), proceeded rapidly; the final marine temperature fluctuations that led to overall decline began in the Middle Maastrichtian. Ecological theory predicts that most of the extinctions that resulted from these changes should have also been graded among organisms with differing ecological strategies and tolerances. Some groups should have been relatively unaffected; others should have been affected only by the highest stress levels at the end of the Cretaceous; and still others should have become extinct long before the end of the Cretaceous, either because of minor environmental changes associated with the onset of the terminal regression or because of earlier and less intensive periods of change in oceanic environments and ecological structures. For the few major groups (for example, calcareous microbiota) that underwent widespread extinction at—or, very near to—the Cretaceous-Tertiary boundary, it is probable, therefore, that the magnitude and diversity of the converging factors of environmental decline, rather than either any single factor or the rate of decline, led to their abrupt demise.

Furthermore, the kinds and patterns of environmental changes that converged at the end of the Cretaceous and are credited with having caused the terminal Cretaceous extinction event are not unique; similar changes of a lesser magnitude also characterized other parts of the Cretaceous. For example, the end of the Albian was marked by a major global regression, a drop in marine temperatures (Kauffman, 1977, figure 7), and an extensive oceanic anoxic event (Fischer and Arthur, 1977); these changes combined to produce widespread extinction. The Turonian-Coniacian boundary zone and that of the Cenomanian-Turonian were each marked by a relatively abrupt rise in epicontinental marine temperatures and/or salinity that was followed, first, by an extensive anoxic event during peak transgression and global eustatic rise, and subsequently, by a temperature decline during regression. These stage boundaries are greatly accentuated by a closely spaced series of relatively abrupt extinctions, which reflect the high stress that is associated with multiple shifts in the environment over a short period of time. For many marine groups (for example, the ammonites), the intra-Cretaceous crises had a more profound effect on their extinction than did the terminal Cretaceous event. For others, the events of the terminal Cretaceous environmental decline were more profound than, but still similar to, events that occurred several times *during* the Cretaceous. Therefore, the biological-environmental patterns of intra-Cretaceous extinction events that have been so well preserved in many parts of the world can be studied in detail and can serve as natural models for the interpretation of the rarely preserved record of the terminal Cretaceous extinction.

When interpreting terminal Cretaceous extinction patterns, it is important to consider one other factor. The extinction of a group at, or near, a peak in its diversity infers a biologic catastrophe (as happened, for example, to radiolitid and hippuritid rudists, to coccolithophorids, and to planktonic foraminifera; see: Figures 8–1C and 8–2B). The extinction of groups near the end of their evolutionary decline, when they are represented by only a few generalized taxa (as was, for example, the case with inoceramids and ammonites; see Figure 8–1B), is probably of little consequence, even in those cases where the group's extinction happens to coincide with a more dramatic extinction event (see Figure 8–2B). The elimination of evolutionarily "old" stocks during a mass extinction is probably *caused by* the biological crisis (for example, the breakdown of the ecological structure) rather than being causal to—or, even, reflective of—primary environmental changes. The importance of these considerations is born out in Figures 8–1B, 8–2B, and 8–5 to 8–9, which depict Cretaceous mollusks in different stages of their evolutionary history at the time of the terminal Cretaceous extinction event.

The ammonites have been widely cited as a major group that suddenly became extinct at the end of the Cretaceous. The neocatastrophist view of the Ammonoidea (Figure 8-1A; Schindewolf, 1962) as important components of Cretaceous faunas right up to their "abrupt" terminal Cretaceous extinction has been fostered by undifferentiated straight-line plots—such as Figure 8-1A—that lack stratigraphic data below the stage level. Figure 8-1B, on the other hand, depicts the same event, but with generic diversity among superfamilies plotted against stages from the Devonian through the Cretaceous. As a result, this plot makes it clear that the greatest abundance and diversity of ammonites occurred between the Triassic and the Middle Cretaceous; it further shows that the few surviving lineages were on an evolutionary decline at the time of their apparently abrupt Late Maastrichtian extinction. This rendering of the data certainly lessens the impact of the so-called extinction event. Figure 8-5 more accurately plots against substage divisions the origin, extinction, and diversity levels of those ammonite genera that occurred in the Cretaceous. The plot shows that these evolution and extinction events were episodic; further, it indicates that these episodes bore a general relationship to global eustatic fluctuations, which produced transgressive-regressive pulses. Six of eight major radiation events correlate with transgressions because these are periods of expanding ecospace, niche diversification, ameliorating climates, and low levels of stress other than that due to competition and predation. All major extinction episodes occurred between peak transgression and peak regression. Regressions, on the other hand, were times of increasing stress due to diminishing ecospace, the loss of habitats, temperature decline, and widespread environmental deterioration. The terminal Cretaceous ammonite extinction fits this pattern: a major regression, oceanic oxygen restriction, and rapid temperature fluctuations were in progress. The most important aspect of Figure 8-5 is the demonstration that, when plotted at this level, all but one of the intra-Cretaceous extinction events (namely, that of the Upper Valanginian) were *greater than*—and, apparently, just as abrupt as—the terminal Cretaceous event. A thorough study of the greatest Cretaceous ammonite extinction event—the one at the end of the Albian—as well as an examination of its causes, should shed much light on the Maastrichtian extinction.

Finally, in areas where the Late Cretaceous-Paleocene boundary sequence is reasonably complete, some very detailed data are available on the nature of ammonite extinction throughout the Maastrichtian. Wiedmann (1964, figure 4) documented a gradual loss of ammonite genera and species through the Campanian-Maastrichtian sequence in Spain, where the complete boundary sequence has been preserved in marine pelagic strata at Zumaya and elsewhere. His data seem to indicate that morphologically more specialized and/or geographically more restricted taxa (stenotypes?) became extinct early in this final evolutionary decline and that more generalized and widespread genera (for example, Pachydiscidae) disappeared later in the Maastrichtian. The history of the disappearance of the ammonites that Wiedmann observed at Zumaya is especially revealing (pers. comm. in 1977): geographically more restricted warmwater lineages (stenotypes?) disappeared near the Middle-Upper Maastrichtian boundary; moderately diverse cosmopolitan (eurytopic?) ammonites ranged through the early Upper Maastrichtian and then gradually began to drop out; the few surviving groups, all of which were cosmopolitan generalists, became smaller (dwarfed?) in the later Maastrichtian and gradually disappeared in the highest 75 meters of the section. Just below the boundary, only a single cosmopolitan pachydiscid species remained. These data suggest that, throughout the Late Cretaceous in Tethys, ammonite extinction was graded, with ecological specialists disappearing first, and generalists last. If this sequence is representative, then the extinction of ammonites in pelagic Tethyan settings was graded and, possibly, was related to changes in marine temperature and chemistry during the Campanian-Maastrichtian and to increased predation from fishes and reptiles, rather than having been related to some terminal Cretaceous environmental disaster.

A survey of various fossil groups that show significant levels of extinction at the end of the Cretaceous reveals four major extinction patterns (demonstrated by families of Bivalvia in Figures 8-6 through 8-9). These patterns are related, in part, to the evolutionary "stage," or antiquity, of each group (and, thus, to the degree to which each had been already stressed *during* the Cretaceous by competition, the evolution of new predators, the loss of principal habitats, and so on); the patterns are also related, in part, to each group's tolerance for the large-scale, episodical environmental changes that characterized Cretaceous history. Of these four extinction patterns, only one records a dramatic (that is, catastrophic) Maastrichtian extinction of diverse groups at a peak in their radiation; and this pattern characterizes less than 25% of the lineages that become extinct in the latest Cretaceous. The other three patterns are far more subtle examples of long-term, graded and/or low-level Cretaceous extinctions, have quite different implications, and represent the great majority of taxa that died out by the end of the Cretaceous: that is, these three patterns represent the *normal* extinction picture.

*Extinction pattern #1* characterizes geologically old marine stocks that had pre-Mesozoic origins and evolutionary peaks and underwent evolu-

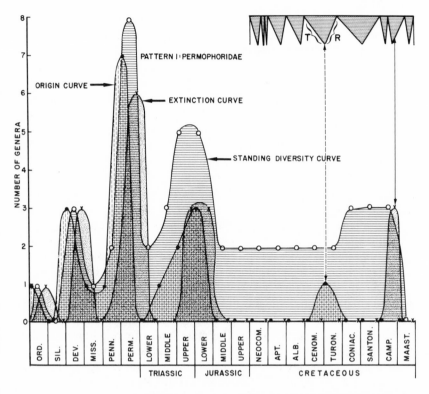

Figure 8-6. Cretaceous extinction pattern #1 among Bivalvia. The family Permophoridae reached its evolutionary peak in the Late Paleozoic and had its major extinction episode in the Late Permian. The Mesozoic evolution of this "archaic" family involved infrequent radiation and extinction events of a low to moderate magnitude; the family was on the evolutionary wane throughout the Mesozoic; and its final extinction near the Campanian-Maastrichtian boundary predated the end of the Cretaceous by 4 to 5 myr. Yet, this family has been portrayed as having become abruptly extinct during the terminal Cretaceous biotic crisis.

Pattern #1 is typical of the extinction histories of many "old" macroinvertebrate lineages that became extinct near, or at, the end of the Cretaceous.

Data has been modified from Cox et al. (1969). Taxonomic units are genera plotted against substages (the points at the bottom). Transgressive-regressive curve for the Cretaceous (upper right) is after Kauffman (1977a).

Note a pattern that is common to nearly all of the lineages that have been analyzed (see Figures 8-5 and 8-7 through 8-9): the single Cretaceous radiation peak (indicated by the vertical dashed line) occurred at a transgressive maximum and the terminal Cretaceous extinction event (indicated by the solid vertical line) occurred during a regressive minimum.

tionary decline through the Mesozoic. The decline was mainly due to competitive exclusion and replacement by more advanced organisms. During the Cretaceous, these groups reached the last stages of their evolutionary decline and were represented by a few generalized lineages, a generally small population size, and—in some cases—a limited dispersion that involved ecological refugia. Many of these groups became extinct at, or near, the Cretaceous-Tertiary boundary; but their final extinction involved from one to (at most) a few species and seems to have been related more to widespread deterioration of the marine ecosystem (for example, the community or food-chain structure) than to the primary environmental causes for this deterioration.

The predominantly Paleozoic family Permophoridae characterizes this pattern among Bivalvia (Figure 8–6). During the great Mesozoic radiation of infaunal bivalves, this group of primitive, semi-infaunal to shallow infaunal, nonsiphonate, suspension-feeding bivalves was competitively replaced by more advanced siphonate forms (Stanley, 1968). Mesozoic survivors of the group have mainly occurred as endemics in biogeographically disjunct centers; some have occurred in brackish marginal marine environments, which could have been possible temporary refugia for this normally marine group. During the Upper Cretaceous, new adaptive forms that arose in these habitats finally drove off the remaining Permorphoridae; some rare Myoconchinae possibly survived into the Maastrichtian, but none are known from the Late Maastrichtian. The extinction of the Permophoridae was mainly related to competitive exclusion; the environmental changes of the latest Cretaceous regression expanded the diverse marginal marine habitats, where relict Permophoridae was living; these changes cannot be viewed, therefore, as being primarily responsible for the demise of Permophoridae.

*Extinction pattern #2* characterizes groups that arose in the latest Jurassic and Early Cretaceous, reached their peak radiation and diversification during the Middle Cretaceous, and were into evolutionary decline at the time of the Maastrichtian extinction. In some groups, Maastrichtian representatives were still moderately diverse, and the extinction event was significant; but, in most cases, only a few generalists represented these lineages at the time of their extinction. Cretaceous ammonoid superfamilies (Figures 8–1B and 8–5), the Inocermidae, and the rudist family Requieniidae (Figure 8–7) show this pattern. In all cases, the major evolutionary decline in the later Cretaceous can be attributed mainly to ecological factors: predation, changes in habitat, and competitive exclusion. During the Upper Cretaceous, the ammonites were subjected to increasing predation levels; this situation, coupled with the general deterio-

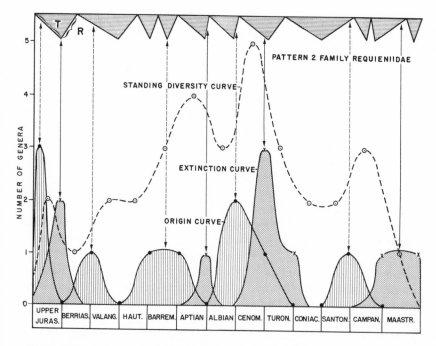

Figure 8–7. Cretaceous extinction pattern #2 among Bivalvia. The rudist family Requieniidae reached its evolutionary peak in the "middle" Cretaceous, when it became a dominant member of the Tethyan "reef" community. Competitive exclusion by more advanced rudists produced a major extinction episode in the latest Cenomanian and, subsequently, lesser extinction events. The family was on the evolutionary wane for 27 myr, until the Late Maastrichtian, when the final extinction of the generalized surviving stocks occurred. Although many rudist families had—like the Requieniidae—experienced their major extinction and were on the evolutionary decline long before the end of the Cretaceous (compare Figures 8–2B and 8–8), the rudists have generally been depicted as a group that died out abruptly during the terminal Cretaceous extinction event (see Figure 8–1A).

Data, after Cox et al. (1969), has been modified and plotted as the number of genera that originated or became extinct per stage (shown at the bottom). The eustatic curve at the top (after Kauffman, 1977a) shows widespread transgressions (T) and regressions (R) with essentially synchronous global peaks.

Unlike other lineages, the Requieniidae lack a clear correlation between radiation events (indicated by the vertical dashed lines) and eustatic history—only three of the five radiation events are associated with transgressive pulses—as well as lacking a clear correlation between extinction events (indicated by the solid vertical lines) and eustatic history—only two of the four extinction events are associated with regression. This lack of correlation may reflect the complicated sea-level history of the Cretaceous rudists' primary habitats in island arc systems and shallow carbonate shelves of such tectonically active areas as the southern Mediterranean platform and the Caribbean.

ration of their favored midwater to epibenthic habitats during the Late Cretaceous regression, probably triggered their decline. The final extinction of the ammonites seems to have been related to marked changes in water chemistry and temperature during the Maastrichtian. The Inoceramidae, predominantly epibenthic bivalves, largely fell prey to diversifying crabs and bottom-feeding vertebrates during the Late Cretaceous.

After their Late Jurassic origins, the rudist family Requieniidae (Figure 8–7) rapidly radiated into reefoid carbonate flat environments to become an important component of "middle" Cretaceous rudist frameworks (Kauffman and Sohl, 1974). But their coiled, recumbent to semi-recumbent valves made them awkward framework builders, and they were largely replaced in Late Cretaceous reef environments by the more erect, more rapidly growing, efficiently clustered radiolitid and hippuritid rudists. These events led to the steady decline of requieniids throughout the late Cretaceous (see Figure 8–7); only one generalized group (*Bayleia*) probably survived beyond the Lower Maastrichtian. Even so, the fossil record of this Cretaceous rudist and many others does not extend through the last 1 to 1.5 myr of the Maastrichtian. The extinction of the Requieniidae and other primitive rudists in the Maastrichtian (see Figure 8–2B) was, therefore, due largely to competitive exclusion by more advanced rudists, and the extinction occurred *before* the biological-environmental "crisis" at the end of the Maastrichtian. Only the abrupt extinction of the Radiolitidae, Hippuritidae, and certain Caprinidae in the Middle to early-Upper Maastrichtian bears significantly on the problem of the terminal Cretaceous extinction event and its causes.

*Extinction pattern #3* is characterized by seemingly-abrupt and widespread extinction among diverse organisms during a Maastrichtian peak in their radiation. This pattern has been commonly envisioned for most taxa that did not survive the terminal Cretaceous biotic crisis, including such groups as the ammonites and the inoceramids that had actually experienced evolutionary decline throughout the Late Cretaceous and were only represented in the Maastrichtian by a few generalized taxa. Cretaceous marine groups that do show extinction pattern #3 include most of the planktonic foraminifera, calcareous nannoplankton, larger benthic foraminifera, scleractinian hermatypic corals, many ostreid and trigoniid bivalves, and advanced rudists, as well as nerineid and actaeonellid gastropods. The great majority of the lineages that followed pattern #3 were primarily stenothermal (?) inhabitants of upper neritic and shallow shelf habitats in the tropical-subtropical Tethyan realm. It is among these that we should seek evidence for the primary causes for the terminal Cretaceous extinction event.

The radiolitid rudists characterize extinction pattern #3 (Figure 8–8);

they had Early Cretaceous origins and diversified through a series of discrete radiation events that were tied to eustatic rise and epicontinental transgression during the Late Cretaceous. One of four radiation events occurred early in the Albian transgression, two others were at the peaks of transgressive pulses (the Cenomanian and the Coniacian-Santonian), and the fourth took place just following the peak of the Upper Campanian transgression (see Figure 8-8). During this diversification, radiolitids became important components of many different types of rudist frameworks and of most stages in their succession (Kauffman and Sohl, 1974). At the peak of their Maastrichtian radiation, the family became extinct within a short time span (Figures 8-2B and 8-8). The exact timing of this event—the only major extinction in radiolitid history—is questionable because, due to widespread regression in the latest Cretaceous, Upper Maastrichtian carbonate shelf deposits have rarely been preserved. Even in those areas—such as Jamaica—where marine carbonate shelf faunas and facies characterize slightly younger Middle Maastrichtian deposits, both Caribbean and Mediterranean Tethyan sequences seem to show widespread extinction of diverse rudists, including most radiolitids, near the Middle-Upper Maastrichtian boundary. Some rudists—the more generalized radiolitids among them—survived into the Late Maastrichtian, but virtually all rudists appear to have disappeared before the boundary event. The disappearance of so many rudists in the late Middle Maastrichtian is remarkable, however, and may be the first reflection of the widespread environmental decline that led to the terminal Cretaceous crisis; this disappearance generally corresponds to the first stage of marine regression, oceanic oxygen decline, and the rapid fluctuations and regional decline in temperature that culminated at the end of the Maastrichtian.

Of the four extinction patterns that are under discussion, pattern #3 is probably the only one that reflects the primary causes for the terminal Cretaceous biotic crisis. One of these causes was certainly the abrupt drop in oceanic and epicontinental marine temperatures during the Late Maastrichtian, as well as the effects of these temperature declines on the stenothermal groups that had been inhabiting Tethys. The hardest hit of these groups seem to have been those that were inhabiting the upper part of the tropical marine water column: such groups as shallow-water benthos and permanent planktonic-nektonic biota or, temporarily, planktotrophic larvae of the shelf benthos. The great majority of benthic shelf invertebrates living in the tropics have long-lived planktotrophic larvae.

*Extinction pattern #4*   is characteristic of many groups that were important components of the Cretaceous biota up to the end of the Maastrichtian and underwent only slight to moderate levels of extinction as a result

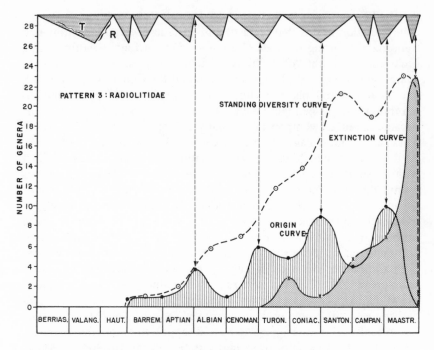

Figure 8–8. Cretaceous extinction pattern #3 among Bivalvia.

The dominant, structurally advanced, Late Cretaceous rudist family Radiolitidae—together with the Hippuritidae (Figure 8–2B)—had replaced more primitive Lower Cretaceous families in reef habitats and had achieved their maximum evolutionary development at the point of their dramatic Late Cretaceous extinction.

Data has been modified from Cox et al. (1969) and plotted as the number of genera that originated or became extinct per stage (shown at the bottom). Although this coarse plot suggests that the major extinction event was near—or, at the end of—the Late Maastrichtian, detailed field data show that the major disappearance of Radiolitidae actually occurred near the Middle-Upper Maastrichtian boundary and that probably few groups survived to the end of the Maastrichtian. The eustatic curve at the top (after Kauffman, 1977a) shows global transgressions (T) and regressions (R).

Note that all radiation events (indicated by vertical dashed lines) seem to have occurred during transgression or near peak transgression and earliest regression; also note that the main extinction event (the solid vertical line) occurred with a latest Cretaceous regressive pulse ($R_9$ near the Middle-Upper Maastrichtian boundary, if the data were to be plotted in detail; $R_{10}$ in the cruder plot used here, based on published ranges).

Extinctions that follow patterns such as this one are the only "catastrophic" events near the end of the Cretaceous and are directly reflective of environmental factors that led to the biological crisis. Interpretation of the terminal Cretaceous extinction event lies in explaining such patterns, which are almost wholly restricted to warm-water organisms.

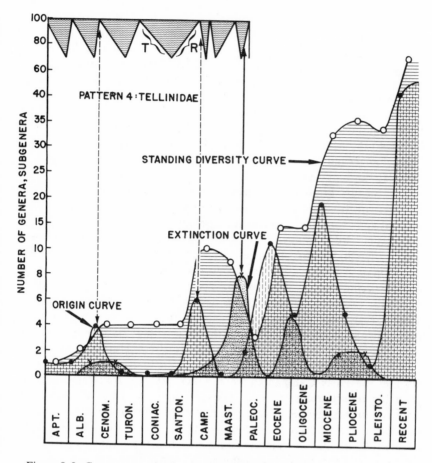

Figure 8-9. Cretaceous extinction pattern #4 among Bivalvia.

The Tellinidae are a widespread family of largely eurytopic bivalves that had their origins in the Mesozoic, a moderate—but significant—extinction event near the Cretaceous-Tertiary boundary, their main radiation in the Cenozoic, and their highest diversity in recent times. Many genera and lineages crossed the Cretaceous-Tertiary boundary, especially in the temperate realm.

Data from Cox et al. (1969), Afshar (1969), and Boss (1966, 1968, 1969a, 1969b) has been plotted as the number of genera and subgenera per stage or epoch (shown at the bottom). The eustatic curve that shows Cretaceous transgressions (T) and regressions (R) is after Kauffman (1977a).

Note that both major Cretaceous radiations (indicated by vertical dashed lines) took place during transgressive pulses and that the only major extinction event (indicated by the solid vertical line) is associated with the terminal Cretaceous regression. This pattern typifies the patterns of successful Cenozoic groups that underwent their initial major radiations in the Cretaceous.

of the crisis (Figure 8–9). Typically, these groups became quickly reestablished in the Paleocene and continued their radiation to even higher levels of Cenozoic diversification. Many groups that showed this extinction pattern are represented today by trophic and/or habitat generalists, and/or they have habitat preferences for temperate climatic shelf zones and deeper marine environments. The fact that these groups experienced relatively low levels of extinction at the end of the Cretaceous may well have been due to their greater tolerance for the environmental events that led up to the Maastrichtian biotic crisis; and/or it may have been that these changes had had a greater impact on tropical marine habitats than they had on temperate ones.

The bivalve family Tellinidae typifies this extinction pattern. The members of this family are infaunal deposit-feeders in shallow to deep tropical and temperate marine environments. Given their broad habitat range, their stable food resources, and their broad environmental tolerance, they should have been less stressed than the more stenothermal plankton or Tethyan suspension-feeding benthos by terminal Cretaceous marine events that involved oxygen decline, rapid temperature fluctuation, widespread regression, and the destruction of the oceanic plankton. Their larvae may also have had shorter planktonic stages that were deeper in the water column—as is the case with some modern Tellinidae—and, therefore, were less susceptible to stress. During this family's Cretaceous history, radiation events during the latest Albian-Cenomanian and the Early Campanian were coincident with transgressions that probably reflected eustatic rise. The only major Cretaceous extinction occurred during the great Maastrichtian regression. Despite their being environmental and trophic generalists, nearly two-thirds of the tellinid genera that are known from the Cretaceous became extinct as a result of the Maastrichtian crisis; significantly, the ones that did were mainly warm-water forms. This fact is an important measure of the magnitude of the terminal Cretaceous environmental deterioration. Paleogene radiation of the Tellinidae was rapid, however, and of considerable magnitude; by the end of the Eocene, tellinid diversification had reached proportions that exceeded their Cretaceous levels. The Tellinidae are now at the known peak in their evolutionary history (Figure 8–9).

*Conclusions*  Groups of marine organisms that showed significant extinction or that disappeared altogether at the end of the Cretaceous had diverse extinction histories that fall into the four general patterns which have been described above. Of these four patterns, only one (#3) shows a relatively abrupt Maastrichtian extinction of lineages at a peak in their evolution, and this pattern represents less than one-fourth of the taxa that disappeared by the end of the Cretaceous. Yet, pattern #3 is the scenario that has commonly been envisioned for most of the terminal Cretaceous extinctions. Furthermore, where pattern #3 does apply, the major extinction events—with the exception of the one for the calcareous microplankton—predate the end of the Maastrichtian by a million years or more. The other three extinction patterns either reflect ecological factors of competition and predation that decimated groups long before the end of the Cretaceous, or else these other patterns involved more eurytopic groups whose radiation was only temporarily interrupted by the environmental crisis at the end of the Cretaceous (Figure 8–9). In the latter cases, as well as in groups that show pattern #3, tropical-subtropical organisms of either shallow shelf or upper pelagic habitats were largely involved. Temperate extinction history was more graded, was of a lesser magnitude, and involved fewer groups. These data suggest that the Late Cretaceous biotic crisis was caused by environmental factors that affected mainly shallow tropical habitats.

Examining the fabric of extinction involves sorting out and interpreting the individual patterns of evolution shown by those groups that were involved in the extinction event. This critical step allows one to focus on extinctions that were caused by major changes in the environment and, thus, on a clearer interpretation of these factors; in addition, it places into perspective those extinctions that were the secondary results of ecosystem shock and, as a result, permits a more objective evaluation of the terminal Cretaceous extinction event.

## THE PRESERVED RECORD OF MARINE MACROFAUNAL EXTINCTION AT THE END OF THE CRETACEOUS

In a few widely scattered areas of the world, stratigraphic sequences preserve most, or all, of Late Maastrichtian through Danian marine history. Most of these sequences are pelagic deep-water ones that contain few macrofossils other than burrows. The details of stratigraphy, sedimentary environments, and biologic history (mainly of microbiota) have been well documented only in the following sections: the Gubbio area of Italy; Zumaya, Spain; an area of northwestern Europe that includes the Netherlands, Denmark, and southern Sweden; the Crimea, Union of Soviet Socialist Republics; the Tampico area of Mexico; southern Haiti; and near Recife in northeastern Brazil. Numerous deep-sea cores have provided additional details of the boundary event, but only the northwestern European shelf sequence has yielded a large macrofauna. When attempting an interpretation of the nature of the terminal Cretaceous extinction, however, the history of the sparse macrofauna in more offshore pelagic sequences is of equal interest. That history is reviewed here, as is the preserved record in such critical, but less complete, boundary sequences as those of Tethyan carbonate platform environments (rudist facies).

The absence, or the paucity, of shelled macrofossils in such deep-water pelatic carbonate sequences as those from Gubbio, Zumaya, and Tampico has been variously attributed to dissolution below ACD (Araganite Compensation Depth) and CCD (Calcite Compensation Depth), as well as to the general ecological restriction of macrobenthos in deeper marine habitats. Neither hypothesis is wholly supported by existing evidence. Preservational aspects of the microbiota in the same sediment rule out extensive dissolution at many of these sites. Furthermore, deep-water organic-rich muds today support one of the most diverse known benthic macrofaunas, especially shelled mollusks (Sanders, 1969). Even below the ACD, the thick periostracum of many mollusks protects them from dissolution for some time after death (Kauffman, 1976). The abundance and diversity of trace fossils in many pelagic sediments and most studied boundary sequences rule out the hypothesis of severe environmental-ecological restriction of the shelled benthos being a *normal* aspect of this facies. The paucity of shelled macrofaunas in the Late Maastrichtian of many sequences (for example: Gubbio, Zumaya, and Tampico) should be viewed as a significant reflection of the terminal Cretaceous extinction event and the large-scale environmental factors that caused it.

## TETHYAN PELAGIC SEQUENCES

*Gubbio, Italy* An essentially complete sequence of tropical pelagic carbonates spans the Maastrichtian-Danian interval at Gubbio in Italy. This sequence has been chosen as the type section for the Late Cretaceous geomagnetic reversal time scale (Alvarez et al., 1977; Roggenthen and Napoleone, 1977; Lowrie and Alvarez, 1977); the lithostratigraphy, sedimentology (Arthur and Fischer, 1977), biostratigraphy, and evolutionary history of the planktonic calcareous microbiota has been thoroughly studied (Premoli Silva, 1977; Monechi, 1977, 1979).

The Maastrichtian-Danian contact, as defined by the abrupt extinction of virtually all of the Cretaceous calcareous planktonic microbiota, lies within a limestone sequence. A thin clay lies less than a meter above the contact. No boundary unconformity has been reported by Arthur and Fischer (1977), and Premoli Silva (1977), cited a complete planktonic zonal sequence from the Upper Maastrichtian through Paleocene time; but other authors (for example, Gamper, 1977, citing Luterbacher and

Premoli Silva, 1964) have suggested that, due to an apparent paraconformity at the boundary, the highest part of the Late Maastrichtian *Abathomphalus mayaroensis* zone is missing. No macrofossils have been reported either from the Upper Maastrichtian or from the lowest Danian (Arthur and Fischer, 1977; Premoli Silva, 1977, who specifically notes the lack of ammonites). Yet, the foraminifera did not show extensive destruction of calcite due to dissolution or diagenesis; some dissolution of coccoliths was noted (Premoli Silva, 1977, p. 372; Monechi, 1979). These observations suggest that, if benthic and midwater macrofaunas with calcite and (possibly) aragonite skeletons were actually present at Gubbio, then at least some elements of them should have been preserved.

The occurrence of abundant and diverse trace fossils throughout the Cretaceous-Tertiary boundary sequence at Gubbio was noted by Arthur and Fischer (1977, p. 370); Fischer (pers. comm. in March of 1979) noted that:

> Within the Campanian through Paleocene portion of the Scaglia Rossa, original laminations are virtually absent, and though some burrow-mottling is present, limestones and marls are basically homogeneous; bioturbation eliminated the details of depositional history.

Obviously, a benthic shelly macrofauna was not excluded from the Gubbio section by anoxic and/or $H_2S$-enriched substrates or by immediately overlying bottom waters. Elsewhere in the Cretaceous, such pelagic sequences as these, as well as sequences that lack extensive bioturbation, do preserve benthic bivalves—such as *Inoceramus*—along with their epibiont-endobiont communities of worms, barnacles, bryozoans, oysters, pteriacean and pectinacean bivalves (Kauffman, 1978b, 1978c): groups that seem to have had a high tolerance for oxygen depletion during the Cretaceous. The lack of macrofauna in the Upper Maastrichtian at Gubbio is, therefore, most probably a reflection of earlier environmental decline *within* the marine water column; this decline probably then spread throughout the midwater habitats and, eventually, by the time of the Late Maastrichtian, throughout the epibenthic habitats as well. It was this earlier environmental decline that ultimately led to the terminal catastrophe among the warm-water, upper pelagic, calcareous microbiota at the Cretaceous-Tertiary boundary.

*Zumaya, Spain* The complete pelagic boundary sequence at Zumaya, located between Bilbao and San Sebastian on the north coast of Spain, has yielded more definitive evidence than any other deep-water pelagic sequence concerning the Upper Maastrichtian extinction event. Percival and Fischer (1977, p. 1) have described "an essentially uninterrupted sed-

imentary sequence" of deep-water flysch that transgresses the Cretaceous-Tertiary boundary and comprises 75 meters of latest Maastrichtian
"purple marls" (*Abathomphalus mayaroensis* zone) that are conformably
overlain by pyritic "boundary shale" with a thickness of 25 to 35 centimeters. Planktonic foraminifera (*A. mayaroensis* zone) and calcareous
nannoplankton extend 10 centimeters up into the boundary shale and
disappear abruptly at the Cretaceous-Tertiary boundary (Percival and
Fischer, 1977, p. 7). The upper 15 to 28 centimeters of the boundary shale
has been assigned to the lowest-Paleocene planktonic zone of *Globigerina
eugubina*. Gray Paleocene limestones and shales (zone of *Globorotalia
pseudobulloides*) conformably overlie the boundary shale.

Among the pelagic microbiota at Zumaya, the fabric of extinction has
been well documented by Herm (1965) and by Percival and Fischer
(1977) and can be summarized as follows:

1) Normal, diverse assemblages of oceanic Tethyan planktonic foraminifera and calcareous nannoplankton characterize the latest Maastrichtian to within a few meters of the contact.

2) A few meters below the top of the Maastrichtian, Herm (1965)
documented a decrease in the size of planktonic foraminifera, the selective extinction of some globotruncanid species, the disruption of pelagic
community structure, and an increase in the relative abundance of calcareous bethonics; the marine nannoflora remained normal.

3) The continued decline of the planktonic foraminiferal fauna was
associated with the introduction among the nannoflora and dinoflagellates of some ecological generalists and opportunists (so-called "disaster
species")—especially Braarùdospheres and Thoracospheres—about
10,000 years before the terminal extinction (Percival and Fischer, 1977).
The presence of these disaster species, which are normally excluded from
complex microbiotic assemblages of ecological specialists, first occurs in
the upper few centimeters of the purple marls and indicates the onset of
stress conditions and the widespread disruption of the pelagic ecosystem.
During the deposition of the lower half of the boundary shale, disaster
species gradually became numerically more dominant, while the normal marine nannoflora diminished in numbers. At the same time, benthonic calcareous foraminifera became dominant over planktonics
(Herm, 1965).

4) The remaining Cretaceous calcareous planktonic foraminifera and
nannofossils abruptly disappear about 15 centimeters above the base of
the boundary shale, presumably due to their extinction at the end of the
Cretaceous.

5) "Persistent species" of Braarudospheres and Thoracospheres (Per-

cival and Fischer, 1977) become numerically dominant in the basal Paleocene, but their diversity is low. Planktonic foraminifera were apparently initially sparse and very small; upsection, they increase in diversity. This assemblage, having a temporal span of about 1 myr, extends through 6 meters of strata in the upper boundary shale and lower "Danian Limestone" (Percival and Fischer, 1977, figure 2).

6) The rapid diversification of the Early Danian nannoflora and planktonic foraminifera followed, as did the emergence of the typical Paleocene pelagic microbiota.

Macrofossils are known from the Maastrichtian at Zumaya up to the basal few centimeters of the boundary shales; the presence of these macrofossils provides a unique opportunity to study the extinction event among benthic and midwater larger invertebrates in a pelagic Tethyan sequence. Wiedmann (1964; and in prep.) has studied the Zumaya ammonites, and Kauffman (in prep.) the Zumaya bivalves. Jost Wiedmann (pers. comm. in 1977) has kindly provided the following outline of shelled macrofossil distribution in the Maastrichtian of Zumaya:

Only one small (dwarfed?) pachydiscid ammonite was found in the basal boundary shale at 10 to 15 centimeters below the Cretaceous-Tertiary contact. Below this find, in 75 meters of purple marls, there irregularly occurs a moderately diverse cosmopolitan assemblage of 10 ammonite species that belong to *Diplomoceras, Scaphites,* and *Pachydiscus;* all specimens are small for adults, having possibly been dwarfed due to ecological stress; upsection in this interval, ammonite diversity and abundance decrease. In marls 75 to 100 meters or more below the contact, the same general ammonite assemblage is represented by normal-sized adult individuals. This marl is underlain by a few meters of unfossiliferous limestone, below which is a thick limestone sequence that bears the first shelled macrobenthos in the lower part of the Maastrichtian sequence. This limestone unit is characterized by two biofacies: the upper few meters with abundant inoceramid bivalves (a plexus of *Trochoceramus radiosus, T. somaliensis,* and *T. zitteli* ) and a distinct warm-water ammonite assemblage; below this, similar ammonites are associated with a more normal marine tropical benthic fauna that is composed mainly of echinoids and mollusks. The Zumaya sequence was intensely bioturbated at most levels across the Maastrichtian-Danian boundary (Crimes, 1973; Percival and Fischer, 1977). Although burrows are somewhat less common in the boundary zone, they are continuously represented, thus ruling out severe anoxic or $H_2S$ poisoning of benthonic sediments at the Cretaceous-Tertiary contact.

The succession of marine events at Zumaya that can be inferred from these data is as follows:

1) The late-Late to early-Middle Maastrichtian warming of oceanic paleotemperatures (Kauffman, 1977a, figure 7; 1979, figure 2) favored the development of normal marine benthic biotas, midwater biotas, and pelagic Tethyan, as well as cosmopolitan, biotas above the ACD and/or the CCD.

2) The virtual disappearance of the more stenotopic tropical benthos—which was due, possibly, to oxygen depletion in the lower water column with continued late-Middle Maastrichtian warming—allowed more eurytopic inoceramid bivalves to colonize free substrates.

3) The elimination of the Tethyan midwater ammonites and all except the burrowing soft-bodied macrobenthos occurred near the Middle Maastrichtian-Late Maastrichtian boundary, which was associated with the expansion of benthic and/or midwater oxygen minimum zones during a transgressive (eustatic) peak and oceanic temperature high (Kauffman, 1977a, figure 7).

4) The initial Late Maastrichtian cooling (for example, Douglas and Saving, 1973, figure 3) was associated with an influx of cosmopolitan temperate ammonites in normal populations.

5) Ecological stressing of the entire marine biota was associated with widespread environmental decline, for example: the latest Maastrichtian regression, the destabilization of marine temperatures, and the widening of the surface to bottom gradient (Boersma and Shackleton, 1979), as well as the expansion of anoxic and low-oxygen zones. This ecological stress would cause stunting of ammonites and a gradual decline in their diversity, subtle decreases in the density of benthic soft-bodied burrowers, and the first changes in the planktonic foraminiferal assemblages (Herm, 1965).

6) The extinction of the last generalized lineage of ammonites, a decrease in the density of benthonic burrowing organisms (trace markers), and the introduction of abundant pelagic Braarudospheres and Thoracospheres occurred about 10,000 years before the final extinction of the pelagic calcareous microbiota.

*The Western Dinarides, Yugoslavia*  Pavlovec and Pleničar (1979) have provided an outline of the stratigraphic and biologic succession across relatively complete marine and marginal-marine boundary sequences that bear macrofossils in the western Dinarides. Unfortunately, the details of this sequence—which might allow patterns of extinction to be resolved—are still unpublished. The sequence suggested by their data, from oldest to youngest, is:

1) rudist-bearing Maastrichtian carbonates of shallow-water origins; 2) very shallow-water to intertidal Maastrichtian ("youngest Cretaceous") carbonates without rudists, showing local exposure and karstification; this probably represents the terminal Cretaceous regressive peak ($R_{10}$, Kauffman, 1979); 3) transgression and the deposition of shallow-water marine to lagoonal "oldest Tertiary" deposits.

This sequence sounds similar to that of the Spanish Pyrenees (Garumnien), but seems to have more dominantly marine–marginal-marine boundary sequences. Knowledge of the precise age, diversity, and ecological distribution of the rudists will be critical to our hypothesis: that Tethyan rudists underwent their main extinction long before the end of the Maastrichtian stage. The sparse Yugoslavian data suggest this pattern.

*Northwestern Brazil*   Mabesoone, Tinoco, and Coutinho (1968) have reported an essentially complete sequence of Santonian through Paleocene strata between Recife and João in northeastern Brazil. The Cretaceous-Tertiary boundary is represented by a "sharp faunistical break" (Mabesoone, Tinoco, and Coutinho, 1968, p. 161) in a "continuously deposited" sequence of micritic pelagic limestones and calciclastics. No boundary clay has been noted. The faunal break is mainly expressed by foraminifera, although numerous macrofossils have been reported both above and below the boundary. This warm-temperate to subtropical sequence would seem to be an ideal place to study the Cretaceous extinction event; unfortunately, except for the foraminiferal data, the nature of the faunal boundary has been only generally reported. An analysis of the foraminiferal lists that have been presented by these authors suggests the existence in this region of a major paraconformity—or, a subtle disconformity—at the Cretaceous-Tertiary boundary: the sequence omits both the highest Maastrichtian planktonic foraminiferal zone of *Abathomphalus mayaroensis* and the Lower Danian zone of *Globigerina eugubina;* the lists cite no foraminifera that are characteristic of either zone.

Despite these omissions in the sequence, there is an interesting distribution of macrofossils and environmental events through the Middle Maastrichtian transgressive sequence (*Globotruncana contusa, G. stuarti* assemblage) below the contact, as well as through the Lower Paleocene

(*Globigerina pseudobulloides* zone) above it; this distribution has been reported as follows:

1) The lowest part of the Middle Maastrichtian Gramame Limestone is characterized by shallow-shelf clastic limestones with a diverse subtropical molluscan fauna that includes heavy-shelled bivalves such as *Cucullaea* and *Veniella*, the gastropod *Turritella*, large echinoids, and abundant vertebrates.

2) Middle Middle Maastrichtian phosphatic offshore limestones mark continued transgression and contain an even more diverse, but smaller, subtropical assemblage of thinner-shelled mollusks that includes such bivalves as *Lucina, Cardium,* and *Plicatula,* as well as gastropods such as *Cypraea, Xenophora, Anchura,* and *Turritella;* vertebrates are abundant. This assemblage becomes greatly depleted at the top of the unit.

3) A fine-grained pelagic limestone extends to the Cretaceous-Tertiary contact (Middle Maastrichtian below Middle Danian) and contains a distinct bivalve assemblage of *Inoceramus* (*sensu lato*) and *Pecten,* diverse gastropods as in the underlying beds, below, the echinoid *Hemiaster,* crustaceans, and diverse vertebrates. This assemblage, which suggests some environmental deterioration of the benthic zone, becomes less diverse upwards in the section. The progressive disappearance of bivalves (*Inoceramus* last) is complete about 5 meters below the contact where the sediment starts to become pyritiferous; echinoids drop out 1.25 meters below the top, and ammonites disappear 1 meter from the top. The assemblage depicts successive extinctions: the first, of the benthic specialists; the next, of the benthic generalists; and the last, of the midwater macrofauna. The top meter of the Cretaceous is pyrite-enriched and barren of shelled macrofauna, although microfossils (including benthic foraminifera) are abundant to the top, where widespread extinction abruptly occurs (emphasized by the paraconformity). Planktonics dominate below the boundary, and benthonic foraminifera above it. Bioturbation continues uninterrupted across the boundary zone. The basal Paleocene also lacks shelled macrofauna.

This sequence of biotic events is very similar to those in complete Maastrichtian-Tertiary pelagic boundary sequences (for example, the one at Zumaya), except that in this sequence the uppermost "barren" zone is more restricted due to the unconformity.

*Tampico Area, Mexico*   The most complete Central American Cretaceous-Tertiary boundary sequence lies within the Mendez Formation near Tampico in Mexico. This sequence is also predominantly a deepwater Tethyan pelagic limestone succession. Diverse planktonic foraminiferal assemblages transgress the boundary (Gamper, 1977). Gamper has reported that in this sequence—unlike similar sequences in Gubbio, Italy, and Zumaya, Spain—the planktonic boundary in the uppermost Maastrichtian is transitional, and that the *Abathomphalus mayaroensis* foraminiferal assemblage is gradually replaced by the *Globigerina eugubina* assemblage; size reduction among planktonics, however, does occur across this interval. Gamper (1977, p. 23) has suggested that the Cretaceous-Tertiary boundary should be raised to the top of the *A. mayaroensis-G. eugubina* "interval-zone." Since Gamper has neither given sampling intervals nor reported degrees of bioturbation (a possible source of planktonic foraminiferal mixing), it is not possible to evaluate the seeming absence of a biological crisis among the planktonic microbiota at the end of the Maastrichtian in this area. The field observations of C. C. Smith of the United States Geological Survey (pers. comm. in March of 1979) support the apparent total lack of macrofossils and bioturbation in at least the upper 25 meters of exposed Upper Maastrichtian. The terminal Cretaceous crisis may be represented only by a size reduction in planktonic foraminifera, but here it has obviously been preceded by the elimination of benthic and midwater macrofaunas long before the terminal Cretaceous extinction event.

## THE MACROFAUNAL RECORD OF THE EXTINCTION EVENT ON CARBONATE PLATFORM FACIES IN TETHYS

Some of the most spectacular extinctions that have been reported from the latest Cretaceous involved tropical and subtropical invertebrates of shallow platform environments; this deserves special attention here. Broadly eradicated during the terminal Cretaceous extinction event were, in particular, the radiolitid and hippuritid rudists, as well as many groups of large oysters, trigoniid bivalves, nerineid and actaeonellid gastropods, scleractinian hermatypic corals, larger irregular echinoids, and larger foraminifera. In many cases, the extinction of these Tethyan forms generally correlated with the times when they were at peaks in their evolutionary development. Figure 8–2B depicts the widely accepted concept of the abrupt catastrophic extinction of all rudists at the very end of the Maastrichtian; other groups are viewed as having had similar histories. The rock record does not, however, provide good evidence for this interpretation.

Maastrichtian shelf carbonates that contain rudist and coral frameworks and diverse shallow-water tropical invertebrates are best known from the Caribbean (especially Jamaica and Puerto Rico; see Kauffman and Sohl, 1974), and from the Mediterranean (that is, the Spanish and French Pyrenees: the "Garumnian" facies; see Liebau, 1973). These shelf sequences are typical of the Tethyan realm. As would be expected of shallow platform environments during the latest Cretaceous eustatic drawdown that produced large scale marine regression ($R_{10}$; Kauffman, 1977a), the Caribbean and Mediterranean sequences are stratigraphically incomplete and grade upward from shelf carbonates into more nearshore clastics; both sequences are terminated by a major disconformity at the Cretaceous-Tertiary boundary. In the Garumnian, a thick sequence of marginal-marine to nonmarine sediments further separates rudist-bearing Cretaceous marine carbonates from the Paleocene marine sequence. In neither area is there any evidence for Upper Maastrichtian, fully marine strata (for example, *A. mayaroensis* Zone) representing reef platform facies. The final 2 myr of evolutionary history among Maastrichtian tropical carbonate platform biotas is apparently missing in these sections and in all other Tethyan ones that have been studied. There apparently is no direct fossil evidence that supports a catastrophic latest-Maastrichtian extinction among the carbonate platform benthos; their disappearance can neither be compared to, nor correlated with, the near-catastrophic demise of tropical calcareous microplankton in those pelagic Tethyan environments that have been described from Spain, Italy, Mexico, and Haiti. There are, however, some significant extinction patterns preserved in Tethyan Maastrichtian platform facies that have an important bearing on an interpretation of the terminal Cretaceous event.

Liebau (1973) has studied the Garumnian regressive sequence of the southern Spanish Pyrenees in an uninterrupted arrangement of shelf carbonates and lagoonal-paludal deposits. In this sequence, rudists were once thought to have extended up to, and even into, the Danian (Liebau, 1973, p. 3). He has documented this regressive sequence as follows:

1) early-Early Maastrichtian pelagic marine marls and limestones that contain abundant foraminifera, ostracods, and rare rudists;

2) middle-Early Maastrichtian interbedded lagoonal and marine carbonate platform facies with low diversity rudists biostromes and small frameworks that are dominated by *Praeradiolites* and *Hippurites* spp. This is the *last* appearance of abundant rudists in the sequence;

3) nonmarine to marginal-marine brackish facies, at the top of which is a thin marine incursion characterized by the ostracod *Limburgina* and local biostromes of rudists composed of only two species of *Praeradiolites* (Liebau, 1973, figure 1) of probable Middle Maastrichtian age;

4) nonmarine terrestrial and lacustrine facies with charophytes, ostracods, and mollusks that are similar to those in underlying Middle Maas-

trichtian facies. Liebau has tentatively placed these beds in the upper part of the exposed Middle Maastrichtian. The youngest diverse carbonate platform biota in this sequence is, therefore, of middle-Early Maastrichtian age; the last, restricted marine fauna with rudists is of Middle Maastrichtian age. There is no good marine record in this classic area for the final 2.5 myr of the Maastrichtian Stage.

Kauffman and Sohl (1974) have studied the Tethyan rudist frameworks and associated biotas on Late Cretaceous Caribbean carbonate platforms. Their subsequent work (in preparation) has centered on the Campanian-Maastrichtian sequence and on the Cretaceous-Tertiary boundary zone, the most complete known record of which occurs in Jamaica. The following Maastrichtian-Paleocene sequence, taken from the Jerusalem Mountain Inlier in the northwestern part of that island, is here cited in ascending stratigraphic order:

1) Middle Maastrichtian (*G. gansseri* zone) carbonate platform facies that contain diverse tropical mollusk-echinoid assemblages, rudist-dominated biostromes, and complex frameworks comprised mainly of *Titanosarcolites* and overgrowing layers of diverse, small radiolitids;

2) marls and silty shales of shallow shelf origin that contain a diverse, normal-marine mollusk-echinoid fauna, but have no rudists, large hermatypic corals, or large oysters, and only very few larger foraminifera; in older Caribbean strata, all of these taxa are typically found in Tethyan carbonate platform facies;

3) fine-grained limestone ("Oyster Limestone") that is similar to the limestone in underlying carbonate platform facies, yields an assemblage of tropical marine mollusks and echinoids that are dominated by clusters of large oysters (marine *Lopha*), but contains neither rudists nor stromatoporoids nor hermatypic corals; both units 2 and 3 are considered to be Middle Maastrichtian in age, based on the occurrence of *G. gansseri* zone planktonic foraminifera in unit 2 marls, that are just below the oyster limestone;

4) disconformity: to the extent that the Upper Maastrichtian and Paleocene sequence had ever been deposited, it was removed by this major erosional event during the terminal Cretaceous eustatic drawdown and regression;

5) in western Jamaica, the oldest Paleogene strata are the shallow carbonate platform marls of the Eocene "Ham Walk Facies" that contain a tropical molluscan-echinoid assemblage.

From these Tethyan carbonate platform sequences as well as from many others, it would appear that the main biological crisis for rudists, hermatypic corals, stromatoporoids, larger foraminifera, and many other elements of the tropical benthos may have occurred near the Middle-Upper Maastrichtian boundary, and not in the latest-Maastrichtian. In particular, the seemingly rapid disappearance of most groups of rudists and associated framework builders *within* the late-Middle Maastrichtian platform carbonate sequence of Jamaica may be significant and may mark an earlier extinction episode. Other tropical bivalves and echinoids extend stratigraphically higher than the rudists and corals, but in tropical platform facies, these others also largely disappeared before the latest Maastrichtian. It is noteworthy that this stratigraphic level (the Middle-Upper Maastrichtian boundary zone) is approximately the same as the one at which the tropical benthic macrobiota became greatly restricted and eventually disappeared in both the Zumaya (Spain) and the northeastern Brazil sequences that have previously been discussed.

Some rudists did extend into the Late Maastrichtian, but those that are coeval with sediments of this age are characterized by low diversity (rarely more than 3 to 5 spp.) and by a dominance of apparent ecological generalists (eurytopes) among the radiolitids and/or hippuritids. No major Upper Maastrichtian rudist frameworks are known, but local paucispecific biostromes and low mounds have been recognized (for example, van de Geyn, 1940, for the Maastricht area in the Netherlands). The "Upper Maastrichtian" rudist "reefs" reported from the Anatolian Basin of Turkey (Norman, Gökcen, and Senalp, 1979, and references therein) are poorly dated (generally cited only as "Maastrichtian"), and occur mainly as cemented flysche blocks in basinal deposits of younger age. The broad expanses of Late Maastrichtian shelf chalk and marl in northern and northwestern Europe contain only scattered rudists as a rule, mainly small epibionts on other shells (*Gyropleura* and related generalized Requieniidae), or isolated eurytopic Radiolitidae, Hippuritidae, or Caprinidae in small populations, which could apparently tolerate more Warm Temperate marine climates (Coates, Kauffman, and Sohl, in press).

It can be argued that the paucity of Upper Maastrichtian rudists, hermatypic corals, and other framework builders, the absence of significant reef structures, as well as the great decline in typical Cretaceous carbonate platform benthos among the echinoderms, mollusks, and larger foraminifers is merely a reflection of the terminal Cretaceous marine regression caused by eustatic fall of as much as 300 to 660 meters (Hancock and Kauffman, 1979). This fall would have drained most of the shallow tropical platform environments before the end of the Maastrichtian. No doubt this major regression did leave few shallow-water subtropical and tropical sites extant by the latest Maastrichtian, and thus it enhances the image of an earlier extinction event among reef and reef-associated macrofaunas. But it can not explain the decline in rudists, hermatypic corals, larger foraminifers, actaeonellid and nerineid gastropods, and many groups of larger tropical bivalves in shallow water carbonate platform settings that survived into the Late Maastrichtian, nor can it explain the absence of tropical shallow-water elements in pelagic sequences marginal to the remaining Upper Maastrichtian carbonate platforms (for example, in Zumaya, Spain; Mexico; Brazil), even in debris flows. Finally, Upper Maastrichtian regression cannot account for the loss of the reef biota *within* Middle Maastrichtian tropical platform settings (for example, Jamaica). There is sufficient evidence to suggest that the major extinction of the tropical shelf macrobiota began near the end of the Middle Maastrichtian, and that this first pulse of extinction was greater for these groups than for any others at the end of the Maastrichtian.

In attempting to explain a Middle to Late Maastrichtian crisis for tropical organisms of shallow-water habitats, it should be kept in mind that major factors of environmental decline were already operative at that time. The terminal Cretaceous regression was well under way in many parts of the world (Kauffman, 1979a, figure 2), and by the early-Late Maastrichtian was everywhere broadly affecting shallow shelf and platform habitats, diminishing ecospace and the number of habitats, increasing competition, and producing more variable turbidity, salinity, and water movement in surviving shallow-water Tethyan habitats.

Some oxygen isotopic data based on Pacific foraminifera (Douglas and Savin, 1973; summarized in Kauffman, 1979a, figure 2) indicate a major drop in marine surface temperatures through the earliest to middle Late Maastrichtian coincident with the demise of the carbonate platform assemblage in Tethys. Other isotopic data from diverse benthic and pelagic organisms (Kauffman, 1979a, figure 2), especially from Atlantic foraminifers (Boersma and Shackleton, 1979) and Danish pelagic carbonate analyses (Buchardt and Jørgensen, 1979), suggest highly variable Late Maastrichtian and boundary zone temperature patterns. Various organisms with different habitats show distinct trends. While many benthic and midwater groups show upward temperature excursions of a few degrees centigrade in the earliest-Late Maastrichtian (collected data in Kauffman, 1979a, figure 2), some planktonic foraminifer data (Douglas and Savin, 1973) indicate a coincident temperature decline that suggests a decrease in the temperature gradient through the oceanic water column. In contrast, Boersma and Shackleton (1979) have suggested a general broadening of the vertical Atlantic temperature gradients across the Cretaceous-Tertiary boundary zone, but variably expressed at different sites by a) overall temperature decline of 1 to 3 °C, which is greater in benthic than in surface waters; or b) at one site a 3 °–C rise in surface temperatures as opposed to a 2°-C drop in benthic temperatures. Buchardt and Jørgensen (1979) have noted major excursions in oxygen/carbon isotopic data across the Cretaceous-Tertiary boundary in Denmark, which, if interpreted wholly as temperature, indicate a 12°-C fluctuation with a net loss of 4°-C within a few hundred thousand years, and diminished oceanic oxygen levels. Although there exists debate over these conflicting interpretations, these interpretations collectively indicate rapid, significant temperature excursions through the Late Maastrichtian and across the boundary of certain tropical and North Temperate sites, factors that would have markedly stressed stenothermal Tethyan taxa well *before* and at the end of the Maastrichtian.

Finally, data from pelagic sequences suggest spreading oxygen depletion in the world's tropical oceans and demise of the stenotopic benthic and midwater biotas, beginning at about the middle-Upper Maastrichtian boundary (for example: at Zumaya, Spain, previously discussed; also Boersma and Shackleton, 1979). These are some of many environmental factors which, in concert, could have broadly stressed the tropical shelf biota at the end of the Middle Maastrichtian and started the terminal Cretaceous biotic crisis on its way nearly 2 myr before the end of the period.

Additional complete or almost complete marine Cretaceous-Tertiary boundary sequences within the Tethyan realm, proven by calcareous microbiotas, but for which no comprehensive macrofossil data are yet available, are reported from: the western Carpathians, Czechoslovakia (Gasparikova and Salaj, 1979); at the Nahal Avdat Section, Negev, Israel (Romein, 1979); in the lower Indus Basin, Pakistan (Kureshy, 1979) where a sharp change in planktonic and larger foraminifera contrasts sharply with the gradual, low-level turnover among benthic foraminifera, with many species shared by the Maastrichtian and Danian; and at two localities (El Kef and Hedil) in Tunisia (Perch-Nielsen, 1979). Subsequent study of these sequences may yield valuable information concerning the Cretaceous macrofaunal extinction pattern.

## THE NORTH TEMPERATE SEQUENCES

*Russia*   Naidin (1979) has summarized the biological evidence for relatively complete Maastrichtian-Danian boundary sequences in the USSR. Sequences that bear abundant macrofossils of marine origin are to be found in the Crimea and in Mangyshlak. In general, the macrofossil assemblages resemble those elsewhere in northern Europe (for example, Denmark, Netherlands), and the extinction history of the macrofauna is similarly developed: that is, it is graded and moderate in terms of extinction levels, most of which involve genus and species level terminations. Several extensive works on the marine invertebrates are currently in press or in preparation. Preliminary results suggest:

1) a sharp change in calcareous microplankton at the Cretaceous-Tertiary boundary.

2) slight to moderate boundary changes among benthonic foraminifera at various locations, with 13 of 27 genera and 14 of 46 species of Maastrichtian benthonics passing into the Danian;

3) no marked changes among supraspecific groups of scaphopods, gastropods, nautiloids, and marine turtles at the boundary;

4) moderate change in the bryozoan assemblage similar to that in Denmark, with the Danian assemblage somewhat more allied to Paleocene faunas that to those of Maastrichtian;

5) gradual demise of brachiopods of Mesozoic affinities during the Cretaceous, the greatest during the Maastrichtian in Russia: Danian brachiopods are therefore much less diverse, and are for the most part taxonomically distinct from those of the Maastrichtian;

6) gradual decline of the Cretaceous Bivalvia with extinction of numerous lineages during the Late Cretaceous. These Bivalvia underwent their greatest extinction during the Maastrichtian, when about 200 genera including the rudists, inoceramids, and many oysters disappeared. Among oysters in the Crimea and southern Russian platform, 8 of 17 Maastrichtian genera and 4 of 50 species passed into the Danian;

7) near extinction of many typical Cretaceous invertebrates like the ammonites, belemnites, and inoceramid bivalves that were sparsely represented by a few taxa in the Maastrichtian, just prior to their extinction;

8) gradual to moderate changes among echinoids across the Cretaceous-Tertiary boundary. Extinction of typical Cretaceous irregular echinoids spread over the Late Cretaceous, and by the end of the Maastrichtian, 33% of the families, 67% of the genera, and all Cretaceous species died out; but many Maastrichtian lineages passed over the boundary with only species-level changes;

9) a major species-level change in ostracods at the boundary;

10) extinction of marine barnacles, several Cretaceous stocks extinct at the boundary and in the Danian, but the Danian assemblage is essentially a depauperate Maastrichtian assemblage;

11) gradual demise of larger marine reptiles during the Cretaceous, sparse and not diverse at the point of their extinction in the Maastrichtian.

Thus, the Russian data, that is summarized for temperate (north-central USSR) and marginal Tethyan (Crimean) boundary sequences, show characteristic extinction patterns. The dramatic Maastrichtian warm-water extinction is manifest in the rudistid, ostreid, and other bivalves, the ostracods, brachiopods, certain gastropods and echinoids, and among the calcareous microplankton. But among more typically temperate groups of mollusks, irregular echinoids, barnacles, bryozoans, benthic foraminifera, and reptiles, the extinction was slight to moderate for common groups, and terminal mainly for characteristic Cretaceous cephalopod, bivalve, and reptile groups, that were already on the evolutionary wane and sparsely represented at the beginning of the Maastrichtian. Publication of the detailed studies of macroinvertebrates from specific sites (Tethyan as opposed to Temperate) in Russia will probably shed much light on the fabric of the terminal Cretaceous extinction event in these biogeographically distinct realms.

*Stevns Klint and other regions, Denmark* The most intensely studied Cretaceous-Tertiary boundary sequences in the North Temperate realm occurs in the vincinity of Stevns Klint, southeastern Zealand, Denmark. There, the Upper Maastrichtian comprises several meters of white, bioturbated, pelagic (coccolithic) chalk, which probably represent middle to outer shelf depth environments, and which contain a diverse pelagic and benthonic biota. Macrofossils are well preserved up to the boundary. An erosional disconformity with relief of up to 1 meter marks the top of the Cretaceous; the high parts of this erosion surface comprise local bioturbated hardgrounds. Berggren (1960, 1962) has documented the fact that only the upper part of the Upper Maastrichtian zone of *A. mayaroensis* and part or all of the basal Danian *G. eugubina* planktonic foraminifer zone are missing in the boundary diastem (see also Gamper, 1977, figure 2). The abundance of macrofossils and microfossils of diverse benthonic and midwater groups in the Upper Maastrichtian-(Stevnsian)-Danian sequence of Denmark are particularly important to interpretation of the terminal Cretaceous extinction event (summarized in Rasmussen, 1965; papers in Birkelund and Bromley, eds., 1979).

A thick montmorillonitic clay unit (The Fish Clay or Fiskeler) measuring 0 to 10 centimeters discontinuously overlies the Maastrichtian chalks (Skrivekridt), mainly concentrated in depressions on the disconformable surface. This unit contains reworked chalk clasts, abundant fish remains (reflecting a mass mortality event?), and a limited microbiota. The fish clay grades upward into nodular mollusk-rich bioturbated chalk (the Cerithiumkalk) or coccolithic limestone of Lower Paleocene age (*G. daubjergensis* zone), 0 to 50 centimeters thick, with a disconformable top; this grades laterally into a clay layer mainly containing echinoids (*Brissopneustes*)—the so-called "dead layer" (Berggren, 1960). Fifty to seventy-five meters of bryozoan-rich calcarenites, calcilutites, and cocco-lithic limestones overlie the Cerithiumkalk, which completes the Paleo-cene sequence in this area. Only about 0.5 to 0.75 myr of time seems to be missing at the boundary disconformity at Stevns Klint; this disconformity diminishes downslope into the Danish Basin, where the boundary is es-sentially complete, and is marked at best by an indistinct, iron-stained "firmground." In some sections, continuous chalk sedimentation proba-bly occurred across the Cretaceous-Tertiary boundary.

The Cretaceous-Tertiary boundary succession in Denmark has a greater diversity of well-studied biotic groups than is found in any other essentially complete boundary sequence in the world. It uniquely com-bines a rich benthic self-upper slope biota with a diverse, warm water planktonic microbiota, which makes it the most important data base available for interpretation of the Cretaceous-Tertiary extinction event. Brotzen (1960), Berggren (1960, 1962), Rasmussen (1965), and authors in Birkelund and Bromley (1979), among others, have extensively docu-mented biotic change across this boundary in Denmark.

The extinction of the calcareous planktonic microbiota in Danish se-quences, as obtained elsewhere in the world, is relatively abrupt, but is exaggerated by the boundary hiatus. Berggren (1960, 1962) first docu-mented this event among planktonic foraminifera. Typical Cretaceous planktonics become wholly extinct at the boundary. But the boundary event is nowhere so dramatic as in tropical Tethyan sequences like those in Gubbio, Italy (Premoli-Silva, 1977) and Zumaya, Spain (Herm, 1965). Stenestad (1979) documented four Upper Maastrichtian foraminiferal zones in Denmark, in ascending order, the *Pseudouvigerina cimbrica* Zone, *P. rugosa* Zone, *Pseudotextularia elegans* Zone, and *Stensioeina es-nehensis* Zone. The middle *P. elegans* Zone represents the last Maastrich-tian transgressive peak, and is associated with a flood of warm water planktonic and calcareous benthonic foraminifera and a marked decrease in benthic agglutinated forms. Berggren (1960, 1962) has similarly noted a marked increase in abundance and diversity of planktonic foraminifera (especially keeled Tethyan forms), certain calcareous benthics, and os-tracods in the upper 10 to 15 meters of the Skrivekridt, as compared to underlying Maastrichtian beds. He interpreted this as a migratory event without linking it to any cause. It is significant that this correlates closely with a temperature rise in oceanic waters documented in the early Upper Maastrichtian by isotopic analysis of planktonic foraminifera and to the major eustatic rise (transgressive pulse $T_{10}$) of the Cretaceous (Kauffman, 1977a, figure 7; Stenestad, 1979). The same warming pattern was noted at Zumaya. But beginning at the top of the *P. elegans* Zone, about 0.5 myr before the end of the Cretaceous, and extending through the *S. esnehensis*

Zone, Stenestad (1979, pp. 102–107) recorded a major *ecologically controlled* change in foraminiferal faunas associated with the terminal Cretaceous regressive event (eustatic fall): "foraminiferal faunas demonstrate the withdrawal of the sea through rapidly changing assemblages, decreasing diversity, and diminishing planktic-benthic index" (Stenestad, 1979, p. 107). A major increase in agglutinated forms characterized the latest Maastrichtian foraminiferal assemblages. Stenestad noted almost complete extinction of the depauperate Cretaceous foraminiferal fauna by the end of the Maastrichtian, and its rapid replacement by ecologically similar, but taxonomically distinct, Paleocene assemblages, that were dominated by agglutinated species (1977). Brotzen (1960) has interpreted the change differently; he noted as well a decline in diversity and a shift to benthic foraminiferal communities dominated by arenaceous forms, with few planktonics, near the end of the Maastrichtian. But he cited 21.6% survival of benthic species across the boundary and a major shift to foraminiferal assemblages dominated by calcareous benthics in the Early Paleocene. In either case, it is obvious that decline of the benthic and pelagic foraminifer assemblages began well before the end of the Maastrichtian in the North Atlantic, that it was largely controlled by eustatic drawdown producing widespread regression and environmental shift, and that the Cretaceous-Tertiary boundary extinction was much less dramatic in Denmark and it involved many fewer planktonic taxa, than the Tethys.

Calcareous nannoplankton further shows an abrupt extinction event at the end of the Maastrichtian in Denmark and surrounding areas (Gartner and Keany, 1978; Gartner, 1979; Perch-Nielsen, 1979). Perch-Nielsen, Ulleberg, and Evensen (1979), however, have clearly shown a small scale grading of the extinction, with declining diversity and abundance of the Upper Maastrichtian nannoplankton assemblage within 3 to 4 meters of the boundary, and progressive introduction of typical Danian coccoliths, as well as the so-called "disaster species" (*Thoracosphaera* spp., *Acanthosphaera* spp.) of Percival and Fischer (1977) within 20 meters of the boundary. The extinction sequence thus resembles that of Zumaya, Spain reported by Percival and Fischer (1977). Gartner and Keany's (1978) and Gartner's (1978) interpretation of this boundry as reflective of two freshwater injection events from the Proto-Arctic ocean, for example, two Cretaceous-Danian boundary sequences as seen in the Ekofisk (North Sea) well, is rejected by Perch-Nielsen, Ulleberg, and Evensen (1979) on the basis of the graded extinction "boundary" and on sedimentologic evidence indicating that the boundary sequence is structurally repeated and complicated by debris flow events in the Ekofisk (and probably other) wells. No such repetition has been noted in exposed shelf or basin margin sequences preserving the Cretaceous-Tertiary boundary. The freshwater injection theory is further rejected on the basis that:

1) The extinction event is less dramatic among the collective biota in the temperate North Atlantic, and especially in Denmark and surrounding areas, than it is toward, and especially in, the Tethyan (tropical) realm: the reverse trend would be expected with a freshwater injection through the Viking Graben-Danish Basin-North Sea region.

2) Only selective calcareous taxa show "dramatic" extinction at the boundary, whereas many marine groups with equal salinity requirements today are not greatly affected by the terminal Cretaceous extinction event (see papers in Christensen and Birkelund, 1979; Birkelund and Bromley, 1979).

3) There is considerable doubt that the Proto-Arctic sea was fresh water or even severely brackish (Tappan, 1979).

Equally important is the lack of a dramatic Maastrichtian-Danian extinction event among many noncalcareous elements of the marine microbiota in Denmark and surrounding areas, and throughout most of the world (Tappan, 1979), especially figures 2, 4 for dinoflagellates and silicoflagellates). In Denmark, Hansen (1979) noted:

1) gradual addition to Tertiary dinoflagellates throughout the Maastrichtian;

2) no major change in dinoflagellates at the Cretaceous-Tertiary boundary, but a major turnover in the Upper Danian and Selandian; and

3) the Maastrichtian and Lower Danian assemblages of dinoflagellates *both* dominated by species of *Spiniferites,* reflecting "global paleoenvironmental (for example: climatic) changes" throughout this extended boundary sequence. This stands in marked contrast to the extinction history of the *calcareous* pelagic microbiota in the same boundary sequences in Denmark, Spain, Italy, and elsewhere. These observations fit those on dinoflagellates (Bujak and Williams, 1979) and siliocoflagellates (Bukry and Foster, 1974; Bukry, 1975) for most of the world; these groups did *not* show massive terminal Cretaceous extinction in the same marine systems where the nannoplankton and planktonic foraminifera were decimated.

Macrofaunal changes during the Late Maastrichtian and across the Cretaceous-Tertiary boundary in northwestern Europe stand in sharp contrast to coeval extinction in the Tethyan realm (for example, Zumaya, Spain; Caribbean Islands). The major oceanic warming trend near the Middle-Upper Maastrichtian boundary is reflected in the Skrivekridt of

Denmark by the influx of abundant and diverse keeled planktonics, Tethyan ostracods and bryozoans, calcareous nannoplankton, and in the diversification of the macrobenthos among large warm-water echinoids, bryozoa, brachiopods, and mollusks. Burrows, including large *Thalassinoides,* are common throughout the Skrivekridt and Danian limestones. Liebau (1978) has similarly documented two major waves of diversification of Cytheracean ostracods, and earliest Upper Maastrichtian subtropical pulse, and a slightly younger early-Upper Maastrichtian tropical pulse, in the type Maastrichtian of the Netherlands. But following this warming pulse, the Northwest European sequences does *not* show the abrupt decrease in macrofaunal diversity during the early-Upper Maastrichtian that occurs at Zumaya and other Tethyan sections. Instead, the Upper Maastrichtian biotic decline in Denmark is gradual for most groups; some show little change during this interval (papers in Birkelund and Bromley, 1979). In part this obviously reflects shelf as opposed to basin contrasts in biotas and paleoenvironments; in part, however, it probably reflects important differences in the extinction event between tropical and temperate environments.

In tropical Tethys, as best reflected in the Caribbean (for example, Jamaican) sequence and the "Garumnien" sequence of the Spanish Pyrenees (Liebau, 1973), the shelf biota was widely decimated during the "terminal" Cretaceous extinction event, "perhaps beginning as early as the late-Middle Maastrichtian. Rudistid bivalves, hermatypic corals, nerineid and actaeonellid gastropods, trigoniid and many larger ostreid bivalves, larger foraminifera, and such disappeared rather abruptly at a peak in their radiation and were not replaced in the Paleocene by related analogs. The cumulative magnitude of the Tethyan macrofaunal extinction closely paralleled that of the pelagic microbiota, even though it occurred earlier in the Late Maastrichtian and was graded over 1 to 2 myr.

In North Temperate Denmark and the Netherlands, however, evolutionary decline of the macrofauna was gradual and variable during the Maastrichtian (for example, Surlyk and Birkelund, 1977; papers in Birkelund and Bromley, 1979) and only moderate levels of macrofaunal extinction occurred at the Cretaceous-Tertiary boundary. The ammonites, belemnites, and inoceramid bivalves were represented by low diversity and rare individuals in the Upper Maastrichtian, and their terminal Cretaceous extinction was not dramatic. The latest-Cretaceous and Danian molluscan, bryozoan, and echinoid faunas are remarkably similar, reflecting the low extinction levels. Ravn (1903) compared faunas of the Skrivekridt with those of the Danian and, although his taxonomy is dated, his data are significant. Sixty-one percent of the molluscan and brachiopod genera and 35% of the species found in the Upper

Maastrichtian cross into the Danian. Comparable numbers result in comparing only the Cerithiumkalk with the Skrivekridt, although reworking may be a problem here. Rosenkrantz (1939, 1960) has noted that the macrofaunal break between the Stevnsian and Danian was "not of importance." Brotzen (1960, p. 6) documented continuation of the important echinoid lineage of *Tylocidaris* across the boundary with only species level changes. Rosenkrantz (1924) has noted Maastrichtian *Echinocorys* in the Cerithiumkalk. Graveson (1979) has regarded only species level changes within the continuing regular echinoid lineages across the boundary in Denmark. The irregular echinoids, on the other hand, show a marked break at the boundary, with over half the Maastrichtian genera and all but one species disappearing (Asgaard, 1979). The important group of spatangoid echinoids are an exception, and Stokes (1979) has reported no major change in spatangoid genera and continuation of lineages across the Maastrichtian-Danian boundary. Rasmussen (1979) has noted little change in crinoid, asteroid, and ophiuroid genera across the Cretaceous-Tertiary boundary of Denmark, and remarked that species of each genus that occur on either side of the boundary are closely allied. The major evolutionary change in these echinoderms occurred later, at the Danian-Selandian boundary (Paleocene). Surlyk (1979) has stated that, although the Danian brachiopods of northwestern Europe are not well known and in general seem to be specifically distinct from those of the Maastrichtian, several latest Maastrichtian brachiopod species pass into the lowest few meters of the Danian, and commonly, the brachiopod assemblages on either side of the boundary are morphologically, ecologically, and generically similar.

Even more impressive is Voigt's (1959) Bryozoa data (Figure 8–3) which show that 73% of the genera occuring in the Upper Maastrichtian extend into the Danian, and 53% into the Montian. Hakansson and Thomsen (1979) have restudied the Bryozoa across the boundary and recorded only a modest change from cheilostome-dominated to cyclostome-dominated assemblages mainly reflecting ecological shifts. Seventy-five percent of the cyclostome species cross the Cretaceous-Tertiary boundary; 20% of Maastrichtian cheilostome species also occur in the Danian, and boundary extinctions are restricted mainly to species and genera.

Thus, modern studies of the best preserved and most diverse Late Maastrichtian-Early Danian boundary biota of the North Temperate realm reveals, at best, modest levels of extinction at lower taxonomic levels among most of the important groups and similar ecosystems developed on either side of the boundary. This is true even where a hiatus of about 0.5 myr is developed across the Cretaceous-Tertiary boundary, which should accentuate the record of the terminal Cretaceous extinction event. As elsewhere, only the calcareous planktonic microbiota shows dramatic extinction among common higher taxa at the boundary. This stands in sharp contrast to the extinction event in the Tethyan (=Tropical) realm in either shelf edge (Brazil) or pelagic facies (Gubbio, Zumaya, Tampico).

*Other North Temperate boundary sequences*   N. F. Sohl (U.S. Geological Service) and I have noted a temperate zone pattern of extinction similar to that in Denmark when we compared shelf faunas of the Middle Maastrichtian "Monmouth" Formation with those of the overlying Middle Danian Brightseat Formation in Maryland, despite the wider boundary hiatus. Many lineages cross the boundary with only species-level change. Miller (1956) has disputed this hiatus, when he cited "continuous deposition" in glauconitic sand facies across the Cretaceous-Paleocene boundary. Olsson (1960) and his students presented foraminiferal data in support of a continuous latest–Creataceous-Danian boundary sequence as well. But subsequent studies by Minard et al. (1969) and Minard, Owens, and Sohl (1976) have failed to support these findings, and have identified a boundary disconformity which cuts out the Late Maastrichtian and lowest-Paleocene. Preliminary observations concerning the large macrofaunas found on Greenland (Rosenkrantz, 1970) also suggest strong similarities between the Cretaceous and the Danian mollusks. In North and South Dakota, the marine Lower to Middle (?) Paleocene Cannonball Formation separates from late-Early Maastrichtian marine strata of the Fox Hills Sandstone by 122 to 267 meters of brackish water and nonmarine beds of the lower Lance Formation (Stanton, 1921), and probably by one or more major disconformities; 2 to 30 myr of marine evolutionary history are missing. The mid-Temperate marine faunas of these epicontinental Maastrichtian and Lower Paleocene strata, however, are so similar (except for microbiota) that Stanton (1921) considered the Cannonball fauna to be of Late Cretaceous age, and he cited 40% of the species as being shared with the Late Cretaceous Pierre-Fox Hills fauna, and only one species with the "Eocene." Cvancara (1966) has noted that 60% of the Cannonball bivalve species also occurred in marine Lower to Middle Maastrichtian strata of the area. Vaughan (1921) has stated that the Cannonball corals were neither typically Cretaceous nor Tertiary.

*General observations*   The pertinent observations that emerge from this review of the most completely preserved and best studied Cretaceous-Tertiary boundary sequences are as follows:

1) TROPICAL:   Tethyan and marginal Tethyan extinctions are abrupt for the calcareous microbiota at the very end of the Maastrichtian, but are more graded, though widespread, among the microbiota. The pattern of macrobiota extinction, to the limited extent that it is known, begins with abrupt decimation of the warm water benthos including many rudists, hermatypic corals, large trigoniid and ostreid bivalves, nerineid and actaeonellid gastropods, and large foraminifera near the middle-Upper Maastrichtian boundary. This is followed by Upper Maastrichtian decline and eventual extinction of more eurytopic macroepibenthos (for example, inoceramid bivalves), lowering of diversity and ecological stressing of midwater ammonites, and finally extinction of the last remaining (pachydiscid) lineage of ammonites within a few meters of the Upper Maastrichtian-Tertiary boundary.

Burrowing infaunal organisms, especially polychaete and other worms and small arthropods, do not seem to have been greatly affected by the extinction event. Bioturbation is intense and spans the boundary in most sequences. Burrowing decreases slightly to moderately in some sections at the Cretaceous-Tertiary boundary, and is rarely absent.

2) TEMPERATE:   In North Temperate settings, extinction at the Cretaceous-Tertiary boundary is, at best, only moderately defined among the macrobenthos; about 50% of lineages representing diverse taxa and ecological strategies in the Late Cretaceous cross the boundary with only species-level evolutionary breaks. Nor is extinction dramatic among other common elements of the fauna, except for Calcareous microplankton (planktonic foraminifera and nannoplankton). Typical Temperate Zone Cretaceous groups that became extinct at the boundary (ammonites, inoceramid bivalves, and such) were largely well on their way toward evolutionary decline at the time of their disappearance, and were represented only by a few generalized (eurytopic?) lineages and species in the Late Maastrichtian. Some groups, like the Bryozoa and Ostracoda, actually diversify within lineages across the boundary. Sparse data from South Temperate sequences strongly suggest a similar extinction pattern. This situation stands in marked contrast to the Tethyan record of the biological crisis and suggests a partial temperature control on Cretaceous extinctions.

3) TO SUM UP,   data on the Cretaceous-Tertiary extinction event are at best sparse, especially among the macrobiota, and reflect the relatively few sections in the world that preserve complete or nearly complete boundary sequences. These are predominantly pelagic deep-water se-

quences with rare or no macrofossils. In most exposed Cretaceous-Tertiary boundary sequences, at least the Late Maastrichtian and Early Paleocene are missing in marine facies due to regressive nonmarine intercalations and/or to major disconformities/paraconformities at the boundary. In most areas the stratigraphic gap is greater. A minimum of 1 myr of marine evolutionary history is missing in over 90% of the world's exposed sequences at the Cretaceous-Tertiary boundary.

As a result, only generalized hypotheses concerning the pattern of macrobiotic extinction, at best, can be formulated from sparse data such as those available at Zumaya. The testing of these hypotheses must involve the study of more fully preserved major marine extinction events. Fortunately, three such extinction events—similar to that of the Late Maastrichtian but of lesser magnitude—are completely preserved in Cretaceous marine strata at many world localities. These events, the extinctions at the Albian-Cenomanian, the Cenomanian-Turonian, and the Turonian-Coniacian stage boundaries, provide valuable data for the interpretation of the terminal Cretaceous biological crisis.

## INTRA-CRETACEOUS EXTINCTION EVENTS

Many Cretaceous stages have been divided by significant biological discontinuities, that represent widespread extinction and disjunct evolutionary events, since the time given for their inception by d'Orbigny (see Hancock, 1977, p. 9). Some stage contacts are unusually sharp and can be precisely correlated throughout the world; these are marked by abrupt extinctions below, and relatively rapid replacement by new biotas above the boundary, commonly within one to a few meters of continuously deposited marine sediments. Barren, or poorly fossiliferous intervals, that represent some kind of an environmental crisis may be associated with the contact zone. Somewhat inconsistant extinction levels between macrofaunal elements and calcareous planktonic microbiota in these contact zones have sometimes led to controversy over the precise placement of stage boundaries. It is common for major environmental fluctuations involving sea level, transgressive-regressive pulses, temperature, salinity, and oxygen to coincide with Cretaceous stage boundaries. In many respects, therefore, certain intra-Cretaceous stage boundaries are somewhat lesser representatives of the terminal Cretaceous extinction event, with many of the same characteristics. They can serve as useful models for the interpretation of the Maastrichtian biotic crisis.

Fortunately, the most abrupt intra-Cretaceous extinction events are well preserved in many parts of the world and can be studied in detail.

These events occur in the "Middle" Cretaceous at the Albian-Cenomanian, Cenomanian-Turonian, and Turonian-Coniacian boundaries. The record is well preserved in epicontinental basins, along continental margins, and, in part, throughout deep ocean basins (the Turonian sequence is the exception). Significantly, this was also a time of extremely active plate tectonism, sea floor spreading, rapid eustatic fluctuations in sea level, and major changes in oceanic and epicontinental marine temperatures and oxygen levels (Kauffman, 1977a; 1979a). These conditions led to episodic stress on marine ecosystems of a magnitude that approached situations during the Middle and Late Maastrichtian, with predictably severe effects on the biota. The sequencing of these tectonic, oceanographic, and climatic changes relative to the patterns of "Middle" Cretaceous extinction can provide important clues to the biological and environmental fabric of the terminal Cretaceous event.

Representative, well exposed, stratigraphically continuous, richly fossiliferous, and extensively studied sequences across "Middle" Cretaceous stage boundaries occur in central and western Europe, northern and western Africa, and in the gulf Coast and Western Interior regions of North America. Complete stage boundary sequences have closely similar biological characteristics in all areas. The North American Interior sequence will be used to document three intra-Cretaceous extinction events in this paper, each of which has well documented global expression.

The most complete stage boundary sequences in North America lie near the center of the Western Interior seaway between north Texas and Alberta; they have been especially well studied in Kansas (Hattin, 1962, 1964, 1965, 1975, 1977, 1978; Scott, 1970a,b), Oklahoma (Kauffman, Hattin and Powell, 1977), central and eastern Colorado (Scott and Cobban, 1964; Eicher, 1965, 1966: Kauffman, Powell, and Hattin, 1969; Kauffman; 1969, 1977a,b; Eicher and Worstell, 1970), and Wyoming (Cobban and Reeside, 1952; Cobban, 1958, Cobban and Scott, 1972: Eicher, 1960, 1967; Reeside and Cobban, 1960; and so on). Detailed studies of biological and environmental events across these boundaries reveal patterns of response that shed considerable light on the nature of the terminal Cretaceous extinction event.

*The Albian-Cenomanian boundary zone*   The boundary between the Lower and Upper Cretaceous is one of the severest extinction events in the Cretaceous, which involves taxonomically and ecologically diverse organisms. In particular the ammonites (Figures 8–1B, 8–5) belemnoids, many tropical groups of bivalves (but not rudists), scleractinian corals, irregular echinoids, articulate brachiopods, and larger foraminifera were affected by this biological crisis (summary and references in Kauffman,

1979a). Warm water taxa were primarily involved in the Albian-Cenomanian extinction event, which was further characterized by environmental decline in shallow marine and upper pelagic settings caused by a major Late Albian regression (eustatic fall) and decline in marine temperatures of up to 5 °–C (Kauffman, 1977a, figure 7; 1979a). The parallels to the Upper Maastrichtian crisis are striking.

In the central and northern part of the Western Interior basin the Albian-Cenomanian contact is largely contained in black, thinly and evenly laminated shale sequences with thin interbedded sandstones; many sections lack evidence for a disconformity, paraconformity, or hiatus at the lower-Upper Cretaceous boundary. Eicher (1960, 1965, 1967) studied the stratigraphy and microfaunas of these sequences, Reeside and Cobban (1960) and Cobban and Scott (1972) studied the ammonites, and Kauffman (1975) and Reeside (1923), among others, the bivalves. In this region, the Late Albian sequence begins with a transgressive disconformity and/or basal sandstone (Cheyenne, Plainview, "First Cat Creek" sandstones) and reaches a first transgressive peak in dark organic-rich black shales and silty to sandy shales (Kiowa, Glencairn, Skull Creek, Thermopolis, Joli Fou formations/members), which contain a low (north) to high (south) diversity benthic biota. A middle-Late Albian regression is represented by thin sandstones like the "Dakota," Muddy and Viking formations/members. A major latest-Albian through Early Turonian transgression ($T_6$, Kauffman, 1977a, figure 7) follows with black silty organic-rich shales (Shell Creek-Mowry, Aspen shales) that contain abundant volcanic ash beds and sideritic concretions, and which extend without apparent depositional break across the Albian-Cenomanian boundary. These boundary shales have, at best, a sparse benthic biota of arenaceous foraminifers, eurytopic bivalves, and trace fossils. Throughout much of Alberta and irregularly as far south as central Colorado, a thin but widespread "Fish Scale Zone," probably reflecting a mass mortality event, lies at the Albian-Cenomanian contact. The overlying Lower Cenomanian shales (lower Graneros and Belle Fourche shales) are black, finely and evenly laminated (non-bioturbated), and organic-rich and contain abundant dense ironstone concretions and bentonite beds, and are nearly devoid of macrofauna; sparse benthonic arenaceous foraminifera and inoceramids occur at scattered levels (Eicher, 1965, 1967). Normal marine pelagic and benthonic faunas first occur in dark upper Graneros and Belle Fourche shales, in the basal Middle Cenomanian.

There is virtually no overlap of microfaunal and macrofaunal species between the Late Albian and the Lower to Middle Cenomanian sequences, and there is rare generic overlap, mainly of epifaunal bivalves and agglutinated foraminifera. The Albian-Cenomanian extinction event is ecologically and taxonomically widespread, but is not dramatic in terms of abrupt termination of numerically dominant taxa at the boundary. Most typical Albian taxa became extinct with epicontinental regression in the middle-Late Albian.

The sequence of biological and environmental events that led up to and across the stage boundary in this region is as follows:

1) Early to middle-Late Albian eustatic rise and transgression correlated with a warming peak in oceanic and epicontinental waters ($T_5$ of Kauffman, 1977a, figure 7); dark muds were deposited offshore in moderately shallow marine epicontinental settings that graded to carbonates in Tethyan platform settings. During transgression, progressively more diverse macro- and microfaunas developed in the southern part of the basin (for example, Kiowa fauna of Scott, 1970a,b, 1977) that included warm water mollusks, rare hermatypic colonial corals, and planktonic foraminifera in Kansas that reflect a significantly northerly shift in marginal Tethyan influence. To the north, simpler temperate inoceramid–ostreid– ammonite assemblages dominated, planktonic foraminifers successively dropped out, and benthic microfaunas remained moderately diverse (Eicher, 1960). Warm normal marine southerly waters grading northward to cooler, possibly more brackish seas are implied.

2) Late Albian regression produced widespread shoaling, loss of habitats, diminishing ecospace, and more brackish water conditions in the basin. This is reflected by lowering diversity among the benthic macro- and microbiota and relatively abrupt loss of marginal Tethyan faunas throughout the Western Interior seaway; oceanic marine temperatures dropped sharply (Kauffman, 1977a, figure 7).

3) Renewed transgression (eustatic rise) associated with intense volcanism that produced dark organic and silica-rich shales with sparse benthic faunas, diminished upsection. Sediments, biotas, and low bioturbation levels indicate widespread oxygen restriction and other chemical deterioration.

4) A widespread mass mortality among fish (fish scale marker zone) occurred at the lower-Upper Cretaceous boundary in the northern and central portions of the basin as oxygen levels dropped to near anoxic conditions, and possibly a major oxygen overturn occurred in the basin.

5) Widespread anoxic to low oxygen conditions in the basin and possibly through much of the water column prevailed during most of the Early Cenomanian, as is reflected in the predominance of dark, thinly and evenly bedded organic-rich shales and rare, scattered occurrences of ammonites, benthic inoceramid bivalves, and arenaceous foraminifera

(Eicher, 1965, 1967). Abundant large dense ironstone concretions suggest low benthic oxygen and widespread reducing conditions.

6) Relatively abrupt influx of diverse Temperate Zone mollusks and microbiotas near the base of the Middle Cenomanian was associated with declining vulcanism, increasing benthic circulation and oxygenation, increasing water temperatures, and continued transgression. This succession of events, which took 4 to 5 myr, closely compares with that which characterized the Late Maastrichtian-Danian boundary zone.

*The Cenomanian-Turonian boundary zone*    This stage boundary is marked worldwide by a seemingly abrupt change in calcareous benthic and planktonic foraminifera, calcareous nannoplankton, scleractinian corals, and many genera of irregular echinoids and ammonites (Kauffman, 1979a, and references therein). The primary groups involved in the extinction inhabited warm temperate to tropical paleoenvironments. The extinction and replacement of the Cenomanian biota by that of the Turonian appears to have taken place within a half million years or less (catastrophic). In oceanic and marine epicontinental sequences, where the boundary is preserved in continuously deposited strata, the abrupt change in fauna normally takes place in less than 2 meters of stratigraphic section; Figure 8-10, based on the Western Interior of North America, is typical for many parts of the world (see subsequent discussion). In some cases only a boundary horizon separates the two biotas.

The change in marine biota near the Cenomanian-Turonian boundary generally corresponds with a transgressive maximum (one of the largest in the Cretaceous: Kauffman, 1979a, figure 2), rising oceanic temperatures, a sharp rise in epicontinental marine temperature (Coates, Kauffman, and Sohl, in press) and/or salinity (for example, Figure 8-10; Scholle and Kauffman, 1977), and a widespread anoxic event in the world's oceans which probably encroached on the epicontinental seas (Frush and Eicher, 1975; Schlanger and Jenkyns, 1976; Kauffman, 1977a, 1979a). Many of these same environmental factors, on a somewhat larger scale, merged at the time of the terminal Cretaceous extinction event. The principal differences between these two events is the correlation of extinction with a transgressive peak at the Cenomanian-Turonian boundary, and with a regressive peak at the Maastrichtian-Danian boundary. Both extinctions, however, coincide with a narrow time interval that involves major reversal in sea level trend.

Detailed study of the Cenomanian-Turonian extinction event at several complete sections in Europe and North America reveals a consistent fabric of extinction that involves a sequence of closely spaced biotic and environmental changes, many of them abrupt. The study does *not* reveal a simultaneous mass extinction, however, as some workers have assumed.

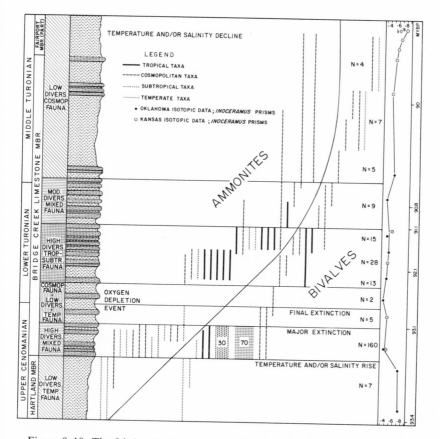

Figure 8–10. The fabric of extinction across the Cenomanian-Turonian boundary in the Western Interior seaway, United States and southern Canada. This stage boundary is one of the most sharply defined in the Upper Cretaceous and is noted for the abrupt extinction of numerous taxa just prior to the Lower Turonian global transgressive peak ($T_6$ of Kauffman 1977a; Figure 7). Stratigraphy (left) and ammonite range data (to left of curved diagonal line) primarily from Cobban and Scott (1972). Bivalve range data modified from Kauffman (1975; to right of curved diagonal line). Oxygen isotope curve, depicting changes in temperature and/or salinity, averaged from Oklahoma and Kansas data of Scholle (Scholle and Kauffman, 1977); note abrupt rise in temperature and/or salinity levels just prior to major latest Cenomanian radiation and subsequent extinction event, and gradual decline in salinity and/or temperature levels following peak transgression and Tethyan radiation in middle Lower Turonian. Radiometric scale (right) after Obradovich and Cobban (1975) and Kauffman (1977a) as modified by application of new decay constants (Obradovich, personal communication, 1978). $N$ numbers are macrofossil diversity at each level, including some taxa whose ranges are not given on this chart; data from Central Colorado, western Kansas, Oklahoma, and western Utah. Patterns differentiate biogeographic-ecologic interpretations at left, and indicate rapid, largely cyclic changes in paleoenvironments and faunas connected with latest transgression and early regression; similar patterns indicate similar environments and faunal response.
See discussion in text and model (Figure 8–13).

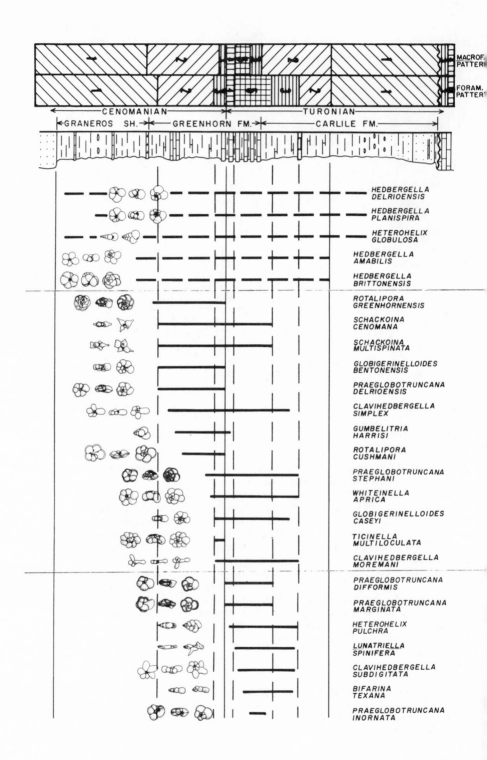

In North America, the Cenomanian-Turonian boundary zone is most fully preserved in essentially uninterrupted offshore sequences of the upper Hartland Shale and Bridge Creek Limestone members of the Greenhorn Formation in central and eastern Colorado (Kauffman, 1969, 1977b; Cobban and Scott, 1972), northwestern Oklahoma (Kauffman, Hattin, and Powell, 1977), and western Kansas, and in the Hartland Shale, Jetmore Chalk, and Pfeifer Shale Members of Central Kansas (Hattin, 1975, 1977; Hattin and Siemers, 1978). Figures 8–10 and 8–11 summarize biological and environmental observations across this boundary. The numbers below refer to the sequence of events indicated by the facies changes on the top side of Figure 8–11, and to similar patterns in Figure 8–10.

1) FACIES 1:  Initial environmental and biological changes connected with the extinction event occurred near the middle-Upper Cenomanian boundary (contact zone between the Graneros Shale and Lincoln Member of the Greenhorn Formation). These changes include: gradual warming of the epicontinental sea; the first influx of abundant calcareous microplankton (Figure 8–11) producing carbonate-enriched sediments; and breakdown of deeper water circulation with resultant oxygen restriction

---

Figure 8–11.  Evolutionary history of planktonic foraminifera across the Cenomanian-Turonian extinction boundary in Western Interior of North America, comparing ecologically generalized, geographically widespread taxa (dashed range zone lines) with more stenothermal Tethyan and marginal Tethyan forms (solid range lines).
Data modified from Eicher (1967, 1969), Eicher and Worstell (1970). Two columns of patterns and numbers at left depict different faunal responses to environmental changes across the Cenomanian-Turonian boundary, with each pattern/number depicting one major environment/fauna relationship. Left hand column taken from Figure 8–10, with pattern 1 added (Mid-Temperate faunas of low to moderate diversity with some cosmopolitan and endemic elements; cool shelf-depth waters with possibly subnormal salinity at times). Second column shows foraminiferal ecological/biogeographic response to the same sequence of environments, as taken from data presented on chart. Note that diversification phase (3) of the biota, with immigration of numerous Tethyan and marginal Tethyan elements associated with temperature and/or salinity increases in the basin (Figure 8–10), begins earlier and lasts longer for planktonic foraminifera than for benthic and mid-water mollusks. Similarly, Tethyan faunal dominance (4) lasts longer for planktonics. Both relationships may reflect more sensitive temperature control on mollusks, or vertical temperature zonation of Western Interior seaway during invasion of Tethyan waters, with surface waters warming earlier and cooling later than lower zones in water column. Stratigraphic column from Eicher (1967), modeled after Pueblo, Colorado section. Note distinct differences in evolutionary rates and extinction patterns across stage boundary. Foraminifera show a less dramatic extinction at the stage boundary than mollusks (Figure 8–10), but a greater extinction at the Lower Turonian termination of Tethyan faunal influence, than do mollusks (Figure 8–10).

(Eicher, 1969; Frush and Eicher, 1975). Within 3 meters of section, the following sequence of events is preserved: a) great restriction of benthic arenaceous foraminifera coupled with abrupt increase in numbers and diversity of, first, eurythermal and second, more stenothermal warm water planktonic foraminifera (Figure 8–11; Eicher, 1969; 1977, figure 1). Temperate Zone benthic macrofaunas remain relatively diverse; and b) marked restriction of the benthic macrofauna to simple *Inoceramus* and oyster-dominated paleocommunities (mostly opportunistic eurytopes), and total elimination of the benthic microfauna; 85 percent of arenaceous species that characterize the Middle Cenomanian in this part of the basin became extinct (Eicher, 1965, 1977). This extinction was due, in part, to benthic facies changes; but calcareous benthonics failed to replace agglutinated foraminifera in the same niches, which suggests that widespread oxygen depletion was a more important cause.

2) FACIES 2:  Warm, stable to slightly rising temperatures (biotically defined) and gradual normalization of surface salinity (isotopically defined) were associated with continuing transgression during most of the Upper Cenomanian (Lincoln-Hartland members of the Greenhorn Formation; Figures 8–10 and 8–11). Benthic oxygen restriction gradually increased, as marked by the continuing absence of a benthic microfauna and even greater restriction of benthic Temperate macrofaunas; infaunal organisms were rare to absent through this 1 myr interval except for brief periods when bioturbation was moderately dense. Some intervals lack benthic biota. In contrast, the planktonic microbiota and ammonite diversity more than doubled during this period (see Figures 8–10 and 8–11).

3) FACIES 3:  An abrupt increase in epicontinental marine temperatures and/or salinity levels (to normal) followed, as is indicated by immigration of Tethyan biotas and oxygen isotope analyses (whole rock and *Inoceramus* prisms) from the upper Hartland and/or Lower Bridge Creek members of the Greenhorn Formation (Figure 8–10; Scholle and Kauffman, 1977). This increase in temperatures probably reflected rapid global eustatic rise, allowing warm Tethyan waters and biotas to circulate widely through the Western Interior (Coates, Kauffman, and Sohl, in press). The impact of this abrupt temperature-salinity increase on the biota was immense. Benthic circulation was reestablished, and Temperate and Tethyan biotas broadly mixed, which resulted in a great diversification of calcareous benthonic foraminifera (maximum diversity for the Middle Cretaceous; Eicher, 1969, 1977) and macrofaunas (140 to 150 spp.; Figure 8–10; Koch, 1978) and a major increase in numbers and diversity among the pelagic microbiota (Figure 8–11). Rapid evolutionary

rates (Kauffman, 1970, 1972, 1978a) and high levels of endemism characterized many lineages in the center of the seaway (Figure 8–10). The pelagic microbiota diversified somewhat earlier than did the benthos (Figure 8–10 in contrast to 8–11).

4) FACIES 4:   An abrupt and massive extinction event that eliminated the great majority of the Cenomanian benthic and midwater biota followed, and caused widespread extinction among the Cenomanian pelagic calcareous microbiota; these events were replaced with an even more diverse tropical pelagic microbiota (Figure 8–11). Benthic foraminifera disappeared, and benthic macrofaunas reduced in number from over 160 to less than 5 species within a meter of section. This extinction event comprises the Cenomanian-Turonian boundary of most workers. The biotic-environmental change implies an abrupt lowering of oxygen in deeper parts of open marine and epicontinental seas that was caused by diminution of the north to south surface temperature gradient, increased horizontal stratification due to temperature and salinity changes, lack of cooling and sinking of poleward waters, and breakdown of deep marine circulation. As a result of this short oxygen depletion event and environmental shock of still rising temperatures and/or salinity (Figure 8–10) on temperate elements of the epicontinental marine ecosystem, a brief benthic to midwater "dead period" ensued. This dead period is characterized by 1 to 2 meters of nearly barren limestone, chalk, and marl with abundant ferruginous nodules, an absence of benthic foraminifera, sparse to moderate bioturbation, and rare macrofauna. In the Western Interior of North America, the sparse macrofauna of this interval, which spans the Cenomanian-Turonian boundary, is comprised almost exclusively of 1 or 2 genera of eurytopic cosmopolitan ammonites and two lineages of inoceramid bivalves; these are of Cenomanian affinities at the base, and of Turonian affinities at the top of the "barren zone." The main turnover in the pelagic microbiota slightly precedes the final extinction of the last few typical Cenomanian ammonite and bivalve lineages in many cases, but concides with it in some sections (condensed?). Thus, the last Cenomanian mollusks disappear *above* the major extinction event that workers have usually taken as the stage boundary (see Smith, 1975). The terminal Cenomanian extinction is, therefore, clearly graded among organisms with different ecological tolerances over 0.5 myr or so of time.

5) FACIES 5:   At peak Early Turonian transgression, isotopic data indicate short-term stabilization of marine temperature and/or salinity levels in Euramerican epicontinental seas (for example, Figure 8–10) and continued rise of oceanic temperatures (Kauffman, 1977a, figure 7; Scholle

and Arthur, 1980, figure 3). Restoration of benthic circulation is evidenced by moderately diverse calcareous benthic foraminifer assemblages (Eicher, 1977, figure 1), diversification of macrobenthos (mainly epifauna; Figure 8–10), and widespread bioturbation in the Western Interior sequence. Planktonic microbiotas reached their maximum diversity with the addition of many new stenothermal, dominantly Tethyan to marginal Tethyan species (Figure 8–11). Moderately diverse benthic and midwater macrofaunas that characterized this phase are markedly different from those of the Late Cenomanian (*S. gracile* Zone) in being dominantly composed of tropical, subtropical, and cosmopolitan lineages (Figure 8–10). These represent pulses of Tethyan waters flooding rapidly into epicontinental areas during peak transgression (Coates, Kauffman, and Sohl, in press).

6) FACIES 3 REPEATS:   another prominent extinction event, involving the majority of the tropical-subtropical macrofauna and much of the benthic calcareous microfauna (Eicher, 1977, figure 1) was associated with initial, low-level decline in epicontinental marine temperatures and salinity, and the onset of regression near the end of the Lower Turonian (Figure 8–10). The extinction in epicontinental areas was relatively abrupt (Figure 8–10) and may mark the attainment of the low lethal threshold temperature and/or salinity for much of the tropical-subtropical macrobiota. Planktonic foraminifera were relatively unaffected by this event (Figure 8–11), which suggests greater tolerance for declining temperatures and salinity. The event may also reflect significant cooling or oxygen restriction in the lower portion of the epicontinental water column prior to the cooling of surface waters during eustatic fall and regression.

7) FACIES 2 REPEATS:   early in the Middle Turonian regression (upper Bridge Creek Limestone Member, Greenhorn Formation and Fairport Chalk Member, Carlile Shale: Figures 8–10 and 8–11) biotic and isotopic data indicate a more significant decrease in epicontinental water temperatures and surface salinity, respectively. This decrease in temperatures and salinity produced a marked decline in benthic and pelagic diversity. The final extinction and/or emigration of all Tethyan and many cosmopolitan lineages of the benthic and midwater macrofauna and of all stenothermal planktonic foraminifera is graded over a few meters of section during this decline (Figure 8–10). Specialized tropical-subtropical planktonic foraminifera disappeared in two waves that represented distinct thermal crisis; the extinction of each wave was relatively abrupt (Figure 8–11).

8) FACIES 1 REPEATS:   low to moderate diversity macrofaunas of warm temperature and cosmopolitan affinities and low diversity (eurythermal?) microplankton assemblages characterized the middle part of the Turonian regression. Biotic and isotopic data, respectively, indicate continued lowering of epicontinental marine temperature and surface salinity (Figure 8–10); increasing restriction of the benthic biota, with episodic elimination of benthic foraminifera (Eicher, 1977, figure 1), suggests an onset of oxygen depletion, at least near the sediment-water interface.

In summary, the Cenomanian-Turonian extinction "event" in the Western Interior of North America is actually a series of closely spaced extinctions that reflect convergence of several major environmental changes linked to eustatic fluctuations in sea level (and thus plate tectonic events); changes in marine circulation, oxygen, and salinity levels; and changes in temperature at various levels in the water column. Biological parameters of competition, stress, evolutionary rates and patterns, and ecological structuring were linked to these environmental changes, but played a minor casual role in the principal extinction events.

Eustatic rise, transgression, gradual and then abrupt warming and normalization of salinity in the seaway, and episodic oxygen depletion events characterized the late Cenomanian and earliest Turonian. These factors produced waves of regional extinction, first among Temperate Zone benthonic foraminifera with widespread oxygen restriction, and then dramatically among diverse Warm Temperate macrofaunas and stenothermal pelagic microbiotas (stage boundary), with a rapid rise in salinity and temperature coupled with the onset of a second anoxic event; and third, among sparse, eurytopic, cosmopolitan Cenomanian bivalve and ammonite lineages (Figure 8–10, 8–11) and residual calcareous microplankton (Smith, 1975; Figure 8–11) during the peak of the anoxic event spanning the Cenomanian-Turonian stage boundary (Figure 8–11). Following this brief, but widespread, period of oxygen restriction, diversification among tropical, subtropical, and cosmopolitan immigrant biotas marked peak transgression with normal marine conditions. Early regression in the Early and Middle Turonian was characterized by lowering temperatures, salinity levels, and, once more, by the onset of low oxygen benthic conditions with increasingly more restricted epicontinental marine circulation. The Cenomanian-Turonian extinction event was completed with moderately abrupt elimination of the Lower Turonian Tethyan macrofauna (Figure 8–10), followed by two waves of extinction among stenothermal planktonic foraminifera (Figure 8–11). Radiometric data suggest that the entire extinction event around the Cenomanian-Turonian boundary zone took about 2.5 myr (Figure 8–10), with the main part between 91 and 93 Ma. The most dramatic event at the top of the *Sciponoceras gracile* ammonite zone, with extinction of over 150 species and numerous genera, probably took only a few thousand years. The latter was associated with the most dramatic temperature increase in the history of the seaway and normalization of salinity, followed closely by a global anoxic event (Figure 8–10).

*The Turonian-Coniacian boundary zone*   The paleoenvironmental setting and the biological response at the well-defined Turonian-Coniacian boundary (fide Kauffman, Cobban, and Eicher, 1976) was almost exactly the same as for the Cenomanian-Turonian boundary event, and involved similar sedimentary facies in the Western Interior Cretaceous seaway of North America. The boundary event is best preserved near the center of this seaway in continuously deposited argillaceous and calcareous sediments of the upper Carlile Shale (Sage Breaks Member and equivalents), the Fort Hays Limestone Member and the lower Smoky Hill Member of the Niobrara Formation (Scott and Cobban, 1964; Kauffman, 1977b). The main extinction event at or near the base of the Niobrara coincided with a warming trend in world oceans (data in Kauffman, 1977a; Scholle and Arthur, 1980), which subsequently peaked in the Coniacian. The extinction closely followed an abrupt rise in epicontinental temperature and/or salinity in the Western Interior (Figure 8–12), and was accompanied by a widespread anoxic event, which was partially causal. Abrupt changes in marine biotas and alternation of discrete Temperate Zone and marginal Tethyan biotas reflect these rapid environmental shifts. All phases of the extinction event occurred near a major transgressive peak ($T_{7a}$ of Kauffman, 1977a), which represents global eustatic rise (Hancock and Kauffman, 1979, figure 4).

Evidence available for interpretation of this extinction is less extensive than for the Cenomanian-Turonian event. The calcareous microbiota has not been well documented for the boundary interval. Eicher's study of Carlile foraminifera (1966) was in an area where upper Turonian Sage Breaks age shales were largely missing along a series of disconformities. Kent's (1967) work and previous studies of the Niobrara foraminifera did not include a detailed study of Lower Coniacian (Fort Hays age) forms across the Turonian-Coniacian boundary zone. Kent (pers. comm. 1979) reported a moderate turnover of planktonic foraminifera without significant loss of diversity, but with a major depletion of benthonic foraminifera in this interval. The calcareous nannofossils of this boundary have been only superficially documented. Nevertheless, careful work on the stratigraphy, depositional environments, and macrofaunas reveals a pat-

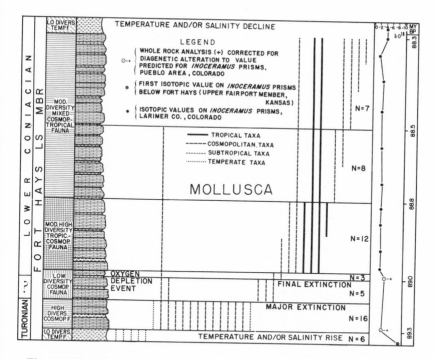

Figure 8-12. Fabric of extinction across the Turonian-Coniacian boundary in the Western Interior of North America.

This is one of the most sharply defined stage boundaries in the Upper Cretaceous, and noted for abrupt extinctions of typical Turonian faunas, even in continuously deposited sequences.

Physical stratigraphy (left) after Scott and Cobban (1964) from Pueblo area of Colorado. Biostratigraphic ranges shown here are for ammonites (Scott and Cobban, 1964; and unpublished data) and bivalves (Kauffman, 1975). *N* numbers refer to macrofossil diversity at each level, including some taxa which do not have range zones shown here (data from Colorado, Kansas, Utah, New Mexico). Generalized oxygen isotope curve indicating temperature and/or salinity changes in the seaway after data presented by Scholle (Scholle and Kauffman, 1977). Radiometric scale (right) after Obradovich and Cobban (1975) and Kauffman (1977a), modified to reflect new decay constants (Obradovich, pers. comm., 1978). Patterns in the left column reflect changing, generally cyclic environments and faunal response across the stage/extinction boundary; similar patterns represent similar environments and faunas on this figure, and on Figures 8-10, 8-11, and 8-13. Note the gradational extinction of first Temperate, then cosmopolitan Cenomanian faunas and finally Tethyan Lower Turonian faunas. Note relationship of diversification and, ultimately, extinction to abrupt temperature and/or salinity increases and subsequently to oxygen depletion. Compare with Figures, 8-10, 8-11, 8-13.

Key to different symbols in species range zones shown in Figure 8-10.

tern of events which is closely comparable to that at the Cenomanian-Turonian boundary (Figure 8–12). The numbering below corresponds to the sequence of patterns (events) along the left side of Figure 8–12, from bottom to top.

1) The upper Carlile Shale (Late Turonian) comprises a transgressive sequence ($T_{7a}$ of Kauffman, 1977a) of basal calcarenites, dark grey finely laminated shales in the middle, grading upward to calcareous shales and marls at the top. Temperate Zone benthic faunas of low to moderate diversity, lacking many infaunal elements, characterize the dark shales. Epibenthic foraminiferal faunas (Eicher, 1966) mainly comprised a few arenaceous species; inoceramid bivalves and their epibionts dominated benthic macrofaunas. The benthic biota indicates reduced oxygen levels at the sediment-water interface, nearly anoxic sediments, and possibly lower than normal surface salinity. Warm Temperate marine water temperatures prevailed during deposition of the Carlile Shale. Rising temperatures gave rise to increasing benthic diversity among macrofaunas, introduction of calcareous benthonic foraminifera, and the first eurythermal planktonics and calcareous nannofossils of the late Carlile (late-Late Turonian) time. Regional extinction of many typical Turonian agglutinated foraminifera accompanied this rise in the early-Late Turonian (Eicher, 1966, figure 2).

2) Isotopic and faunal data from uppermost-Turonian carbonates near the contact of the upper Carlile Shale and the basal Niobrara (Fort Hays Limestone Member; Figure 8–12) suggest a sharp upward shift in epicontinental temperature at this time. Subsequently, diversity among uppermost-Turonian biotas increased significantly with the introduction of more cosmopolitan warm-water mollusks, greater numbers of specialized planktonic foraminifera, and abundant calcareous nannoplankton. Coincidentally, oceanic temperatures rose slightly (Scholle and Arthur, 1980, figure 3). Normalization of epicontinental salinity and benthic oxygenation allowed development of a moderately diverse epifauna and a burrowing infauna (as evidenced by bioturbation). This was the peak macrofaunal diversification during the Late Turonian-Early Coniacian transgression in the Western Interior Seaway (Figure 8–12), insofar as data are available for analysis.

3) The major short-term extinction event which defines the Turonian-Coniacian boundary in the Western Interior occurred in conjunction with the onset of an oxygen crisis in the basin, whose end marked maximum transgression ($T_{7a}$ of Kauffman, 1977a), a peak rise in epicontinental marine temperatures, and apparent normalization of salinity throughout the Western Interior Seaway (biotic data and oxygen isotope determinations; Figure 8–12: Scholle and Kauffman, 1977). This extinction was marked

by: a major depletion of the dominantly temperate and cosmopolitan Turonian macrobiota; large scale changes in the pelagic microbiota; and diminution, but not cessation of bioturbation. All the above occurs commonly within 0.5 meter or less of stratigraphic thickness in continuously deposited calcareous sediments of the lower Niobrara Formation, where it is fully developed. In most cases a single bedding plane marks this abrupt change (Figure 8-12).

4) A meter or less of barren, cyclically deposited limestone and shale with abundant ferruginous nodules overlies the Turonian-Coniacian extinction boundary that represents a basin-wide oxygen depletion event. Reduced bioturbation and rare epibenthic ostreid and inoceramid bivalves of mixed Turonian-Coniacian affinities suggest low benthic oxygen; but pelagic microbiotas (mainly nannofossils) remained abundant and formed the bulk of the carbonate fraction. Kent (1967) did not report calcareous benthonic foraminifera for this barren interval.

5) At peak eustatic rise and transgression (Lower Coniacian, *Inoceramus erectus* zone, middle Fort Hays Limestone; $T_{7a}$, Kauffman, 1977a, b), a moderately diverse warm-water macrofauna and rich calcareous microbiota developed. Bioturbation levels were again high. Normal oxygenation and marine salinities are indicated by the biota and oxygen isotope data. The biota of this peak transgressive phase was a mixture of Tethyan elements (rudistid bivalves, certain ammonites, stenothermal groups of planktonic foraminifera) and cosmopolitan groups, mainly among inoceramid and ostreid bivalves, and ammonites. The rapid influx and subsequent development of Tethyan biotas suggest that epicontinental marine temperatures initially increased sharply, and then remained at a peak for nearly a million years. The return of normal benthic oxygenation implies that basinal current systems changed at peak eustatic rise and maximum transgression despite reduction in the thermal gradient of surface waters from north to south—a situation often called upon to explain breakdown of bottom current systems and deoxygenation of the benthic zone.

6) The onset of eustatic fall and epicontinental regression ($R_{7a}$ of Kauffman, 1977a) is marked lithologically near the center of the Western Interior basin by a gradual change from dominantly pelagic limestone and chalk with secondary shale intervals, to predominantly calcareous shale and shalelike chalk in the upper Fort Hays Limestone and lower Smoky Hill Member of the Niobrara Formation. Isotopic data (Figure 8-12) suggest initially normal, then gradually lowering salinity levels in the seaway; successive biotas suggest a similar temperature trend, although marine temperatures remained in the subtropical to Warm Temperate range. Some restriction of benthic oxygen may be indicated by diminished bioturbation in the marls and shales of the lower Smoky Hill

Member. There was little change in benthic macrofaunas, which dominantly comprised large inoceramid bivalves, their epibionts and endobionts, and oyster beds, except for gradual elimination of stenothermal, dominant Tethyan faunal elements (the final phase of the extinction event: Figure 8–12, top).

7) Marine environments stabilized during the minor; short-lived $R_{7a}$ regression. Warm Temperate conditions prevailed and benthic circulation was moderately restricted. Benthic oxygen was depleted to the point where bioturbation levels became very low, benthic microfaunas were greatly restricted (Kent, 1967), and benthic macrofaunas were limited to oxygen-tolerant inoceramid bivalves and their epi- and endobionts. Pelagic microbiotas remained abundant, but decreased in diversity. This post-extinction phase is represented by the lower shale and limestone unit, and the overlying lower shale unit of the Smoky Hill Member, Niobrara Formation (Scott and Cobban, 1964; Kauffman, 1977b).

## A MODEL OF INTRA-CRETACEOUS EXTINCTION EVENTS

The three most severe and widely correlative extinction events within the Cretaceous Period (Albian-Cenomanian, Cenomanian-Turonian, and Turonian-Coniacian stage boundary zones) are coincident with intervals of active plate tectonism, eustatic fluctuation, and commensurate changes in marine current systems, temperatures, salinity levels, oxygen levels, and the size and diversity of ecological niches. As at the end of the Maastrichtian, *the convergence of diverse major environmental fluctuations within a short period of time is directly correlative with each major extinction event.* Is the sequence of events associated with these extinctions random, or generally consistent from one event to another, allowing predictability for poorly known extinctions like that at the end of the Maastrichtian? If a consistent pattern emerges, does it also reflect a consistent pattern of environmental change? Comparison of the three major intra-Cretaceous extinction events, as they are typically expressed in the Western Interior Cretaceous Seaway provides interesting data which are applicable to these questions. Many events observed in this basin have widely distributed paleogeographic expression in Europe and Africa.

The Cenomanian-Turonian, and Turonian-Coniacian extinction events are remarkably similar, except for the levels of diversity involved (mainly a matter of differential taxonomic research); these two events provide a basis for one model of intra-Cretaceous extinction by which other events might be interpreted (Figure 8–13). Both extinctions were preceded by a period of gradually warming temperatures and benthic

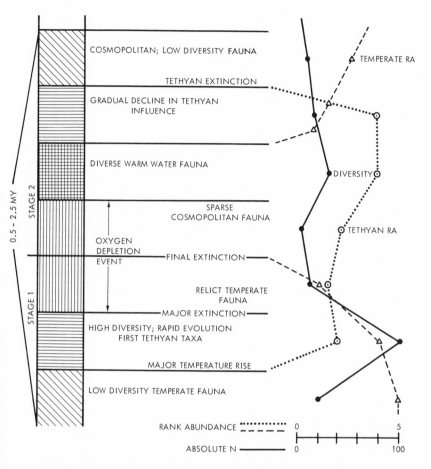

Figure 8–13. Model of Cretaceous extinction events, as generalized from changing faunas, diversity, geochemical, and paleoenvironmental parameters across Albian/Cenomanian, Cenomanian/Turonian, and Turonian/Coniacian Stage boundaries in Western Interior Seaway of North America.

Typical data, from which this model is derived, are presented in Figures 8–10 to 8–12. The model implies close interaction between eustatic fluctuations, transgression and regression, temperature, salinity, oxygen levels, and biological response. Extinctions are clearly gradational, and involve several distinct events crowded into a short time interval. The model shows the relative abundance levels of Tethyan and Temperate rank elements (RA), and species diversity in representative absolute numbers (N), based on such data as presented in Figures 8–10 to 8–12. The patterns in the left column depict separate phases of environmental change, extinction and radiation, and are keyed to the same events shown in Figures 8–10 to 8–12.

oxygen restriction, followed by relatively abrupt temperature and salinity increase, biotic diversification, and re-establishment of benthic circulation. The main extinction level in both events was succeeded by a widespread "anoxic event" during which final extinction of relict eurytopic taxa took place. Subsequently, renewed radiation and extensive mixing of Warm Temperate, Tethyan, and cosmopolitan biotas occurred at peak transgression and temperature-salinity maxima. Short-term but graded extinction of this biota occurred when marine temperatures began to drop. In both cases, extinction events occurred at slightly different times for different types of organisms, which depended upon their primary habitat and degree of ecological tolerance for stress (degree of eurytopy). This is the pattern of extinction that is predictable in light of current evolutionary and ecological theory, but which contradicts the idea of a catastrophic extinction involving only one event. Of primary importance is the correlation of the extinction events at both stage boundaries with convergent environmental fluctuations of sea level, temperature, salinity, oxygen, current circulation, and ecological opportunities characteristic of the mode of transgression or regression.

A model of these two extinctions would involve the following sequence of events, as largely represented by different patterns and labels on Figure 8–13.

1) (not represented in Figure 8–13) Early to middle transgression (eustatic rise) associated with subnormal epicontinental surface salinity, a partially stratified water column, regional oxygen restriction near the sediment-water interface, and low to moderate diversity among Temperate Zone benthic and midwater assemblages; also great restriction of warm-water microplankton.

2) (Figure 8–13, base) Middle to late transgression (eustatic rise), initial normalization of salinity and warming of epicontinental seas (expanding Warm Temperate Zone) associated with moderate diversification of pelagic and midwater biotas and immigration of the first eurythermal subtropical planktonic foraminifera and nannoplankton. Increased oxygen depletion in the lower water column leads to widespread extinction of temperate benthic microfaunas (mainly agglutinated foraminifera), and restriction of benthic microfaunas to trophic generalists and oxygen-tolerant epibenthic eurytopes like the Inoceramidae and their characteristic epi- and endobiont communities. Great reduction of the burrowing infauna results from $O_2$ depletion and $H_2S$ poisoning in deeper offshore sediments.

3) Late transgression (near-peak eustatic rise) produces widespread flooding of Tethyan (tropical) waters into bordering epicontinental seas, abrupt warming and normalization of salinity, and re-establishment of

deep-water circulation and benthic oxygenation. Major diversification and immigration among specialized stenothermal calcareous microplankton occurs in the upper part of the water column. Recolonization of the benthic zone by diverse micro- and macrofaunas also occurs. Ecological mixing of tropical, subtropical, cosmopolitan, and Warm Temperate biotas produces a major burst in diversity and complex ecological restructuring; evolutionary rates increase in response to rising stress levels for many lineages (Kauffman, 1978a). This is the Warm Temperate-marginal Tethyan diversity maximum, and favors evolution of stenothermal organisms.

4) Abrupt, widespread extinction of this diverse biota is associated with onset of a major anoxic event which may widely permeate the deep ocean and expand upward into the deeper portions of epicontinental seas. This is the core of the extinction event (the best stage boundary), and it may affect hundreds of genera and species within a short period of time. The benthic biota is almost totally decimated shortly after attaining a diversity peak during the preceding warming spell; only eurytopic burrowing polychaetes and rare inoceramid and ostreid bivalves remain in some areas. A major evolutionary turnover occurs in the pelagic calcareous microbiota; but planktonic taxa which become extinct during this crisis are quickly replaced by even more stenothermal Tethyan-marginal Tethyan taxa in greater diversity (Figure 8–11). Marine surface temperatures and salinity levels in shallow epicontinental seas reach their peak during this event, which in turn minimizes the oceanic and epicontinental thermal gradients, enhancing the breakdown of deep-water circulation; the marine system becomes widely stratified, and massive oxygen depletion results.

5) Prolongation of widespread oxygen depletion events for nearly 0.5 myr produces final extinction among the few remaining eurytopes from the older Warm Temperate-cosmopolitan biota. The middle and lower parts of the water column become nearly abiotic, although some polychaete bioturbation persists. Toward the end of the anoxic event rare cosmopolitan and subtropical generalists (ammonites, inoceramid bivalves) appear as circulation is slowly restored and approaches peak eustatic rise and maximum transgression. Pelagic diversity remains high during this period (Figures 8–10 to 8–13).

6) Peak transgression (maximum eustatic rise) occurs, during which water temperatures remain high and marine salinity normal. Changes in current circulation and the mixing of water masses brought about by peak eustatic rise fully restores benthic circulation in epicontinental seas. The benthic zone becomes extensively colonized by biotas of moderate diversity. Pelagic biotas remain diverse, with high biomass. Throughout the water column, the biota becomes biogeographically mixed among or-

ganisms of predominantly Tethyan, marginal Tethyan, cosmopolitan and Warm Temperate affinities. The predominance of stenothermal warm-water elements reflects widespread flooding of Tropical surface waters into normally temperate epicontinental seaways during the eustatic peak (Coates, Kauffman, and Sohl, in press).

7) Initial eustatic fall and early regression is associated with a slight decline in epicontinental water temperatures ($12°-C$) and salinity levels. But these slight changes ecologically shock the immigrant stenothermal, subtropical-tropical biotas in temperate epicontinental seas, which leads to waves of extinction and emigration during the initial 1 to 3 myr of regression, first among stenothermal, and subsequently among more eurythermal elements of the macrofauna. The extinction is thus graded over a short period of time, with a resultant decline in diversity.

8) Accelerated decline in epicontinental marine water temperatures and salinity associated with continued regression (eustatic fall) brings about final extinction of the macrofauna with Tethyan affinities, and waves of relatively abrupt extinction or emigration among the stenothermal Tethyan microplankton, which terminates the extinction event in epicontinental seas.

9) By mid-regression (eustatic fall), epicontinental seas return to Warm Temperate conditions, possibly with subnormal surface salinities. Warm Temperate to cosmopolitan biotas of moderate to low diversity predominate. In the benthic zone, oxygen again becomes restricted near the sediment-water interface in epicontinental seas.

In this model (Figure 8–13), five environmental factors converge in epicontinental seas to affect all stages of the extinction event: major changes in sea level, water circulation, marine water temperature, salinity, and benthic to midwater oxygen levels (reflecting circulation). Sea level, temperature, and oxygen changes are coeval in oceanic and epicontinental marine settings. The question remains: Is it the sequence of these events that is important in triggering large-scale extinction, or is it merely the magnitude and convergence of large-scale environmental changes that destabilize marine ecosystems and cause mass extinction? Comparison of the Albian-Cenomanian and the Maastrichtian-Danian extinction histories to this model provides some insight, since both events are obviously different in that they are primarily associated with eustatic fall and near-peak regression rather than with peak transgression, during the Cretaceous.

*Comparison with the Albian-Cenomanian extinction event:* The Albian-Cenomanian boundary zone is characterized by major environmental changes in the marine ecosystem that involve eustatic fluctuations, changing circulation patterns, and, as a result, broad variations in temperature, salinity, and benthic oxygen levels. It thus resembles the major intra-Cretaceous extinction events previously discussed. But here the resemblance ends. The Albian-Cenomanian extinction event was associated with a major latest-Albian eustatic fall, a major *regression* ($R_5$: Kauffman, 1977a), and a peak low in global marine water temperatures (see Scholle and Arthur, 1980, figure 3; Kauffman, 1977a, figure 7). The Cenomanian-Turonian and Turonian-Coniacian extinctions were associated with oceanic warming peaks and eustatic (transgressive) maxima. The Albian-Cenomanian extinction event provides an opportunity to test whether biological trends were similar across major extinction boundaries even though the sequence of events leading to environmental decline were variable. A summary of environmental and biological events across the Albian-Cenomanian boundary zone in the central part of the Western Interior seaway is presented below:

1) Initial Late Albian transgression ($T_5$; Kauffman, 1977a, figure 17) that reflected eustatic rise was associated with a slight rise in global marine temperatures during an overall decline (Scholle and Arthur, 1980, figure 3), subnormal epicontinental salinity, and somewhat restricted benthic circulation, which limited benthic and pelagic diversity.

2) Peak Late Albian transgression was associated with a marked decline in oceanic temperatures (Scholle and Arthur, 1980, figure 3), but a major warming trend in shelf and epicontinental seas, which resulted in migration of reefs well up onto the Texas platform (Coates, Kauffman, and Sohl, in press) and of diverse marginal Tethyan biotas as far north as Colorado. Normalization of salinity occurred in the southern part of the Interior seaway, grading northward to more brackish conditions. Peak diversification of benthic and pelagic biotas resulted, mainly among marginal Tethyan and Warm to mid-Temperate taxa.

3) Initial Late Albian regression (eustatic fall) produced a relatively abrupt epicontinental temperature decline, continued oceanic cooling, and the spread of a brackish water cap over much of the North American Interior seaway. Rapid, widespread extinction of Warm Temperate and marginal Tethyan biotas resulted; this was the major Late Albian extinction event.

4) Maximum Late Albian eustatic fall produced widespread epicontinental regression associated with a near peak temperature decline, widespread brackish conditions, and shoaling and loss of ecospace. Continued lowering of marine diversity in the seaway mainly resulted from the final extinction of warm water elements.

5) Renewed latest Albian transgression (initial $T_6$, Kauffman, 1977a, figure 7) was associated with peak temperature decline in the oceans and epicontinental seas (Kauffman, 1977a, figure 7; Scholle and Arthur, 1980, figure 3), greatly increased volcanic activity in North America, restricted epicontinental salinity, and subnormal benthic to midwater oxygen levels; depauperate upper pelagic biotas and low benthic and lower pelagic diversity resulted; mid-Temperate biotas were largely composed of eurytopic and endemic species.

6) Early $T_6$ (latest-Albian) eustatic rise and epicontinental transgression was associated with a global temperature minimum and great restriction of circulation in epicontinental basins like that of the North American Interior; widespread oxygen depletion resulted in the lowering of diversity and almost complete extinction of the benthic biota, followed by midwater ammonites, as the anoxic zone expanded.

7) Widespread mass mortality (or series of mortality events) among fish and the final extinction of Albian ammonites resulted from maximum oxygen restriction and possibly rapid overturn in the Western Interior epicontinental sea, which correlated with a widespread oceanic anoxic event (Fischer and Arthur, 1978), and marks the Albian-Cenomanian boundary.

8) Continuation of widespread anoxic or low oxygen conditions characterized the Western Interior seaway during Early Cenomanian transgression. Marine surface temperatures and epicontinental salinity remained relatively low. Early Cenomanian faunas in this region comprise generalized arenaceous foraminifera, eurytopic epibenthic inoceramid bivalves, and rare ammonites—all were sparsely and irregularly distributed.

9) Continued (middle) transgression resulted in the warming of marine waters, re-establishment of moderate benthic circulation, and diversification of a Warm Temperate and cosmopolitan biota in the Western Interior Cretaceous seaway; this event took place near the lower-Middle Cenomanian boundary.

Despite major variations in the ordering and magnitude of multiple environmental changes near the Albian-Cenomanian boundary as compared to the Cenomanian-Turonian and Turonian-Coniacian events, the sequence of biological events spanning this extinction is remarkably similar in all three studied boundary zones. In particular:

1) A major diversification of the temperate and cosmopolitan biota, associated with rapidly rising epicontinental marine temperatures, normalization of salinity, good benthic circulation, and near-peak trans-

gression (eustatic rise) occurred well before each main extinction event. Northward migration of subtropical waters and biotas, radiation of Warm Temperate biotas, and broad paleobiogeographic intermixing in the basin combined to create each diversity peak.

2) The major extinction of each biota, especially those that involved warm-water elements, occurred rather abruptly some time before the final extinction of its more generalized elements. This circumstance resulted in great loss of diversity and reduction of the biota to sparse eurytopic, cosmopolitan, and endemic elements; the benthic zone was especially affected. Each major extinction was associated with the rapid onset of oxygen restriction and changes in marine surface temperatures and salinity. In each case, a major regression and temperature decline followed shortly after the main extinction event.

3) A major period of oxygen depletion that reflected a widespread "anoxic event," followed each main extinction event. Early in this oxygen decline, rare surviving eurytopes, endemics, and cosmopolitan taxa from the ancestral biota gradually dies out; the final extinction of the ancestral biota occurred near the middle of the oxygen depletion event, when widespread anoxic conditions set in; this event may be marked by a mass mortality among pelagic macrofaunas, and/or an abrupt changeover of the microplankton. The anoxic events occurred near peak eustatic rise or fall.

4) Following peak oxygen reduction in marine environments, the new biotas of the subsequent stage appeared slowly in response to changing circulation, temperature, and salinity. These biotas comprised low oxygen-tolerant and pelagic generalists, and later, diverse stenotopes. The major radiation of these new biotas, therefore, occurred some time after the final extinction event of the older biota and coincided with the return of more normal marine conditions, especially benthic circulation and oxygenation of the sea floor.

Thus, in the case of the Albian-Cenomanian event, neither the injection of a major regression between maximum diversification and final extinction of the Albian biota, nor a major decline in global marine surface temperatures prior to the end of the Albian greatly altered the basic pattern of the extinction recorded at other middle Cretaceous levels. The most important environmental elements of the three "Middle" Cretaceous extinction events, therefore, seem to be the association of a significant epicontinental warming trend with a transgressive peak, which allowed high diversification of the marine biota (setting them up for the kill?), followed by a rapidly generated shock to the marine ecosystem that involved the convergence of three environmental factors: widespread oxygen depletion, a major eustatic change (reversal from transgression to regression or vice versa), and a major shift in marine water temperatures (up or down). These factors are commonly related; benthic oxygenation may be a factor of the thermal gradient of surface waters and deep water circulation, and both may be related to eustatic rise or fall, rerflecting active and quiet phases, respectively, of plate tectonic activity.

## AN ENVIRONMENTAL INTERPRETATION OF THE CRETACEOUS-TERTIARY EXTINCTION EVENT

There are few relatively complete Cretaceous-Tertiary boundary sequences that are exposed and have been studied in detail. The great majority of these sequences are outer shelf or deeper pelagic facies that are poor in macrofaunas. Except for the microplankton, therefore, data which can be applied to the interpretation of the terminal extinction event are sparse. These data are consistent enough among widely dispersed localities, however, to generate an hypothesis concerning the sequence of events that preceded the final Cretaceous extinction, and to compare it with the model established from well-studied intra-Cretaceous extinctions.

In the Tropical-Subtropical Tethyan realm, the general sequence of events leading up to the terminal Cretaceous extinction is remarkably similar to the intra-Cretaceous extinction model based on more subtropical to warm biotas (Figure 8–13). In summary they are:

1) diversification of the pelagic and benthic biota near the middle-Upper Maastrichtian boundary coeval with a small but relatively abrupt increase in marine surface temperatures, late phases of the final Cretaceous eustatic rise, and widespread epicontinental transgression ($T_{10}$ of Kauffman, 1977a, figure 7).

2) widespread, relatively abrupt extinction of stenotopic elements of the Tethyan macrofauna coincident with maximum transgression, a marine warming peak, and widespread benthic oxygen restriction near or at the middle-Late Maastrichtian boundary. This abrupt extinction probably affected most groups of rudists, hermatypic corals, and larger foraminifers, which characterized Tropical carbonate platform environments. It was the first major extinction event of the Maastrichtian, and included many taxa which are reputed to have survived up to the Cretaceous-Tertiary boundary before their "catastrophic extinction."

3) continued depletion of oxygen extending up to midwater zones during inital regression, while water temperatures were still high. This oxygen depletion resulted in simple benthic communities dominated by

eurytopic, low oxygen-tolerant, inoceramid and pectinid bivalves, and soft-bodied burrowing organisms. Diverse warm water, largely cosmopolitan ammonites and pelagic microbiotas characterized the upper part of the water column.

4) final extinction of the shelled benthos and great restriction of specialized (stenothermal) elements of the midwater ammonite fauna that occurred within a short time period as a result of initial Late Maastrichtian regression, initial lowering of marine temperatures and expansion of the oxygen depletion zone through the water column.

5) onset of a widespread deep water anoxic event essentially killing off all but the pelagic microbiota and most generalized burrowing infauna, which suggests a broad expansion of a midwater oxygen minimum zone.

6) continued eustatic fall, regression and sharp mid-Late Maastrichtian temperature fluctuations leading to overall decline (small to large depending on data used; Kauffman, 1977a, figure 7; Scholle and Arthur, 1980, figure 3) was associated with restoration of circulation and some oxygenation of the midwater zone. Low diversity (<10 spp.) Temperate to cosmopolitan eurytopic ammonite faunas characterized midwater habitats whereas diverse warm-water microbiotas predominated in the upper part of the water column. Shelled benthos remained rare to absent.

7) major deterioration of midwater and benthic environments, probably due to widespread oxygen depletion, broadly stressed burrowers and ammonites through the late-Late Maastrichtian regression. Bioturbation levels diminished; diversity and size (dwarfing?) decreased among remaining eurytopic ammonites. Extinction among the few surviving Late Maastrichtian ammonite lineages went from more stenotopic, biogeographically restricted (temperate) groups, to more eurytopic and cosmopolitan forms last. Deterioration of planktonic foraminifer assemblages began as warm near-surface waters became affected by spread of oxygen depletion and possibly by continued marine temperature fluctuation and/or cooling (e.g. from Ekofisk well; Scholle and Arthur, 1980, figure 3: and North Pacific planktonic foraminifera; Douglas and Savin, 1973).

8) final extinction of the ammonites (1 or 2 species), great depletion of both planktonic foraminifers and calcareous nannoplankton, and introduction of "disaster species" (e.g. braarudospheres) that occurred among the nannoplankton just prior to the regressive Later Maastrichtian maximum. Some data suggest peak lowering of marine temperatures (e.g. Douglas and Savin, 1973) near the end of the Maastrichtian; other data suggest abrupt low-level fluctuations of temperature, or little change (e.g. Scholle and Arthur, 1980, figure 3).

9) final, abrupt, and widespread extinction of the calcareous microplankton which was associated in some areas with a large-scale fish kill

and marks the Cretaceous-Tertiary boundary. This took place at peak eustatic fall and regression, a point of possibly rapid temperature and/or salinity fluctuation, and with the apparent onset of severe oxygen restriction through much of the marine water column. In deeper habitats, depletion continued into the basal Paleocene, and recovery of the ecosystem was slow at first and involved opportunists and eurytopic groups for the most part.

It must be emphasized that macrofossil data in support of this hypothesis are sparse and scattered, and that the sequence of events, therefore, is purposely generalized. The available data do seem to reflect a consistent pattern of macrofaunal extinction events in both shelf and deeper water facies in various parts of the world.

The graded extinction event during the last 1.5 myr of the Maastrichtian followed the same basic pattern of large-scale intra-Cretaceous extinctions, and was basically related to the same environmental processes. Differences between them, that involve details of sequencing or magnitude of any step leading to extinction, relate to variations in the strength and direction of environmental changes during individual extinction events (i.e. magnitude and direction of eustatic fluctuations, temperature shifts, and oxygen depletion at any one time). The greater magnitude of the terminal Cretaceous extinction event reflects the greater magnitude of regression, oxygen depletion, the abruptness of temperature fluctuation, and regional decline at the end of the Maastrichtian, and their close convergence as factors that led to deterioration of the marine environment. Of particular importance in this comparison is the large-scale sequence of: a) diversification of the warm-water biota followed by the major extinction of the macrofauna in conjunction with peak eustatic and temperature rise, and widespread oxygen depletion about 1.5 myr before the end of the Cretaceous; b) graded decline and extinction of more tolerant benthic and midwater macrofaunas with increasing oxygen restriction, eustatic fall, and abrupt changes in Late Maastrichtian marine temperatures; and c) final abrupt extinction of the pelagic warm-water microbiota and (regionally) fishes associated with maximum Late Maastrichtian temperature decline, regression, and oxygen restriction at the end of the Cretaceous. This sequence compares closely with those at the Albian-Cenomanian, Cenomanian-Turonian, and Turonian-Coniacian boundaries, previously discussed. These observations suggest a causal mechanism for Cretaceous marine extinction which is inherent in the relationship between plate tectonic activity, paleoceanography, and paleoclimatic responses; they do not support the idea of extinction that mainly result from extraterrestrial causes, or even abnormal earth-bound causes.

Clearly, they do not support the idea of a catastrophic extinction affecting a wide ecological spectrum of the marine biota at any one instant in time during or at the end of the Cretaceous.

Significantly, the terminal Cretaceous extinction event as preserved in the North Temperate realm (Denmark especially) does not show comparable levels of biological decline in the Late Maastrichtian as is shown by Tethyan shelf and deep-sea sections, or by epicontinental seas connected directly with Tethys (i.e. the Western Interior seaway of North America) (Kauffman, 1979b). But the *pattern* of environmental and biological decline is broadly similar to that of more equatorial waters (see previous discussion), further suggesting a global system of environmental change effecting large-scale extinction; this was initially and most severely manifest in the tropics and deep equatorial ocean sites, and subsequently in more poleward shelf areas.

Interpretation of the Cretaceous-Tertiary extinction event, and many smaller Cretaceous extinctions before it, now seems possible because of great similarities in the patterns of environmental change that led to extinction. I propose the following model for these extinctions:

1) A major eustatic sea level rise generated by active plate tectonism and elevation of topographic features on the seafloor is accompanied by global transgression, amelioration and warming of the world climate, and diversification of ecological opportunities. Oceanic temperatures rise, and abrupt-temperature increase and normalization of salinity occurs in shallow epicontinental seas connected to Tethys; epicontinental circulation changes.

2) As a result of the above conditions, widespread diversification of marine faunas takes place as evolutionary rates increase and extensive niche partitioning results. Tropical-subtropical waters spread poleward into shallow epicontinental seas and ocean margins that cause widespread mixing of Temperate, Tethyan, and cosmopolitan biotas in formerly temperate marine environments. Widespread niche and resource competition in this broad zone of mixing disrupts original ecosystem structure, that initially causes a significant increase in diversity as niches are temporarily shared by similarly adapted organisms or partitioned through competition, as new niches are occupied, and as endemism increases in the "disrupted" zone. Warming and amelioration of global climates result in lowering of temperature gradients across the world's oceans.

These trends favor the diversification of stenothermal organisms, and result in widespread dominance of most pelagic and benthic shelf habitats by Tethyan, marginal Tethyan, and Warm Temperate biotas. This is an important event relative to any subsequent extinction because large portions of the marine ecosystem are "set up for the kill" if diversification of warm water groups is rapidly interrupted by massive environmental deterioration and breakdown of ecosystem structure. Convergence of major environmental changes such as rapid temperature fluctuation and/or reduction, marine oxygen depletion, and eustatic fall that leads to widespread regression and loss of ecospace, would broadly shock any marine ecosystem dominated by stenotopic organisms. It would especially affect stenothermal groups that comprise the great bulk of the tropical, subtropical, and warm temperature biota. It is specifically this element of the Cretaceous biota that is most dramatically involved in both intra-Cretaceous and terminal Cretaceous extinction events.

3) Rapid change in one or more major environmental factors extensively shocks the marine biota/ecosystem during a period of widespread diversification of warm water biotas and restructuring of large ecological units. Widespread extinction results within a short period of geological time, involving diverse, mainly warm-water taxa. As a result, the marine ecosystem breaks down. In the case of the Cenomanian-Turonian, Turonian-Coniacian, and Maastrichtian-Paleocene extinction events, this initial shock was apparently caused by relatively rapid and widespread oxygen depletion in the world's ocean and/or deeper portions of epicontinental seas. These anoxic events resulted from lowering of the thermal gradient across the ocean surface associated with: rising sea level, rising marine temperatures and possibly salinity; decreased rates of poleward cooling and sinking of "heavy" waters; breakdown of deep water circulation; and widespread stagnation. The effect would have been enhanced if oceanic circulation was dominated by equatorial flow confined to warm waters of the Tethys seaway. Waves of extinction among benthic microfaunas, stenotopic endobenthic macrofaunas followed by epibenthic macrofaunas and ammonites resulted from this "shock." Only poorly structured, low diversity, benthic communities comprised of eurytopic, opportunistic elements remained (e.g. polychaetes). Broad expansion of the oxygen minimum zone(s), with severe oxygen restriction (but not anoxic conditions) that affect the benthic zone below it, is implied by the pattern of extinction. In the case of the Albian-Cenomanian boundary, abrupt temperature decline and rapid eustatic fall combined initially to shock the ecosystem.

In all studied cases, this environmental shock, which accounts for the major portion of the extinction event (greatest number of taxa) occurred 0.5 to 2.0 myr before final extinction of the biota in question. Surviving macrofaunal and benthic microfaunal lineages were represented by relatively few taxa, usually generalized forms; these were clearly on the evolutionary wane at the time of their final extinction. Only low to moderate levels of extinction characterize the pelagic microbiota at this stage.

4) Coincident with the initial stress to the marine ecosystem, or shortly after it begins, one or more additional major environmental change is superimposed on the first event. This overlap leads to accelerated environmental deterioration and greatly magnifies the shock to the marine ecosystem *before* it has had time to restructure in the face of the first major change. Wholesale collapse of large parts of the ecosystem is now possible. It is the magnitude of biological stress resulting from numerous, superimposed sources of environmental deterioration that constitutes the most important factor in causing widespread extinction. In the Cretaceous, eustatic fall, regression, loss or restriction of diverse ecological niches, and marine temperature decline became superimposed on widespread marine oxygen restriction in three of four examples involving Tethys and its margins; this includes the terminal Cretaceous extinction event. At the Albian-Cenomanian boundary, and in the terminal Cretaceous event of North Temperate realm shelf areas, widespread oxygen depletion was superimposed on combined factors of rapid temperature fluctuation, eustatic fall, and loss of shallow marine habitats. The ordering of environmental events was somewhat different in these two cases, but the ultimate effect was the same—widespread extinction.

During this phase of the "crisis" in the model, already depleted warm water macrofaunas including the shallow shelf biota situated above the oxygen minimum zone undergo their final extinction, which suggests a different cause. Benthic macrofaunas of more temperate shelf environments also show significant decline, as do some of the pelagic calcareous microbiota and burrowing infauna. The entire marine system becomes stressed, including for the first time, very shallow water and upper pelagic communities. This suggests an increasingly more important role of rapid temperature fluctuation and/or decline in the upper water column. The greater effect of this extinction event on stenothermal Tethyan and marginal Tethyan organisms than on more eurythermal, Temperate Zone taxa supports this contention.

In the case of the terminal Cretaceous extinction event, the stenothermal, shallow, warm-water biota apparently became caught in a "stress squeeze." Continued depletion of oxygen in the deep ocean caused upward migration of the oxygen minimum zone(s) until it encroached onto shelf areas. At the same time, eustatic fall and extensive marine regression drove shallow-water biotas toward island platform and shelf edges, and forced a geographic constriction of Tethys and its marginal influence. Broadly declining and rapidly fluctuating global temperatures apparently cooled near-surface waters in some regions to the point where the pelagic calcareous microbiota became thermally stressed. Such cooling would also have broadly affected remaining elements of the shallow benthic macrofauna, and especially those of the tropics and subtropics, through thermal stress to their long-lived, stenothermal, planktotrophic larvae (assumed from living analogs). *The dramatic Maastrichtian extinction of ecologically distinct planktonic micro-organisms and tropical macrobenthos like rudists and hermatypic corals can therefore be explained by stress to the one habitat they share in common—the upper layers of the tropical-subtropical oceans.* Differences in the timing of extinction for various groups of tropical-subtropical organisms may simply reflect different tolerances for thermal stress, with macrofaunal elements (through their shallow habitats and/or planktotrophic larvae) reacting to lower levels of temperature stress than the pelagic microfauna; in nearly all examples studied, the Tethyan pelagic microbiota have their principal extinction after that of the shallow water Tethyan macrofauna.

5) A final biotic crisis results from peak levels of environmental deterioration involving multiple factors. Both widespread oxygen depletion and rapid regional temperature decline and/or fluctuation play important roles in bringing this about during the Cretaceous, broadly stressing marine biotas and generally reaching the pelagic microplankton last during the Cenomanian-Turonian (Figure 8–11), Turonian-Coniacian, and terminal Cretaceous extinction events. In the last case, the demise of the calcareous microplankton was "catastrophic," within a few tens of thousands of years as all factors of marine environmental deterioration converged on the pelagic zone, and the biota of the entire water column (especially in Warm Temperate to tropical regions) became stressed. For the marine ecosystem, already in a state of collapse from removal of major biotic components in most environments, the abrupt disappearance of the plankton in the latest Cretaceous was the final shock, and reflected temporary loss or restriction of the base of the marine food chain. In conjunction with other factors of environmental deterioration, this probably accounted for the final extinction or restriction of surviving lineages of trophic specialists, and represented ecologically diverse groups like the rudist, ostreid, trigoniid, and inoceramid bivalves, the ammonites, corals, and even marine reptiles and fishes, most of which had already been greatly depleted during earlier stages of the terminal Cretaceous extinction.

The greater effect of this final event on Tethyan biotas than on temperate biotas may partly reflect greater depletion of the stenothermal tropical plankton (as represented in the fossil record mainly by planktonic foraminifera and calcareous nannofossils) than of the more eurythermal Temperate Zone plankton. A last major anoxic event (Fischer and Arthur, 1977), due to great spread of the oxygen minimum zone into shallow waters and/or to major overturn of deep anoxic waters in world oceans, may also have triggered the final collapse of the warm water pelagic ecosystem and the terminal Cretaceous extinction event. This anoxic event is possibly represented by organic-rich "boundary clays," "fish beds" (mass mortality?), restricted bioturbation, and considerable depletion or temporary disappearance of the benthic biota—all characteristics of the Cretaceous-Tertiary boundary zone in various parts of the world.

In conclusion, the "Terminal Cretaceous Extinction Event" or "Biotic Crisis" was neither broadly catastrophic nor unique (Kauffman, 1979b). In terms of the types and magnitude of environmental changes associated with the extinction, and the general pattern of biological response, the terminal event was very similar to major extinctions at stage boundaries *within* the "Middle" Cretaceous. The extinction corresponded to massive environmental deterioration during the last 1.5 myr of the Maastrichtian that resulted when several large-scale changes in the marine environment became superimposed. These collectively intensified the shock to the ecosystem at a rate that exceeded the ability of the marine biota to readapt, and to restructure and "stabilize" ecological units. Physicochemical environmental factors initially depleted the marine biota through a series of smaller extinctions, each of which predominantly affected a different part of the biota; these factors were primarily: widespread oxygen depletion in the lower and middle parts of the water column; eustatic fall, regression, and loss of ecospace; and rapid temperature fluctuation leading regionally to decline. Of these, oxygen depletion played a major role in early phases of extinction, and temperature was a major factor later in the event. Biological factors, especially widespread breakdown of ecological structure, competition during buildup of stress and diminution of ecospace, and ultimately collapse of a large portion of the food chain with abrupt extinction of warm water microplankton, all played a major role in terminal extinction. Clearly, warm water (Warm Temperate to Tethyan) organisms were more strongly affected than Temperate Zone organisms during large-scale extinctions; each extinction was more widespread and more dramatic and had a longer history in Tethyan and marginal Tethyan biotas than elsewhere. In part this relates to more numerous stenothermal, stenotopic ecotypes in Tethys than toward the poles; in part, it relates to greater exposure of Tethyan benthic biotas, through their long-lived planktotrophic larvae, to environmental stresses in the uppermost part of the water column. It may also relate to greater magnitude of environmental deterioration in more equatorial seas, especially oxygen depletion, because of a dominance of equatorial ocean circulation. Tethys was basically an open circum-global sea in the Cretaceous.

The terminal (boundary zone) Cretaceous extinction event mainly involved calcareous microplankton of warm water environments, and was probably the collective result of a stress squeeze on the pelagic biota resulting from upward migration of the oxygen minimum zone toward the surface at the same time that marine environments were being dramatically reduced in size and diversity by eustatic fall and extensive regression, and surface temperatures were rapidly fluctuating. Collapse of the marine ecosystem, and specifically of the base of the food chain with widespread extinction of the plankton, simultaneously killed off the last remnants of the typical Cretaceous macrofauna, most of which were already well into their evolutionary decline as a result of earlier environmental shocks.

This hypothesis needs extensive testing, but it best fits the observed fossil record of the Cretaceous-Tertiary extinction event. It does not require forces beyond those that are inherent in the paleoceanography, climatic changes, and plate tectonic activity of the Cretaceous Period. It is the magnitude and timing of normal environmental events that seems to have determined whether large scale extinction took place in the marine realm.

How do these diverse observations fit the many recent theories that have been generated for the Cretaceous extinction event? In many cases not very well. Too commonly these theories are based on limited data, either drawn from a detailed analysis of a single biotic group or paleoceanographic factors; in some cases popular hypotheses ignore the massive, detailed biological and paleoenvironmental data which are presently available and which collectively bear on the problem of the terminal Cretaceous extinction. To consider the principal hypotheses for mass extinction:

1) Catastrophic terminal extinction of the Cretaceous marine biota, as exemplified by elements of the calcareous microplankton, has been credited to abrupt lowering of the surface salinity of the world's oceans down to depths of 100 meters. According to this theory, the salinity change was caused by a massive injection of fresh water from the proto-Arctic ocean through narrow passageways newly created by plate tectonic movements (Gartner and Keany, 1978; Gartner, 1979; Thierstein, 1979). This theory presupposes an Arctic Ocean wholly isolated from marine systems so that it freshened significantly during the Late Cretaceous. The North Sea-Danian-Viking graben System is viewed as the principal North Atlantic conduit for this fresh water injection.

Principal arguments against this theory are: lack of evidence for Cretaceous isolation of the Arctic Ocean (Tappan, 1979); the likelihood that such a fresh water wedge would be oceanographically trapped in North

Temperate Atlantic and Pacific gyres and thus would not spread to many parts of the world where the extinction event nevertheless occurred; and lack of biological evidence for such a massive salinity change among those organisms which should be most affected—North Temperate shallow shelf and intertidal taxa thought to have been stenohaline, based on their modern analogs. A broad biotic analysis of the Cretaceous extinction event, considering all major groups, suggests that Maastrichtian extinction was most severe among tropical-subtropical biotas and decreased toward the poles. Ironically, the best studied, most diverse marine biota which shows only slight to moderate Maastrichtian extinction levels characteristic of the North Temperate realm occurs in Denmark (papers in Birkelund and Bromley, 1979, and references therein), near the mouth of the proposed fresh-water injection channel; the Danish biota should show the greatest, not the least evidence for catastrophic extinction if this theory were valid. North Temperate lineages which survive the Cretaceous extinction event and pass on into the Paleocene (40 to 60% among different major taxa) include many stenohaline shallow water groups which should have been most affected by massive desalination of the world's oceans.

2) Worsley (1971, 1974) has proposed that CCD (Calcite Compensation Depth) rose rapidly to the ocean surface during the latest Cretaceous, and apparently created a physiological crisis and caused widespread destruction among diverse calcium carbonate secreting organisms. It is true that tropical to Warm Temperate carbonate-producing invertebrates showed the highest level of extinction during the Maastrichtian (for example, compare the extinction history of planktonic foraminifera and coccolithophorids with that of dinoflagellates and radiolarians). But three arguments make this hypothesis untenable: a) Lack of a mechanism for such a rapid rise of CCD and for having it selectively affect carbonate producing organisms in warm water zones; b) most carbonate-producing organisms have protective organic membranes over the shells during and after formation which only disappear after death, allowing dissolution; this allows carbonate-shelled organisms (for example, mollusks) to live below CCD in abundance today; and c) the record of Cretaceous carbonate-secreting organisms across the Maastrichtian-Danian boundary shows that only a few selected groups became dramatically extinct while many others, with the same shell composition, crossed the boundary at the lineage or generic level without significant evolutionary or calcification crises. These include both deep water and shelf taxa.

3) Massive chemical changes and increase in toxic trace elements in the world's oceans that led to rapid, widespread extinction has been proposed by various authors (for example, Cloud, 1959). These changes have

been variously attributed by most authors to large modifications in continental source materials due to tectonic uplift, erosion events, and/or widespread marine regression; to salinity changes (Gartner and Keany, 1978; Thierstein, 1979): to volcanic activity and oxygen restriction (Tappan, 1968; Fischer and Arthur, 1978; this paper). No geochemical studies of latest Cretaceous sediments or connate waters show such a change in toxic elements (but see Alvarez et al., 1979), nor is it reflected by the patterns of biotic decline previously described. Many other large Cretaceous regressions show no major extinction event. Massive salinity changes have been previously discussed and rejected (see 1 above). Detailed documentation of Cretaceous volcanism relative to tectonic and eustatic events (as in Kauffman, Cobban and Eicher, 1976) shows scattered large regional Maastrichtian events, but in general suggests that the latest Cretaceous was volcanically passive compared, for example, to the active volcanism of the Late Albian, Middle to Late Cenomanian, Early to early-Late Campanian, and so on; yet no major extinction event can be correlated with these earlier volcanic episodes. On the other hand, large-scale oxygen restriction can be documented from organic-rich Cretaceous-Tertiary boundary sequences (Tappan, 1968; Fischer and Arthur, 1978; and herein) and this event, possibly associated with lesser chemical changes of other types, was contributory to massive Late Cretaceous extinction.

4) Nutrient and essential mineral starvation of the world's oceans (Bramlette, 1965), or selective mortality of the microplankton (Tappan, 1968; Kauffman, 1977d) leading to breakdown of the oceanic food chain has been widely cited as a cause for the terminal Cretaceous extinction of diverse organisms. Although these factors may have contributed to terminal Cretaceous extinction of some trophic specialists, they were not primary causes. Careful documentation of the boundary event point out: a) that the extinction of most organisms occupying elevated positions on the food chain largely *preceded* the elimination of organisms at the base of the food chain; they were probably the very last to go in the extinction event; b) further, although there was a significant replacement of dominant calcareous microplankton at the boundary by disaster species such as braarudospheres, there is little evidence for a significant change in microplankton biomass across the boundary. This suggests continuation of primary productivity in the base of the food chain throughout the Cretaceous-Tertiary extinction event.

5) Regression and resultant changes in circulation, cooling, and loss of habitats in the world oceans have been cited by Newell (1967) as primary causes for extinction at the Cretaceous-Tertiary boundary, among others. Whereas these changes strongly contributed to the extinction (previous

discussion) by spatially restricting or eliminating habitats and thermally shocking stenothermal taxa, they are not, overall, the sole causes. The terminal Cretaceous extinction also strongly involved oceanic warm-water plankton whose ecospace was not drastically affected by the regression. Further, only certain regions show significant cooling at this boundary (for example, Douglas and Savin, 1973), while other areas show rapid small-scale temperature fluctuations (Buckardt and Jørgensen, 1979) and still others manifest little significant temperature change (Boersma and Shackleton, 1979). Yet the terminal Cretaceous extinction event affected all marine areas to some extent, despite their thermal and relative eustatic history.

6) The idea that the terminal Cretaceous extinction resulted from a "greenhouse effect" (McLean, 1978), in which the earth's surface rapidly heated up or cooled down (depending upon $CO_2$ production or Earth heat-flow levels in various theories) with development of a dense atmospheric cloud which increased filtration of the sun's rays or hindered heat escape, has been raised several times. In general, increased Maastrichtian volcanic activity (Feldman, 1977) is called upon to create the cloud and it can now be shown that, compared to intense earlier Cretaceous volcanic episodes that caused no major extinction, the Maastrichtian was relatively quiet. It is unlikely ash production worldwide at the end of the Cretaceous was sufficient to have had any significant atmospheric effect. More recently Alvarez et al. (1980) have rejuvenated a kind of greenhouse theory by speculating on a global dust cloud generated by a terminal Cretaceous meteor impact which effectively blocked the sun's rays for many months or possibly one or more years. Not only is there no direct evidence of a meteor crater, tidal wave deposits and erosion surfaces that would have resulted from an oceanic impact, or the dust layer itself in the sedimentary record, but the catastrophic effect such volcanic or meteoric dust clouds would have had on broad leafed deciduous land plants, possibly producing a mass mortality on land, is not recorded in the fossil record. Rather, many lineages of deciduous plants cross the Cretaceous-Tertiary boundary with little or no change.

7) Massive radiation due to a supernova or solar flare has been widely cited as the cause for massive terminal Cretaceous extinction (e.g. Russell and Tucker, 1971; Papers in K-Tec symposium, 1977); such an event would be enhanced by the breakdown of the Van Allen Belt with Late Maastrichtian magnetic reversals. Appealing as this may be to explain catastrophic extinction, it is not widely supported by biological evidence. Many of the most exposed marine organisms (intertidal, shallow subtidal invertebrates) show the least change across the boundary. Similarly tropical plants, with largely unprotected reproductive cells, show much less

change across the boundary than do temperate plants with more protected germ cells. No extensive genetic damage or bizarre mutations are manifest in species found in and just above the boundary zone. Finally, tropical and subtropical marine organisms are much more dramatically affected by the extinction event than are more poleward species. Yet, radiation entering the earth when the magnetic field is active is focused on the poles. The major events of tropical-subtropical extinction are spread over the Upper Maastrichtian, which is bounded by, but which does not contain, magnetic reversals. Further, Alvarez et al. (1979) argued against a supernova from the standpoint of iridium levels found at the Cretaceous-Tertiary boundary.

8) Alvarez and his colleagues (1979, 1980) have demonstrated that complete Cretaceous-Teritary boundary zones in various parts of the world contain abnormally high iridium and other rare trace element levels within a highly restricted stratigraphic interval coincident with the dramatic terminal Cretaceous extinction of calcareous microplankton. These authors attributed this to a large meteor (10 kilometer diameter) impact on the earth's oceanic surface and consider this event and subsequent oceanic disruption and atmospheric screening of the sun's rays by a thick dust cloud as causal for the terminal Cretaceous biotic catastrophe. There analytical data leaves little doubt that some extraterrestrial event did happen at this boundary to produce such iridium and other trace-element anomalies, and that it coincides closely with the final microplankton extinction and "fish clay" (mass mortality?) zone, where it is developed. But was this the principal cause of Cretaceous extinction, or merely the "straw that broke the camel's back"? Detailed biological evidence, which shows that most of the terminal Cretaceous extinction was over by the time of the final calcareous microplankton catastrophe, strongly suggests the latter. Diverse evidence further casts doubt on the meteor impact theory and suggests, instead, that the earth might have been impacted by a comet, or passed through a comet's tail or similar source for these rare elements (Hsü, 1980). There is no impact crater, no deposit reflecting global tidal waves "several kilometers high" which would have been generated by such an impact into the ocean, and no widespread mass mortality event among shallow marine organisms or among land plants that would have been shaded out by a global dust cloud emplaced for several months to several years, according to the theory, as a result of the impact. Whatever the source of the iridium peak at the Cretaceous-Tertiary boundary, and the effect that the event producing it possibly had on the world's oceanic calcareous microplankton, it was largely after the fact of widespread extinction, an accident of space and time that coincided with massive natural environmental decline on earth, and pushed it to an exceptional end.

In conclusion, diverse evidence strongly suggests that Terminal Cretaceous extinction was graded over 1 to 5 myr in the marine realm (Kauffman, 1979b) and was primarily the result of massive environmental deterioration resulting from relatively rapid, large-scale superimposed changes in sea level, water chemistry (especially oxygen), ocean temperature, circulation, climate, niche size and diversity, and resultant biologic effects of increased competition and broad destructuring of ecological units. The extinction was enhanced by some extraterrestrial event near the terminal phase of biotic decline. The temporal overlap of many large-scale environmental factors is necessary for major extinction (affecting diverse organisms with varying ecological plans) to take place; many of these environmental factors are interrelated, others are chance occurrences. But the fact that in many Cretaceous extinction events similar earthbound factors seem to occur in a general order helps to explain the main biological pattern of extinction during the terminal Cretaceous event.

## ACKNOWLEDGMENTS

This research was supported by a grant from the Smithsonian Research Awards program (grant no. 89191016). Numerous colleagues contributed to my thoughts through discussion and by making available to me unpublished data. In particular, I would like to thank Drs. Norman F. Sohl, Peter Scholle, Joseph E. Hazel, and Charles C. Smith of the U.S. Geological Survey, Dr. Jost Wiedmann of the Universität Tübingen, Drs. Alan H. Cheetham and Richard H. Benson of the U.S. National Museum, Dr. Harry C. Kent of Colorado School of Mines, and Dr. Don L. Eicher of the University of Colorado. Finally, I would like to express my sincere gratitude to the editors for their patience and understanding, and to Laurel Smith and Lawrence B. Isham of the U.S. National Museum for assistance in drafting.

## LITERATURE CITED

Afshar, R., 1969. Taxonomic revision of the superspecific groups of the Cretaceous and Cenozoic Tellinidae. *Geol. Soc. Amer. Mem.,* 119:1–215.

Alvarez, L. S., W. Alvarez, F. Asaro, and H. V. Michel, 1980. Extraterrestrial cause for the Cretaceous-Tertiary extinction. *Science,* 208:1095–1108.

Alvarez, W., L. W. Alvarez, F. Asaro, and H. V. Michel, 1979. Anomalous iridium levels at the Cretaceous/Tertiary boundary at Gubbio, Italy: negative results of tests for a supernova origin. In: W. K. Christensen and T. Birkelund (eds.), *Cretaceous-Tertiary Boundary Events. Copenhagen Univ., Proc. Sympos. "Cretaceous-Tertiary Boundary Events"*, vol. 2, p. 69.

Alvarez, W., M. A. Arthur, A. G. Fischer, W. Lowrie, G. Napoleone, I. Premoli Silva, and W. M. Roggenthen, 1977. 5. Type section for the late Cretaceous-Paleocene geomagnetic reversal time scale. *Geol. Soc. Amer. Bull.*, 88:383-389.

Arkell, W. J., W. M. Furnish, B. Kummel, A. K. Miller, R. C. Moore, O. H. Schindewolf, P. C. Sylvester-Bradley, and C. W. Wright, 1957. Part L, Mollusca 4, Cephalopoda Ammonoidea. In: R. C. Moore (ed.), *Treatise on Invertebrate Paleontology* (New York: Geol. Soc. Amer.), pp. L1-L489.

Arthur, M. A., and A. G. Fischer, 1977. Upper Cretaceous-Paleocene magnetic stratigraphy at Gubbio, Italy. 1. Lithostratigraphy and sedimentology. *Geol. Soc. Amer. Bull.*, 88:367-371.

Asgaard, U., 1979. The irregular echinoids and the boundary in Demark. In: T. Birkelund and R. G. Bromley (eds.), *The Maastrichtian and Danian of Denmark. Copenhagen Univ., Proc. Sympos. "Cretaceous-Tertiary Boundary Events"*, vol. 1, pp. 74-77.

Berggren, W. A., 1960. Biostratigraphy, planktonic foraminifera and the Cretaceous-Tertiary boundary in Denmark and southern Sweden. *21st Int. Geol. Congr. Proc. (Copenhagen)*, sec. 5, pp. 181-192.

―――, 1962. Stratigraphic and taxonomic-phylogenetic studies of Upper Cretaceous and Lower Tertiary planktonic foraminifera. *Stockholm Contrib. Geol.*, 9 (2):103-129.

Birkelund, T., and R. G. Bromley (eds.), 1979. *The Maastrichtian and Danian of Denmark. Copenhagen Univ., Proc. Sympos. "Cretaceous-Tertiary Boundary Events"*, vol. 1, pp. 1-210.

Boersma, A., and N. Shackleton, 1979. Some oxygen and carbon isotope variations across the Cretaceous/Tertiary boundary in the Atlantic Ocean. In: W. K. Christensen and T. Birkelund (eds.), *Cretaceous-Tertiary Boundary Events. Copenhagen Univ., Proc. Sympos. "Cretaceous-Tertiary Boundary Events"*, vol. 2, pp. 50-53.

Boss, K. J., 1966. The subfamily Tellininae in the Western Atlantic: the genus *Tellina* (part 1). *Johnsonia*, 4 (45):217-272.

―――, 1968. The subfamily Tellininae in the Western Atlantic: the genera *Tellina* (part II) and *Tellidora. Johnsonia*, 4:273-344.

―――, 1969a. The subfamily Tellininae in the Western Atlantic: the genus *Strigilla. Johnsonia*, 4:345-366.

―――, 1969b. The subfamily Tellininae in South African waters (Bivalvia, Mollusca). *Harvard Univ. Mus. Comp. Zool. Bull.*, 138 (4):81-162.

Bramlette, M. N., 1965. Massive extinctions in biota at the end of Mesozoic time. *Science*, 148:1696-1699.

Bramlette, M. N., and E. Martini, 1964. The great change in calcareous nannoplankton fossils between the Maastrichtian and Danian. *Micropaleontology*, 10:291-322.

Brotzen, F., 1959 (1960). On *Tylocidaris* species (Echinoidea) and the stratigraphy of the Danian of Sweden, with a bibliography on the Danian and the Paleocene. *Sver. Géol. Undersökn.*, ser. C, no. 571, Årsbok 54 (2):2-81.

Buckardt, B., and N. O. Jorgensen, 1979. Stable isotope variations at the Creta-ceous/Tertiary boundary in Denmark. In: W. K. Christensen and T. Birke-lund (eds.), *Cretaceous-Tertiary Boundary Events. Copenhagen Univ., Proc. Sympos. "Cretaceous-Tertiary Boundary Events"*, vol. 2, pp. 54–61.

Bujak, J. P., and G. L. Williams, 1979. Dinoflagellate diversity through time. *Mar. Micropal.*, 4:1–12.

Bukry, D., 1975. Silicoflagellate and coccolith stratigraphy, Deep Sea Drilling Project, Leg 29. *Initial Reports of the Deep Sea Drilling Project* (Washington, D.C.: U.S. Govt. Printing Office), vol. 29, pp. 845–872.

Bukry, D., and J. H. Foster, 1974. Silicoflagellate zonation of upper Cretaceous to lower Miocene deep-sea sediment. *U.S. Geol. Surv. Jour. Res.*, 2:303–310.

Byers, C. W., and D. W. Larson, 1979. Paleoenvironments of Mowry Shale (lower Cretaceous), western and central Wyoming. *Amer. Assoc. Petrol. Geol. Bull.*, 63 (3):354–361.

Christensen, W. K., and T. Birkelund (eds.), 1979. *Cretaceous-Tertiary Boundary Events. Copenhagen Univ., Proc. Sympos. "Cretaceous-Tertiary Boundary Events"*, vol. 2, pp. 1–250.

Cloud, P. E., 1959. Paleoecology—retrospect and prospect. *Jour. Pal.*, 33 (5):926–962.

Coates, A. G., E. G. Kauffman, and N. F. Sohl, (in press). Cretaceous migrations of tropical organisms into the North Temperate Realm. In: E. G. Kauffman and D. E. Hattin (eds.), *Cretaceous of the North Temperate Realm* (Strouds-burg, Penn.: Hutchinson & Ross).

Cobban, W. A., 1958. Late Cretaceous fossil zones of the Powder River Basin, Wyoming and Montana. *Wyoming Geol. Assoc. 13th Ann. Field Conf., Pow-der River Basic Guidebk.*, pp. 114–119.

Cobban, W. A., and J. B. Reeside, Jr., 1952. Correlation of the Cretaceous for-mations of the Western Interior of the United States. *Geol. Soc. Amer. Bull.*, 63:1011–1044.

Cobban, W. A., and G. R. Scott, 1972. Stratigraphy and ammonite fauna of the Graneros Shale and Greenhorn Limestone near Pueblo, Colorado. *U. S. Geol. Surv. Prof. Paper*, 645:1–108.

Coogan, A. H., 1969. Evolutionary trends in rudist hard parts. Part N, Volume 2 (of 3), Mollusca 6, Bivalvia. In: R. C. Moore (ed.), *Treatise on Invertebrate Paleontology* (New York: Geol. Sco. Amer.—, pp. N766–N776.

Cox, L. R., et al., 1969. Part N, Volume 1–3, Mollusca 6, Bivalvia. In: R. C. Moore (ed.), *Treatise on Invertebrate Paleontology* (New York: Geol. Soc. Amer.), pp. xxxviii, N1–N1224.

Crimes, T. P., 1973. From limestones to distal turbidities: a facies and trace fossil analysis in the Sumaya flysch (Paleocene-Eocene), north Spain. *Sedimentol-ogy*, 20:105–131.

Cvancara, A. M., 1966. Revision of the fauna of the Cannonball Formation (Pa-leocene) of North and South Dakota. *Michigan Univ. Mus. Pal. Contrib.*, 20 (10):277–374.

Douglas, R. G., and S. M. Savin, 1973. Oxygen and carbon isotope analyses of the Cretaceous and Tertiary foraminifera from the central North Pacific. *In-*

*itial Reports of the Deep Sea Drilling Project* (Washington, D.C.: U.S. Govt. Printing Office), vol. 17, pp. 591–605.

Eicher, D. L., 1960. Stratigraphy and micropaleontology of the Thermopolis Shale. *Yale Univ. Peabody Mus. Nat. Hist. Bull.*, 15:1–126.

———, 1965. Foraminifera and biostratigraphy of the Graneros Shale. *Jour. Pal.*, 39 (5):875–909.

———, 1966. Foraminifera from the Cretaceous Carlile Shale of Colorado. *Cushman Found. Foram. Res. Contrib.*, 17 (1):16–31.

———, 1967. Foraminifera from Belle Fourche Shale and equivalents, Wyoming and Montana. *Jour. Pal.*, 41 (1):16–31.

———, 1969. Cenomanian and Turonian planktonic foraminifera from the Western Interior of the United States. *Proceedings of the 1st International Conference on Planktonic Microfossils*, Geneva, 1967 (Leiden: E. J. Brill), pp. 163–174.

———, 1977. Stratigraphic distribution of Cretaceous planktonic foraminifera, Rock Canyon Anticline. In: E. G. Kauffman (spec. ed.), Cretaceous facies, faunas, and paleoenvironments across the Western Interior Basin. *Rocky Mt. Assoc. Geol. Mountain Geol.*, 14 (3,4):153–154.

Eicher, D. L., and P. Worstell, 1970. Cenomanian and Turonian foraminifera from the Great Plains, United States. *Micropaleontology*, 16 (3):269–324.

Feldman, P., 1977. Astronomical evidence bearing on the supernova hypothesis for the mass extinctions at the end of the Cretaceous; and discussions. In: K-Tec symposium; Cretaceous-Tertiary extinctions and possible terrestrial and extraterrestrial causes. *Canadian Nat. Mus. Syllogeus*, 12:125–152.

Fischer, A. G., and M. A. Arthur, 1977. Secular variations in the pelagic realm. In: H. E. Cook and P. Enos (eds.), *Deep Water Carbonate Environments. Soc. Econ. Pal. Min. Spec. Pub.*, 25:19–50.

Gamper, M. A., 1977. Acerca del limite Cretacico-Terciario en Mexico. *Univ. Nac. Auton. Mexico Inst. Geol. Rev.*, 1 (1):23–27.

Gartner, S., 1979. Terminal Cretaceous extinctions, a comprehensive mechanism. In: W. K. Christensen and T. Birkelund (eds.), *Cretaceous-Tertiary Boundary Events. Copenhagen Univ., Proc. Sympos. "Cretaceous-Tertiary Boundary Events"*, vol. 2, pp. 26–28.

Gartner, S., and J. Keany, 1978. The terminal Cretaceous event; a geological problem with an oceanographic solution. *Geology*, 6:708–712.

———, 1979. The coccolith succession across the Cretaceous/Tertiary boundary in the subsurface of the North Sea (Ekofisk). In: W. K. Christensen and T. Birkelund (eds.), *Cretaceous-Tertiary Boundary Events. Copenhagen Univ., Proc. Sympos. "Cretaceous-Tertiary Boundary Events"*, vol. 2, pp. 103–105.

Gill, J. R., and W. A. Cobban, 1973. Stratigraphy and geologic history of the Montana Group and equivalent rocks, Montana, Wyoming, and North and South Dakota. *U. S. Geol. Surv. Prof. Paper*, 776:1–37.

Graveson, P., 1979. Remarks on the regular echinoids in the upper Maastrichtian and lower Danian of Denmark. In: T. Birkelund and R. G. Bromley (eds.), *The Maastrichtian and Danian of Denmark. Copenhagen Univ. Proc. Sympos. "Cretaceous-Tertiary Boundary Events"*, vol. 1, pp. 72–73.

Hakansson, E., and E. Thomsen, 1979. Distribution and types of bryozoan communities at the boundary in Denmark. In: T. Birkelund and R. G. Bromley (eds.), *The Maastrichtian and Danian of Denmark. Copenhagen Univ. Proc. Sympos. "Cretaceous-Tertiary Boundary Events"*, vol. 1, pp. 78–91.

Hancock, J. M., 1977. The historic development of biostratigraphic correlation. In: E. G. Kauffman and J. E. Hazel (eds.), *Concepts and Methods of Biostratigraphy* (Stroudsburg, Penn.: Dowden, Hutchinson, and Ross, Inc.), pp. 3–22.

Hattin, D. E., 1962. Stratigraphy of the Carlile Shale (upper Cretaceous) in Kansas. *Kansas Geol. Surv. Bull.*, 156:1–155.

————, 1964. Cyclic sedimentation in the Colorado Group of west-central Kansas. In: D. F. Merriam (ed.), Symposium on cyclic sedimentation. *Kansas Geol. Surv. Bull.*, 169:205–217.

————, 1965. Stratigraphy of the Graneros Shale (upper Cretaceous) in central Kansas. *Kansas Geol. Surv. Bull.*, 178:1–83.

————, 1975. Stratigraphy and depositional environment of Greenhorn Limestone (upper Cretaceous) of Kansas. *Kansas Geol. Surv. Bull.*, 209:1–128.

————, 1977. Upper Cretaceous stratigraphy, paleontology, and paleoecology of western Kansas. In: E. G. Kauffman (spec. ed.), Cretaceous facies, faunas, and paleoenvironments across the Western Interior Basin. *Rocky Mt. Assoc. Geol. Mountain Geol.*, 14 (3,4):175–218.

Hattin, D. E., and C. T. Siemers, 1978. Guidebook: Upper Cretaceous stratigraphy and depositional environments of western Kansas. *Kansas Geol. Surv. Guidebook*, ser. 3, pp. 1–102.

Herm, D., 1965. Mikropaläontologisch-stratigraphische Unterschungen im Kreide-flysch zwischen Deva und Zumaya (Prov. Guipuzcoa, Nordspanien). *Deutsch. Geol. Ges. Zeitschr., Jahrg.,* 1963, 15:277–348.

House, M. R., 1963. Bursts in environment. *Adv. Sci.,* 19:499–507.

Hsü, K. J., 1980. Terrestrial catastrophy caused by cometary impact at the end of the Cretaceous. *Nature*, 285:201–203.

Kauffman, E. G., 1969. Cretaceous marine cycles of the Western Interior. *Rocky Mt. Assoc. Geol., Mountain Geol.,* 6 (4):227–245.

————, 1970. Population systematics, radiometrics, and biostratigraphy—a new biostratigraphy. *Proceedings of the North American Paleontological Convention, Section I.,* pt. F, pp. 612–666.

————, 1972. Evolutionary rates and patterns of North American Cretaceous Mollusca. *Int. Geol. Congr., Montreal, 1972, Proceedings,* sec. 7, pp. 174–189.

————, 1973. Cretaceous Bivalvia. In: A. Hallam (ed.), *Atlas of Palaeobiogeography* (Amsterdam: Elsevier Pub. Co.), pp. 353–383.

————, 1975. Dispersal and biostratigraphic potential of Cretaceous benthonic Bivalvia in the Western Interior. In: W. G. E. Caldwell (ed.), The Cretaceous system in the Western Interior of North America. *Geol. Assoc. Canada Spec. Paper,* 13:163–194.

————, 1976. Deep-sea Cretaceous macrofossils: Hole 317A, Manihiki Plateau. *Initial Reports of the Deep Sea Drilling Project* (Washington, D.C.: U.S. Govt. Printing Office), vol. 33, pp. 503–535.

————, 1977a. Geological and biological overview: western interior Cretaceous Basin. In: E. G. Kauffman (spec. ed.), Cretaceous facies, faunas, and paleoenvironments across the Western Interior Basin. *Rocky Mt. Assoc. Geol. Mountain Geol.*, 14 (3,4):75–99.

————, 1977b. Upper Cretaceous cyclothems, biotas, and environments, Rock Canyon Anticline, Pueblo, Colorado. In: E. G. Kauffman (spec. ed.), Cretaceous facies, faunas, and paleoenvironments across the Western Interior Basin. *Rocky Mt. Assoc. Geol. Mountain Geol.*, 14 (3,4):129–152.

————, 1977c. Evolutionary rates and biostratigraphy. In: E. G. Kauffman and J. E. Hazel (eds.), *Concepts and Methods of Biostratigraphy* (Stroudsburg, Penn.: Dowden, Hutchinson, and Ross, Inc.), pp. 109–141.

————, 1977d. Cretaceous extinction and collaspe of marine trophic structures. *Jour. Pal.*, 51, suppl. to no. 2 (3):16 (abstract).

————, 1978a. Evolutionary rates and patterns among Cretaceous Bivalvia. *Roy. Soc. London Phil. Trans.*, sec. B, 284:277–304.

————, 1978b. Benthic environments and paleoecology of the Posidonienschiefer (Toarcian). *Neues Jahrb. Geol. Pal. Monatsh.*, 157:18–36.

————, 1978c. Short-lived benthic communities in the Solnhofen and Nusplingen Limestones. *Neues Jahrb. Geol. Pal. Monatsh.*, 12:717–724.

————, 1979a. Cretaceous. In: C. Teichert and R. A. Robison (eds.), *Treatise on Invertebrate Paleontology: Part A.* (Lawrence, Kansas: Univ. Kansas Press/Geol. Soc. Amer.).

————, 1979b. The ecology and biogeography of the Cretaceous-Tertiary extinction event. In: W. K. Christensen and T. Birkelund (eds.), *Cretaceous-Tertiary Boundary Events. Copenhagen Univ., Proc. Sympos. "Cretaceous-Tertiary Boundary Events"*, vol. 2, pp. 29–37.

Kauffman, E. G., W. A. Cobban, and D. E. Eicher, 1976 (1978). Albian through lower Coniacian strata and biostratigraphy, western interior United States. In: R. A. Reyment and G. Thomel (eds.), Evenements de la Partie Moyenne du Cretace: mid-Cretaceous events: Uppsala 1975–Nice 1976. *Mus. Hist. Nat. Nice, Ann.*, 4:XXIII.1–XXIII.52.

Kauffman, E. G., D. E. Hattin, and J. D. Powell, 1969. Stratigraphic, paleontologic, and paleoenvironmental analysis of the Upper Cretaceous rocks of Cimarron County, northwestern Oklahoma. *Geol. Soc. Amer. Mem.*, 149:1–150.

Kauffman, E. G., J. D. Powell, and D. E. Hattin, 1969. Cenomanian-Turonian facies across the Raton Basin. *Rocky Mt. Assoc. Geol. Mountain Geol.*, 6:93–118.

Kauffman, E. G., and P. A. Scholle, 1977. Abrupt biotic and environmental changes during peak Cretaceous transgressions in Euramerica. *Jour. Pal.*, vol. 51, no. 2, suppl., part III, p. 16.

————, 1979. Biological response to temperature/salinity changes during the Cenomanian and Coniacian transgressions, western interior United States. In: E. G. Kauffman and D. E. Hattin (eds.), *Cretaceous of the North Temperate Realm* (Stroudsburg, Penn.: Dowden, Hutchinson, and Ross, Inc.)

Kauffman, E. G., and N. F. Sohl, 1974. Structure and evolution of Antillean Cretaceous rudist frameworks. *Naturf. Ges. Basel, Verhandl.*, 84 (1):399–467.

Kent, H. E., 1967. Microfossils from the Niobrara-equivalent portion of the Mancos Shale (Cretaceous) in northwestern Colorado. *Jour. Pal.*, 41 (6):1433–1456.

Koch, C. F., 1978. Bias in the published fossil record. *Paleobiology*, 4 (3):367–372.

K-TEC Symposium, 1977. Cretaceous-Tertiary extinctions and possible terrestrial and extraterrestrial causes. *Canada Nat. Mus. Nat. Sci. Syllogeus*, 12:1–163.

Kureshy, A. A., 1979. The Cretaceous/Tertiary boundary in Pakistan. In: W. K. Christensen and T. Birkelund (eds.), *Cretaceous-Tertiary Boundary Events. Copenhagen Univ., Proc. Sympos. "Cretaceous-Tertiary Boundary Events"*, vol. 2, pp. 214–221.

Liebau, A., 1973. El Maastrichtiense lagunar ("Garumniense") de Isona. *Proceedings of the 13th Colloquim on European Micropaleontology*, 3 (87)–28 (112).

————, 1978. Palaobathymetrische und palaoklimatische Veranderungen im Mikrofaunenbild der Maastrichter Tuffkreide. *Neues Jahrb. Geol. Pal.*, 157 (1,2):233–237.

Loeblich, A. R., Jr., H. Tappan, J. P. Beckmann, H. M. Bolli, E. M. Gallitelli, and J. C. Troelsen, 1957. Studies in foraminifera. *U. S. Natl. Mus. Bull.*, 215:1–323.

Lowrie, W., and W. Alvarez, 1977. Upper Cretaceous-Paleocene magnetic stratigraphy at Gubbio, Italy. 3. Upper Cretaceous magnetic stratigraphy. *Geol. Soc. Amer. Bull.*, 88:374–377.

Luterbacher, H. P., and I. Premoli Silva, 1964. Biostratigrafia del limite Cretaceo-Terziario nell' Appennino centrale. *Riv. Italiana Pal. Stratigr.*, 70:67–128.

Mabesoone, J. M., I. M. Tinoco, and P. N. Coutinho, 1968. The Mesozoic-Tertiary boundary in northeastern Brazil. *Palaeogeogr. Palaeoclimatol. Palaeoecol.*, 4 (3):161–185.

McLean, D. M., 1978. A terminal Mesozoic "greenhouse": lessons from the past. *Science*, 201:401–406.

Miller, H. W., Jr., 1956. Paleocene and Eocene, and Cretaceous-Paleocene boundary in New Jersey. *Amer. Assoc. Petrol. Geol. Bull.*, 40 (4):722–736.

Minard, J. P., J. P. Owens, and Sohl, N. F., 1976. Coastal Plain stratigraphy of the upper Chesapeake Bay region. *Geol. Soc. Amer. Field Trip Guidebk.*, no. 7a, pp. 1–61.

Minard, J. P., J. P. Owens, N. F. Sohl, H. E. Gill, and J. F. Mello, 1969. Cretaceous-Tertiary boundary in New Jersey, Delaware, and eastern Maryland. *U. S. Geol. Surv. Bull.*, 1274-H:H1-H33.

Monechi, S., and C. Pirini Radrizzani, 1975. Nannoplankton from Scaglia Umbra Formation (Gubbio) at Cretaceous-Tertiary boundary. *Riv. Italiana Pal. Stratigr.*, 81:45–87.

Naiden, D. P., 1979. The Cretaceous/Tertiary boundary in the USSR. In: W. K. Christensen and T. Birkelund (eds.), *Cretaceous-Tertiary Boundary Events. Copenhagen Univ., Proc. Sympos. "Cretaceous-Tertiary Boundary Events"*, vol. 2, pp. 188–201.

Newell, N. D., 1962. Paleontologic gaps and geochronology. *Jour. Pal.,* 36:592–610.

———, 1967. Revolutions in the history of life. *Geol. Soc. Amer. Spec. Paper,* 89:63–91.

Obradovich, J. D., and W. A. Cobban, 1975. A time scale for the late Cretaceous of the western interior of North America. In: W. G. E. Caldwell (ed.), The cretaceous system in the western interior of North America. *Geol. Assoc. Canada Spec. Paper,* 13:31–54.

Olsson, R. K., 1960. Foraminifera of latest Cretaceous and earliest Tertiary age in the New Jersey coastal plain. *Jour. Pal.,* 34 (1):1–58.

Pavlovec, R., and M. Plenicar, 1979. Cretaceous/Tertiary boundary in the limestone sequence of the West Dinarides. In: W. K. Christensen and T. Birkelund (eds.), *Cretaceous-Tertiary Boundary Events. Copenhagen Univ., Proc. Sympos. "Cretaceous-Tertiary Boundary Events",* vol. 2, p. 185.

Perch-Nielsen, K., 1979. Calcareous nannofossils in Cretaceous/Tertiary boundary sections in Denmark. In: W. K. Christensen and T. Birkelund (eds.), *Cretaceous-Tertiary Boundary Events. Copenhagen Univ., Proc. Sympos. "Cretaceous-Tertiary Boundary Events",* vol. 2, pp. 120–126.

Perch-Nielsen, K., K. Ulleberg, and J. E. Evensen, 1979. Comments on "The terminal Cretaceous event: a geologic problem with an oceanographic solution" (Gartner & Keany, 1978). In: W. K. Christensen and T. Birkelund (eds.), *Cretaceous-Tertiary Boundary Events. Copenhagen Univ., Proc. Sympos. "Cretaceous-Tertiary Boundary Events",* vol. 2, pp. 106–111.

Percival, S. F., and A. G. Fischer, 1977. Changes in calcareous nannoplankton in the Cretaceous-Tertiary biotic crisis at Zumaya, Spain. *Evol. Theory,* 2:1–35.

Premoli Silva, I., 1977. Upper Cretaceous-Paleocene magnetic stratigraphy at Gubbio, Italy. 2. Biostratigraphy. *Geol. Soc. Amer. Bull.,* 88:371–374.

Rasmussen, H. W., 1965. The Danian affinities of the Tuffeau de Ciply in Belgium and the "Post-Maastrichtian" in the Netherlands. *Meded. Geol. Stricht.,* 17:33–38.

———, 1979. Crinoids, asteroids and ophiuroids in relation to the boundary. In: T. Birkelund and R. G. Bromley (eds.), *The Maastrichtian and Danian of Denmark. Copenhagen Univ., Proc. Sympos. "Cretaceous-Tertiary Boundary Events",* vol. 1, pp. 65–71.

Ravn, J. P. J., 1903. Molluskerne i Denmarks Kridtaflejringer III: Stratigrafiske Undersøgelser: *D. Kgl. Danske Vidensk. Selsk. Skrift., ser. 6: Naturvid. Math., Afd. XI.,* 6:339–445.

Reeside, J. B., Jr., and W. A. Cobban, 1960. Studies of the Mowry Shale (Cretaceous) and contemporary formations in the United States and Canada. *U. S. Geol. Surv. Prof. Paper,* 335-1-126.

Roggenthen, W. M., and G. Napoleone, 1977. Upper Cretaceous-Paleocene magnetic stratigraphy at Gubbio, Italy. 4. Upper Maastrichtian-Paleocene magnetic stratigraphy. *Geol. Soc. Amer. Bull.,* 88:378–382.

Romein, A. J. T., 1979. Calcareous nannofossils from the Cretaceous/Tertiary boundary interval in the Nahal Avdat Section, the Negev, Israel. In: W. K. Christensen and T. Birkelund (eds.), *Cretaceous-Tertiary Boundary Events.*

*Copenhagen Univ., Proc. Sympos. "Cretaceous-Tertiary Boundary Events"*, vol. 2, pp. 202–206.

Rosenkrantz, A., 1924. Mødet den 6. Oktober 1924 (untitled communication). *Medd. Dansk. Geol. Foren.*, 6:28–31.

———, 1939. Faunaen i Cerithiumkalken og det haerdnede Skrivekridt i Stevns Klint. *Medd. Dansk. Foren.*, 9:509–514.

———, 1960. South-eastern Sjaelland and Mön, Denmark: guide to excursions nos. A42 and C37, part 1. *Int. Geol. Congr., 21st Session, Norden, 1960, Guidebk.*, pp. 1–17.

———, 1970. Marine upper Cretaceous and lowermost Tertiary deposits in West Greenland. *Medd. Dansk. Geol. Foren.*, 19 (4):406–453.

Russell, D., and W. Tucker, 1971. Supernovae and the extinction of the dinosaurs. *Nature*, 229 (5286):553–554.

Sanders, H. L., 1969. Benthic marine diversity and the stability-time hypothesis. In: G. M. Woodwell and H. H. Smith (eds.), Diversity and stability in ecological systems. *Brookhaven Sympos. Biol.*, 22:71–81.

Schindewolf, O., 1954. Über die möglichen Ursachen der grossen Erdgeschichtlichen Faunenschnitte. *Neues Jahrb. Geol. Pal., Monatsh.*, (1954), pp. 457–465.

———, 1955. Die Entfaltung des Lebens im Rahmen der geologischen Zeit. *Sondersabdr. "Studium Generale"*, 8:489–497.

———, 1962. Neokatastrophismus? *Deutsch. Geol. Ges. Zeitschr., Jahrg.*, 114 (2):430–445.

Scholle, P. A., and M. A. Arthur, 1980. Carbon isotope fluctuations in Cretaceous pelagic limestones: potential stratigraphic and petroleum exploration tool. *Amer. Assoc. Petrol. Geol. Bull.*, 64 (1):67–87.

Scholle, P. A., and E. G. Kauffman, 1977. Paleoecological implications of stable isotope data from Upper Cretaceous limestones and fossils from the U. S. Western Interior. *Jour. Pal.*, 1, suppl. to no. 2:24–25 (abstract).

Scott, G. R., and W. A. Cobban, 1964. Stratigraphy of the Niobrara Formation at Pueblo, Colorado. *U. S. Geol. Surv. Prof. Paper*, 454-L:1–29.

Scott, R. W., 1970a. Paleoecology and paleontology of the Lower Cretaceous Kiowa Formation. *Kansas Univ. Kansas Pal. Contrib.*, art. 52 (Cretaceous 1):1–94.

———, 1970b. Stratigraphy and sedimentary environments of Lower Cretaceous rocks, southern Western Interior. *Amer. Assoc. Petrol. Geol. Bull.*, 54 (7):1225–1244.

———, 1977. Early Cretaceous environments and paleocommunities in the southern Western Interior (parts 1 and 2). In: E. G. Kauffman (ed.), Cretaceous facies, faunas, and paleoenvironments across the Western Interior Basin. *Rocky Mt. Assoc. Geol. Mountain Geol.*, 14 (3,4):155–173; 219–221.

Smith, C. C., 1975. Upper Cretaceous calcareous nannoplankton zonation and stage boundaries. *Gulf Coast Assoc. Geol. Soc. Trans.*, 25:263–278.

Speden, I. G., 1970. The type Fox Hills Formation, Cretaceous (Maastrichtian), South Dakota: part 2, systematics of the Bivalvia. *Yale Univ. Peabody Mus. Nat. Hist. Bull.*, 33:1–222.

Stanley, S. M., 1968. Post-Paleozoic adaptive radiation of infaunal bivalve molluscs: a consequence of mantle fusion and siphon formation. *Jour. Pal.*, 42:214–229.

Stanton, T. W., 1920 (1921), The fauna of the Cannonball marine Member of the Lance Formation. *U. S. Geol. Surv. Prof. Paper*, 128:1–60.

Stenestad, E., 1979. Upper Maastrichtian foraminifera from the Danish Basin. In: T. Birkelund and R. G. Bromley (eds.), *The Maastrichtian and Danian of Denmark. Copenhagen Univ., Proc. Sympos. "Cretaceous-Tertiary Boundary Events"*, vol. 1, pp. 101–107.

Stokes, R. B., 1979. Analysis of the ranges of spatangoid echinoid genera and their bearing on the Cretaceous/Tertiary boundary. In: W. K. Christensen and T. Birkelund (eds.), *Cretaceous-Tertiary Boundary Events. Copenhagen Univ., Proc. Sympos. "Cretaceous-Tertiary Boundary Events"*, vol. 2, pp. 78–82.

Surlyk, F., 1979. Maastrichtian brachiopods from Denmark. In: T. Birkelund and R. G. Bromley (eds.), *The Maastrichtian and Danian of Denmark. Copenhagen Univ., Proc. Sympos. "Cretaceous-Tertiary Boundary Events"*, vol. 1, pp. 45–50.

Surlyk, F., and T. Birkelund, 1977. An integrated stratigraphical study of fossil assemblages from the Maastrichtian White Chalk of northwestern Europe. In: E. G. Kauffman and J. E. Hazel (eds.), *Concepts and Methods of Biostratigraphy* (Stroudsburg, Penn.: Dowden, Hutchinson, and Ross, Inc.), pp. 257–281.

Tappan, H., 1968. Primary production, isotopes, extinctions and the atmosphere. *Palaeogeogr. Palaeoclimatol. Palaeoecol.*, 4 (3):187–210.

———, 1979. Protistan evolution and extinction at the Cretaceous-Teritary boundary. In: W. K. Christensen and T. Birkelund (eds.), *Cretaceous-Tertiary Boundary Events. Copenhagen Univ., Proc. Sympos. "Cretaceous-Tertiary Boundary Events"*, vol. 2, pp. 13–21.

Thierstein, H. R., 1979. The terminal Cretaceous oceanic event. In: W. K. Christensen and T. Birkelund (eds.), *Cretaceous-Tertiary Boundary Events. Copenhagen Univ., Proc. Sympos. "Cretaceous-Tertiary Boundary Events"*, vol. 2, pp. 22–25.

Van Hinte, J. E., 1976. A Cretaceous time scale. *Amer. Assoc. Petrol. Geol. Bull.*, 60 (4):498–516.

Vaughan, T. W., 1920 (1921). Corals from the Cannonball marine Member of the Lance Formation. *U. S. Geol. Surv. Prof. Paper*, 128:61–66.

Voight, E., 1959. La signification stratigraphique des Bryozoaires dans le Crétacé Supérior. *84ᵉ Congr. Soc. Savantes, Sec. Sci.*, pp. 701–707.

Waage, K. M., 1968. The Type Fox Hills Formation, Cretaceous (Maastrichtian), South Dakota: pt. 1, stratigraphy and paleoenvironments. *Yale Univ. Peabody Mus. Nat. Hist.*, pp. 1–175.

Wiedmann, J., 1964. Le Crétacé supérieur de l'Espagne et du Portugal et ses Céphalopodes. *Estud. Geol.*, 20:107–148.

Chapter 9

# CAMPANIAN THROUGH PALEOCENE PALEOTEMPERATURE AND CARBON ISOTOPE SEQUENCE AND THE CRETACEOUS-TERTIARY BOUNDARY IN THE ATLANTIC OCEAN

ANNE BOERSMA

Lamont-Doherty Geological Observatory

## INTRODUCTION

The extinctions and physical-oceanographic phenomena at the Cretaceous-Tertiary boundary have elicited a vast number of astute observations from scientists for half a century. The accumulated literature documenting faunal changes among dinosaurs (Russell, 1975), marine invertebrates (Kauffman, 1973), larger foraminifera (Dilley, 1973), benthonic foraminifera (Beckman, 1960), planktonic foraminifera (Rosenkrantz and Brotzen, 1960), and marine floras (Tappan and Loeblich, 1971; Harker and Sarjeant, 1975; Percival and Fischer, 1977) complements the work on geology and tectonics which shows extensive paraconformities (Gignoux, 1950; Newell, 1963) and hiatuses in the deep sea record (Worsley, 1974) along with an extensive and rapid regression (Termier and Termier, 1969; Vail et al., 1977) from a low-lying continental system (Hallam, 1968) largely covered by marine seaways (Dunbar, 1960). A change in spreading rates along oceanic ridge systems is also documented at this time (Pitman and Hays, 1973). Recent evidence from DSDP (Deep Sea Drilling Project) Legs 39 and 43 shows that the CCD (Calcite Composition Depth) in the Atlantic dropped, then rose just below the boundary (Tucholke et al., in press), then dropped again by earliest Tertiary (Boersma, 1977).

Paleobotanists have speculated on late Cretaceous climate (Whitehead, 1969; Tschudy, 1971); they have recorded that extinctions at the boundary are rare in the land plants (Sloan, 1967; Flessa and Imbrie, 1975). Oxygen isotope analysis of foraminifera in deep-sea sediments (Saito and

Van Donk, 1974; Douglas and Savin, 1975) and of shelf invertebrates (Lowenstam and Epstein, 1954) suggest that the marine temperatures reached a low in the Early Maastrichtian, decreased during the Maastrichtian and *may* have decreased across the boundary (Savin, 1977).

Theories on the events that precipitated the Cretaceous/Tertiary boundary crisis include drastic climate change and decreased marine temperatures (Saito and Van Donk, 1974); supernova (Reid, 1977); magnetic reversals affecting biotic systems (Uffen, 1963); the rapid regression that reduced nutrient supplies (Tappan, 1968); and the rising of the CCD into the photic zone (Worsley, 1974). To date only ideological change has been overlooked as a possibility.

During the last four years, I have been investigating marine paleotemperature and geochemical variations through the Maastrichtian, across the Cretaceous-Tertiary boundary, and through the early Tertiary using the oxygen and carbon isotopic variations in foraminiferal carbonate. To date I have analyzed sediments through the Maastrichtian and Early Tertiary of the Atlantic Ocean at DSDP Sites 152, 356, 357, 384, and in

| Site | Latitude | Longitude |
| --- | --- | --- |
| 20C | 28°31'S | 26°50'W |
| 21 | 28°35'S | 30°35'W |
| 94 | 24°31'N | 88°28'W |
| 95 | 24°09'N | 86°23'W |
| 144 | 09°27'N | 54°20'W |
| 151 | 15°01'N | 73°24'W |
| 152 | 15°52'N | 74°36'W |
| 327a | 50°52'S | 46°47'W |
| 329 | 50°39'S | 46°05'W |
| 356 | 28°17'S | 41°05'W |
| 357 | 30°00'S | 35°33'W |
| 384 | 41°21'N | 51°39'W |
| v26-65 | 26°42'S | 27°51'W |
| V16-56 | 41°21'S | 26°38'E |
| V22-127 | 41°18'S | 26°43'E |
| V16-55 | 40°14'S | 25°15'E |
| V22-126 | 41°01'S | 26°30'E |

Table 9–1. Geographic locations of sites from which we have made isotopic measurements.

assorted Vema cores from the Agulhas Plateau and the Rio Grande Rise (Table 9-1; Boersma et al., in press). I have attempted to produce the estimated temperature and carbon isotope record across the boundary at as close a sample spacing as possible in order to evaluate some of the theories mentioned above.

Because of preservational vagaries and the numerous boundary hiatuses, there are slight stratigraphic gaps in several of my records. The section at Site 384 appears to be the most complete, and I have sampled this at 2 to 10 centimeter intervals. The section at Site 356 contains a 31-meter lower Paleocene interval that includes the basal Tertiary, but lacks the very uppermost Cretaceous. The records at the other sites are less complete, but they can be correlated with Sites 384 and 356 to produce meaningful coverage from almost 50° N to 50° S paleolatitude in the Atlantic Ocean.

## SURFACE DWELLING PLANKTONIC FORAMINIFERA

In order to produce a surface temperature record it has been necessary to establish which species of foraminifera lived closest to the ocean surface. A detailed study of foraminiferal stratification patterns through the Maastrichtian (Boersma, in preparation) has demonstrated that foraminiferal stratification order varied through time so that different species must be used in almost every epoch.

Some stratification arrays for the Maastrichtian and Paleocene are shown in Table 9-2. We used *Rugoglobigerina* or *Pseudoguembelina* in the Maastrichtian, and species of *Guembelitria, Acarinina,* and *Morozovella* during the Paleocene to construct our estimated surface temperature records.

It is apparent in Table 9-2 that there is an unusual situation among the planktonic foraminiferal species of the very earliest Tertiary. The surface zone appears to be dominated by the heterohelicids, to judge by their oxygen isotopic composition. In addition, several species record anomalously negative carbon isotope values. Later in the Paleocene the heterohelicids appear to have been deep-dwelling, while other surface-dwelling species yield more positive carbon isotope values.

## SURFACE TEMPERATURES

*Site 384* There is little change in estimated surface temperature through the samples we analyzed (Figure 9-1). Early Maastrichtian surface temperatures lie near 14° C and increase to the 15 to 16° C range in the later

| SITE/ INTERVAL | FOSSIL | $^{18}O$ | $^{13}C$ |
|---|---|---|---|
| | **Maastrichtian** | | |
| 384/13/3/57 | Pseudoguembelina excolata | −0.87 | +2.21 |
| | Globotruncana contusa | −0.72 | +2.21 |
| | Rugoglobigerina rotundata | −0.68 | +2.76 |
| | Pseudotextularia elegans | −0.60 | +2.12 |
| | Globotruncana arca | −0.60 | +2.17 |
| | Heterohelix striata | −0.61 | +2.00 |
| | Globotruncana stuartiformis | −0.57 | +1.93 |
| | Planoglobulina glabrata | −0.59 | +2.20 |
| | Rugoglobigerina rugosa ? | −0.54 | +2.30 |
| | Racemiguembelina fructicosa | −0.54 | +2.30 |
| | Globotruncana mayaroensis | −0.43 | +1.35 |
| | Globotruncanella citae | −0.35 | +1.81 |
| | Heterohelix globulosa | −0.34 | +1.80 |
| | Gublerina sp. | −0.38 | +1.83 |
| | Globigerinelloides sp. | −0.32 | +2.41 |
| | Planoglobulina multicamerata | −0.20 | +1.52 |
| | Hedbergella monmouthensis | −0.17 | +2.43 |
| | Heterohelix pulchra | +0.27 | +1.68 |
| | **Globigerina eugubina Zone** | | |
| 356/29/3/33 | Guembelitria cretacea | −2.49 | +0.21 |
| | Chiloguembelina morsei | −1.78 | +0.80 |
| | Planorotalites eugubina | −1.36 | +1.50 |
| | **Zone P1a** | | |
| 356/29/2/80 | Guembelitria cretacea | −2.09 | +0.12 |
| | Globoconusa daubjergensis | −1.95 | +0.32 |
| | Subbotina pseudobulloides | −1.43 | +1.76 |
| | Chiloguembelina midwayensis | −1.40 | +0.50 |
| | **Zone P1b, c** | | |
| 384/12/1/81 | Morozovella inconstans | −0.72 | +1.66 |
| | Planorotalites compressa | −0.49 | +0.95 |
| | Subbotina pseudobulloides | −0.28 | +1.35 |
| | Chiloguembelina midwayensis | −0.14 | +1.40 |

Table 9–2. Representative stratification arrays of Maastrichtian and Paleocene Planktonic foraminifera from the Atlantic Ocean.

| SITE/ INTERVAL | FOSSIL | $^{18}O$ | $^{13}C$ |
|---|---|---|---|
| | **Zone P1d** | | |
| 356/26/6/80 | Guembelitria cretacea | −2.90 | +0.11 |
| | Planorotalites compressa | −2.26 | +0.64 |
| | Morozovella trinidadensis | −2.17 | +1.13 |
| | Subbotina pseudobulloides | −1.87 | +0.95 |
| | **Zone P2** | | |
| V26/65 | Morozovella praecursoria | −0.88 | +2.80 |
| | Morozovella cf. uncinata | −0.75 | +2.79 |
| | Acarinina spiralis | −0.73 | +2.76 |
| | Morozovella trinidadensis | −0.42 | +1.99 |
| | Chiloguembelina midwayensis | −0.23 | +1.87 |
| | Planorotalites compressa | −0.19 | +1.92 |
| | Subbotina triloculinoides | −0.04 | +2.01 |
| | Subbotina pseudobulloides | +0.39 | +2.00 |
| | **Zone P3** | | |
| 384/11/2/6 | Morozovella conicotruncata | −1.78 | +2.52 |
| | Morozovella pusilla | −1.11 | +1.54 |
| | Subbotina pseudobulloides | −0.81 | +0.64 |
| | Planorotalites compressa | −0.69 | +0.85 |
| | Chiloguembelina midwayensis | −0.54 | +0.60 |
| | Subbotina triloculinoides | −0.45 | +0.92 |
| | **Zone P4** | | |
| 95/7/5/45 | Acarinina mckannai | −1.12 | +4.43 |
| | Acarinina coalingensis | −1.10 | +3.60 |
| | Morozovella velascoensis | −1.00 | +4.85 |
| | Planorotalites pseudomenardii | −0.88 | +2.22 |
| | Subbotina trilocularis | −0.55 | +2.49 |
| | Chiloguembelina midwayensis | −0.58 | +2.36 |

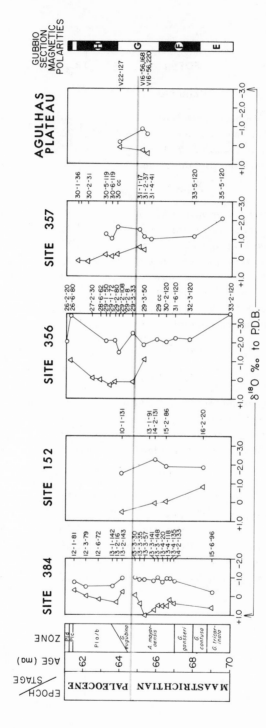

Figure 9–1. Oxygen isotope records of benthic and planktonic foraminifera through the Maastrichtian-Paleocene at five Atlantic sites. 0 = unspecific planktonic foraminifera living in the surface zone; Δ = benthic measurements (see Table 9–3). Time-scale and zonations are from Berggren (1972), Premoli Silva and Bolli (1973), and Luterbacher and Premoli Silva (1962). Results of magnetic measurements at Sites 356 and 384 are questionable. Thus, our data are plotted against the reliable sequence across the boundary at Gubbio (Alvarez and Lowrie, 1977). Hiatuses and stratigraphic gaps are not indicated.

Maastrichtian; the four samples below the boundary and in the boundary zone all register temperatures of 15° C. In Early Paleocene foraminiferal Zone P1, there are similar temperature estimates; latest P1 is missing at this site. The estimated temperature during Zone P2 (Figure 9–2) is about 16° C, the coolest of the post-Danian Paleocene. A temperature maximum near 21° C is recorded in the first million years of Zone P3; then cooler temperatures ranging from 15 to 18° C prevail during the later Paleocene; the 18° C high is reached in Zone P4.

*Site 151/152* Maastrichtian temperatures were derived from only three samples which yielded similar temperature estimates near 20 to 21° C (Figure 9–1). Induration prevented measurements in Cores 13–11, thus the Late Maastrichtian value of 21° C may not be representative of tem-

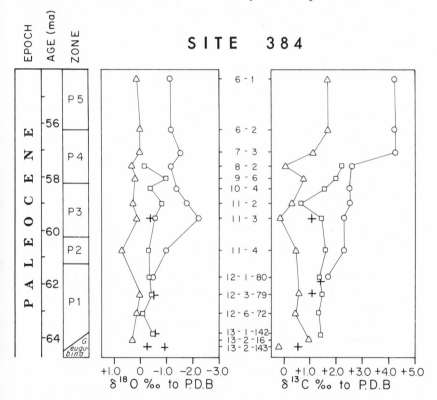

Figure 9–2. Carbon and oxygen isotope records through the Paleocene at DSDP Site 384 in the North Atlantic. 0 = measurements of the planktonic foraminifera *Acarinina* or *Morozovella;* + = *Chiloguembelina* ssp.; □ = *Subbotina* spp.; Δ = benthic measurements (see Table 9–3).

peratures just below the boundary; the earliest Paleocene (early *Globi-gerina eugubina* Zone) was not measured either. Temperatures were near 19° C in the later *G. eugubina* Zone, and decreased to almost 16° C during Zone P1a, b, c. Fossils from Zone P1d were not measured.

A cool value of 16° C in Zone P2 was followed by higher values of 21° C in Zone P3. Unfortunately basal P3 was not present in these samples, so we could not confirm the basal Zone P3 temperature maximum recorded at other sites (Table 9–3). At Site 152, temperatures are slightly cooler in Zone P4 than in later Zone P3; values of 20° C in Zone P4 are then followed by a higher temperature near 23° C in the interval from Zone P5 to P6.

These temperatures at Site 152 in the Caribbean arm of the "Tethys" are in most cases at least 2° C cooler than those at Site 356, which is situated between 35 to 30° C South paleolatitude during this time (Supko et al., 1977). In the *G. eugubina* Zone, *Guembelitria cretacea* records a temperature nearly 5° C higher at Site 356 than at Site 152.

*Site 356*    Oxygen isotope values become significantly less negative between the Late Campanian *Globotruncana calcarata* Zone and the Maastrichtian (*Guembelitria gansseri* Zone; Figure 9–1). The Campanian (33-2-130) value implies ocean surface temperatures close to 30° C at this time. Estimated temperatures decreased gradually through the Maastrichtian, from 20 to 21° C in the latest Maastrichtian (*Abathomphalus mayaroensis* Zone).

The earliest Paleocene sample (29-3–33) was taken just above a visible Cretaceous-Tertiary discontinuity; it comes from the early *Globigerina eugubina* Zone and is excellently preserved. A high temperature of near 24° C is estimated for this sample. Temperature dropped later in the *G. eugubina* Zone to about 21° C. As at all the other localities we have studied, temperatures are relatively low during Zone P1a, b, and c. When we followed the evolution of *Morozovella trinidadensis,* however, we found a temperature rise to nearly 27° C at Site 356, the highest Paleocene temperature estimate we have made so far.

Surface temperature fell during Zone P2 (Figure 9–3) to almost 22° C; from this time on the temperatures at Site 356 have been the warmest we have measured in the Atlantic. A pronounced temperature increase to almost 24° C occurred in earliest Zone P3 and was followed by a fall to nearly 21° C later in Zone P4. At Site 356 the preservation of samples deteriorates markedly in Zones P4 and P5; we consider the analysis for Core 16 (Zone P5) to be invalid for this reason.

*Rio Grande Rise*    Although the preservation of Cretaceous and Paleocene sediments is only moderate at Site 357, the measurements seem to be

| SITE/ INTERVAL | FOSSIL | $\partial^{18}O$ | $\partial^{13}C$ |
|---|---|---|---|
| | **SITE 384** | | |
| | CRETACEOUS | | |
| 14/2/133 | P. excolata | −0.78 | +2.55 |
| | Gublerina sp. | −0.75 | +1.94 |
| | Benthics | +0.39 | +1.37 |
| 13/4/135 | P. excolata | −0.92 | +2.11 |
| | Gavelinella+ | | |
| | Nuttalides | +0.18 | +1.85 |
| 13/4/115 | R. rotundata | −0.93 | +2.85 |
| | Gavelinella | +0.52 | +1.77 |
| 13/4/20 | R. rugosa | −0.97 | +2.64 |
| | Benthics | +0.51 | +0.99 |
| 13/3/141 | R. rotundata | −0.82 | +2.69 |
| | bulk carbonate | −0.05 | +2.55 |
| | Benthics | +0.73 | +1.55 |
| 13/3/57 | P. excolata | −0.87 | +2.21 |
| | Benthics | −1.05 | +1.67 |
| 13/3/35 | P. excolata | −0.87 | +2.50 |
| | H. monmouthensis | −0.32 | +1.79 |
| | Benthics | +0.40 | +1.39 |
| 13/3/30 | R. rotundata | −0.97 | +2.60 |
| | H. monmouthensis | −0.41 | +1.65 |
| | Nuttalides + Gavelinella | −0.05 | +1.63 |
| | PALEOCENE | | |
| 13/2/143 | C. midwayensis | −0.97 | +0.48 |
| | S. pseudobulloides | −0.60 | +1.59 |
| | Benthics | −0.24 | +0.00 |
| 13/2/103 | bulk carbonate | −0.20 | +1.70 |
| 13/1/142 | C. midwayensis | −0.52 | +1.38 |
| 12/3/79 | C. midwayensis | −0.47 | +0.79 |
| | P. compressa | −0.23 | +1.50 |
| | Benthics | +0.00 | +0.55 |
| 12/1/81 | M. inconstans | −0.72 | +1.66 |
| | C. midwayensis | −0.14 | +1.40 |
| | Benthics | −0.32 | −0.16 |
| 11/4/10 | M. uncinata | −0.98 | +2.29 |
| | P. compressa | −0.41 | +1.45 |
| | S. triloculinoides | −0.25 | +1.58 |

Table 9–3. Isotopic data.
(6 pg table)

| SITE/ INTERVAL | FOSSIL | $\partial^{18}O$ | $\partial^{13}C$ |
|---|---|---|---|
| 11/3/30 | M. angulata | -2.22 | +2.29 |
|  | S. triloculinoides | +0.03 | +1.68 |
|  | C. midwayensis | -0.23 | +1.09 |
| 11/2/6 | M. conicotruncata | -1.78 | +2.52 |
|  | C. midwayensis | -0.54 | +0.60 |
|  | S. triloculinoides | -0.45 | +0.92 |
|  | Nuttalides | +0.09 | +0.46 |
| 10/4/54 | M. conicotruncata | -1.38 | +2.50 |
|  | P. ehrenbergi | -0.49 | +1.35 |
|  | S. triloculinoides | -0.39 | +1.56 |
| 8/2/54 | M. conicotruncata | -1.17 | +2.61 |
|  | P. pseudomenardii | -0.31 | +1.91 |
|  | S. triloculinoides | -0.14 | +2.22 |
|  | Benthics | +0.33 | +0.03 |
| 7/3/57 | M. velascoensis | -1.18 | +4.27 |
|  | Benthics | +0.04 | +1.11 |
| 6/2/105 | M. velascoensis | -1.18 | +4.23 |
|  | Benthics | +0.02 | +1.68 |

**SITE 356**

MAASTRICHTIAN

| | | | |
|---|---|---|---|
| 29cc | P. excolata | -2.21 | +2.13 |
| 29/3/50 | P. excolata | -1.89 | +2.56 |
|  | Gavelinella | -1.08 | +1.26 |

PALEOCENE

| | | | |
|---|---|---|---|
| 29/3/33 | Guembelitria | -2.49 | +0.21 |
|  | C. midwayensis | -1.78 | +0.80 |
|  | Benthics | +0.13 | +1.28 |
| 29/3/30 | C. midwayensis | -1.56 | +1.02 |
| 29/3/24 | C. midwayensis | -1.98 | +0.77 |
|  | S. pseudobulloides | -1.47 | +1.60 |
| 29/2/108 | Guembelitria | -1.48 | +0.20 |
|  | S. pseudobulloides | -1.13 | +1.45 |
| 29/2/80 | Guembelitria | -2.09 | +0.12 |
|  | G. daubjergensis | -1.95 | +0.32 |
|  | S. pseudobulloides | -1.43 | +1.76 |
|  | Nuttalides | +0.06 | +1.94 |

Table 9-3. (*continued*)

| SITE/<br>INTERVAL | FOSSIL | $\partial^{18}O$ | $\partial^{13}C$ |
|---|---|---|---|
| 29/1/92 | Benthics | +0.28 | +1.03 |
| 29/1/50 | M. trinidadensis | -2.07 | +0.90 |
| | C. midwayensis | -1.79 | -0.11 |
| | S. pseudobulloides | -1.60 | +0.87 |
| 28/6/62 | Benthics | -0.02 | +0.64 |
| 27/2/30 | Benthics | -0.18 | +0.14 |
| 26/6/80 | Guembelitria | -2.90 | +0.11 |
| | P. compressa | -2.26 | +0.64 |
| | S. pseudobulloides | -1.87 | +0.95 |
| | Nuttalides + Gavelinella | -1.06 | +0.80 |
| 26/2/62 | M. inconstans | -2.14 | +1.98 |
| | P. compressa | -1.86 | +1.97 |
| | S. pseudobulloides | -1.31 | +1.51 |
| | Benthics | -0.14 | -0.29 |
| 25cc | M. uncinata | -2.34 | +1.22 |
| | P. compressa | -1.53 | +0.89 |
| | S. triloculinoides | -1.40 | +1.28 |
| | Benthics | -0.19 | +0.22 |
| 25/4/30 | M. angulata | -2.79 | +1.77 |
| 24/2/80 | M. conicotruncata | -2.34 | +3.26 |
| | C. midwayensis | -1.71 | +0.83 |
| | P. compressa | -1.34 | +0.87 |
| | S. pseudobulloides | -0.96 | +1.19 |
| | Nuttalides | -0.02 | +0.76 |
| 23/6/30 | M. conicotruncata | -2.14 | +3.27 |
| | S. triloculinoides | -1.21 | +1.74 |
| | P. compressa | -1.09 | +1.26 |
| 23/5/30 | M. conicotruncata | -2.59 | +3.24 |
| 22/6/40 | Benthics | +0.42 | +0.46 |
| 22/4/120 | M. conicotruncata | -2.12 | +3.35 |
| | P. compressa | -1.45 | +1.29 |
| | Benthics | +0.51 | +0.78 |
| 20/3/20 | M. conicotruncata | -2.24 | +3.02 |
| | Benthics | +0.22 | +0.60 |
| 20/1/120 | M. conicotruncata | -2.69 | +2.63 |

Table 9-3. (*continued*)

| SITE/ INTERVAL | FOSSIL | $\partial^{18}O$ | $\partial^{13}C$ |
|---|---|---|---|
| 20/1/40 | M. conicotruncata | -2.10 | +2.86 |
| | Benthics | +0.94 | +0.34 |
| 19/5/120 | M. conicotruncata | -2.43 | +3.57 |
| | P. pseudomenardii | -0.68 | +1.33 |
| | Benthics | +0.24 | +1.06 |
| 18/6/119 | Morozovella sp. | -2.65 | +3.03 |
| | Benthics | +0.57 | +1.07 |
| 16/1/112 | Acarinina sp. | -2.59 | +1.33* |
| | Benthics | -1.44 | +0.21* |

### SITE 152

#### MAASTRICHTIAN

| | | | |
|---|---|---|---|
| 16/2/80 | R. rugosa | -1.82 | +3.21 |
| | P. excolata | -1.86 | +2.53 |
| | Nuttalides | -0.82 | +0.74 |
| 15/2/86 | P. excolata | -1.91 | +2.78 |
| | Benthics | -0.02 | +0.98 |
| 14/1/31 | P. excolata | -2.28 | +2.60 |
| | Benthics | +0.06 | +0.41 |
| 13/1/91 | bulk carbonate | -1.13 | +2.07 |

#### PALEOCENE

| | | | |
|---|---|---|---|
| 10/1/130 | Guembelitria | -1.52 | +1.29 |
| | P. eugubina | -1.42 | +1.28 |
| | Benthics | +0.52 | +1.84 |
| 151 11/6/104 | M. uncinata | -1.09 | +1.29 |
| | Benthics | +0.22 | +0.28 |
| 151 10/2/106 | M. uncinata | -1.32 | +2.29 |
| | Benthics | +0.48 | +0.48 |
| 9/1/93 | Nuttalides | -0.04 | +0.46 |
| 8/1/93 | Morozovella sp. | -2.01 | +2.11 |
| 7/2/100 | M. aequa | -2.12 | +3.36 |
| | Benthics | +0.17 | +1.19 |
| 6/2/78 | M. velascoensis | -1.85 | +4.02 |

Table 9-3. (*continued*)

| SITE/<br>INTERVAL | FOSSIL | $\partial^{18}O$ | $\partial^{13}C$ |
|---|---|---|---|
| | **SITE 357** | | |
| | MAASTRICHTIAN | | |
| 31/4/41 | P. excolata | -1.00 | +2.02 |
| 32/2/37 | P. excolata | -1.14 | +2.01 |
| | Gublerina | -0.64 | +1.63 |
| | Benthics | -0.43 | +1.12 |
| 31/1/17 | P. excolata | -1.50 | +2.26 |
| | Benthics | -0.58 | +1.27 |
| | PALEOCENE | | |
| 30cc | Tubitextularia | -1.65 | +0.85 |
| | H. monmouthensis | -1.14 | +1.20 |
| | P. compressa | -1.01 | +1.15 |
| | Nuttalides | -0.19 | +0.84 |
| 30/6/119 | M. inconstans | -1.04 | +1.51 |
| | Benthics | -0.09 | +0.83 |
| 30/5/119 | M. inconstans | -1.26 | +0.54 |
| | Benthics | +0.19 | -0.26 |
| 30/2/31 | S. pseudobulloides | -1.19 | -- |
| | Benthics | -0.17 | +0.82 |
| 30/1/36 | Nuttalides | +0.15 | +0.72 |
| V26-65 | M. praecursoria | -0.88 | +2.80 |
| | Benthics | +0.38 | +0.90 |
| 29/2/39 | M. velascoensis | -1.42 | +3.34 |
| | Gavelinella | +0.03 | +1.10 |
| 29/1/120 | M. angulata | -1.13 | +2.88 |
| | Nuttalides | +0.51 | +1.15 |
| | **SITE 20C, 21** | | |
| 20C<br>6/4/68 | M. velascoensis | -0.90 | +3.79 |
| | C. midwayensis | 0.00 | +1.66 |
| 20C<br>6/3/55 | A. mckannai | -0.92 | +3.37 |
| | Nuttalides | +0.36 | +1.27 |

Table 9-3. (*continued*)

| SITE/ INTERVAL | FOSSIL | $\partial^{18}O$ | $\partial^{13}C$ |
|---|---|---|---|
| 21A | | | |
| 3/6/130 | M. conicotruncata | −0.92 | +2.73 |
| | Nuttalides | −0.06 | +0.69 |
| 21 | | | |
| 2/1/100 | A. nitida | −0.88 | +3.27 |
| | Nuttalides | +0.04 | +0.63 |
| **FALKLAND PLATEAU** | | | |
| 329 | | | |
| 33/9/54 | Acarinina sp. | −1.14 | +3.21 |
| 329 | | | |
| 33/1/44 | Acarinina sp. | −1.68 | +3.54 |
| 327A | | | |
| 10/2/52 | Rugoglobigerina sp. | −0.77 | +2.24 |
| **AGULHAS PLATEAU** | | | |
| V16-56,220 | R. rugosa | −0.58 | +2.59 |
| | Globigerinelloides | −0.04 | +1.99 |
| | H. globulosa | −0.09 | +1.43 |
| | Gavelinella | +0.42 | +1.02 |
| V16-56,168 | R. rugosa | −0.85 | +2.57 |
| | Gavelinella | −0.27 | +0.93 |
| V22-127,160 | G. daubjergensis | −0.17 | +1.59 |
| | C. midwayensis | +0.01 | +1.21 |
| | H. monmouthensis | +0.30 | +1.61 |
| | Gavelinella | +0.11 | +1.84 |
| V16-55,B | Acarinina | −0.20 | +3.25 |
| | Gavelinella | +0.85 | +1.41 |
| V22-126,B | A. mckannai | −0.50 | +3.70 |
| | Planorotalites | +0.22 | +2.36 |
| | C. crinita | +0.35 | +2.31 |
| | Nuttalides | +0.64 | +1.96 |
| | Gavelinella | +0.87 | +1.69 |

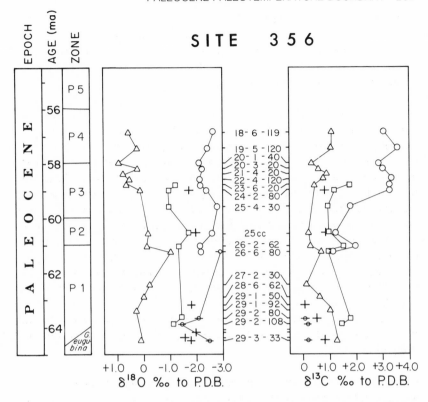

Figure 9-3. Carbon and oxygen isotope records through the Paleocene at DSDP Site 356 in the South Atlantic. Symbols used in this diagram are the same as those in Figure 9-2. In addition, ⊖ = *Guembelitria cretacea.*

reliable. The measurements through the Maastrichtian imply a high temperature during the *Guembelitria tricarinata* Zone, dropping off through the *G. gansseri* Zone and stabilizing near 17 to 19° C in the late *Abathomphalus mayaroensis* Zone (Figure 9-1).

There is a hiatus in sedimentation during the earliest Paleocene. A temperature near 19° C in Zone Pla is registered by an unusual species, *Tubitextularia cretacea,* while *Guembelitria* lives below, at temperatures near 16° C. The presence of *Tubitextularia* in such high abundance is unique to Site 357, although significant amounts of this species also occur on the Aghulhas Plateau. Temperatures then decrease to near 16° C late in Zone P1. As Zone Pld and P2 are missing at this site, sediments from the Zone P1 to P2 intervals at nearby Vema 26–65 were measured. Temperatures there were low, near 15° C (Table 9-3).

Higher temperatures near 17 to 18° C are estimated from samples from mid- and late Zone P3 (earliest P3 was missing). As at Sites 152 and 384, temperatures during Zone P4 were slightly lower, near 16° C. As the Zone P4 samples comes from Site 21, which is located about 400 kilometers to the east of Site 357, temperatures at Site 357 may have been slightly warmer during the same time interval.

*Agulhas Plateau* A few scattered measurements have been made on sediments from the Agulhas Plateau (Figure 9-1). Although not stratigraphically continuous, these sediments are well-preserved and richly fossiliferous. The measurements from V16-56, 220 and 168 centimeters indicate temperatures near 15 to 16° C during the Late Maastrichtian. Temperatures in the earliest Paleocene *Globigerina eugubina* Zone were based, not on *Guembelitria,* but *Globoconusa daubjergensis,* which stratifies just below *Guembelitria* at warmer sites (Boersma, in preparation). These temperatures lie near 13° C. It is possible that warmer temperatures could have been derived from *Tubitextularia cretacea,* however its abundance was too low to permit of measurements. The phyletic maturity of the *Globoconusa daubjergensis* suggests that this sample comes from the later *Globigerina eugubina* Zone.

Cores V16-55 and V22-126 contained sediments from Zones P3 and P4, respectively (Table 9-3). Temperatures fluctuated in these samples from 13 to 14° C.

*Falkland Plateau* Only one surface temperature measurement has been made from the latest Maastrichtian of this area (Table 9-3); temperatures were similar to those on the Agulhas Plateau, near 15° C. No earliest Paleocene was recovered on Leg 36.


BOTTOM TEMPERATURES

Bottom temperature measurements have been made where possible on unispecific samples; in many instances, however, the small size of the specimens, the faunas, or the samples required that several species be mixed for isotope analysis. When possible, only *Gavelinella* and *Nuttalides* were combined; however, multispecific analyses predominate. Accordingly, temperature and, to a greater degree, carbon isotope values are confounded by the specific composition of the sample. As with the planktonic foraminifera, size was held as constant as possible lest variation in test dimension affect isotopic composition.

One means of testing fluctuations in bottom temperatures based on multispecific samples is to compare the temperature record of the deep-dwelling planktonic foraminifera to the bottom temperature record. In the Oligocene at Site 366, we suggested that the parallelism of the two records represented fluctuations in the main body of Atlantic Ocean deep water (Boersma and Shackleton, 1977b). Thus, where possible, we have analysed deeper dwelling planktonic species so as to check the reliability of benthonic analyses.

*Site 384*   In general the bottom temperature record (for an estimated paleodepth near 3500 meters) parallels the surface temperature record through the Maastrichtian and the Paleocene at this site (Figure 9–1). Temperatures in the *Guembelitria tricarinata* Zone were cool, but they rose slightly during the course of the Maastrichtian. Maastrichtian temperatures near 10 to 11 ° C were slightly cooler than those in the Paleocene near 12 ° C. Bottom temperatures just below the Cretaceous-Tertiary boundary dip to near 8 ° C, but they rise again to 10 to 11 ° C just below the boundary. The deep-dwelling planktonic foraminifera show parallel fluctuations in the same samples.

In the earliest *Globigerina eugubina* Zone, the benthics were too minuscule to analyze. In the late *G. eugubina* Zone, temperature was low, but it rose slightly during the remainder of Zone P1a, b, c. Temperature minima near 9 ° C occur during Zone P2 (Figure 9–2), and a high near 12 ° C occurs during the surface temperature maximum early in Zone P3. Bottom temperatures during Zone P4 and P5 vary from 11 to 12 ° C.

*Site 356*   This site has an estimated paleodepth (Supko et al., 1977) near 1000 meters (Figure 9–1). From Figure 9–2 and Table 9–3 we can see that bottom temperatures increased through the *A. mayaroensis* Zone to near 14 ° C just below the boundary.

Bottom temperatures fluctuated near 12 ° C through Zone P1a, b, c, but increased to almost 16 ° C during a surface temperature maximum in Zone P1d. While bottom temperatures generally parallel the surface temperatures during the remainder of the Paleocene, the number of fluctuations at the bottom is slightly greater, up to 3 ° C in later Zone P3 (Figure 9–3). Preservation problems render bottom temperature estimates in Zone P4 suspect, but their comparability to some at other sites suggest that they may be valid.

*Sites 151/152*   The relatively few measurements at these sites suggest that bottom temperature at paleodepths near 3000 meters dropped off from almost 15 ° C in the *Guembelitria contusa* Zone of the Early Maastrichtian to near 12 ° C in the Late Maastrichtian *Abathomphalus mayaroensis* Zone. Bad preservation prevented measurements just below the boundary. A bottom temperature near 14° C was registered in the later *Globigerina eugubina* Zone.

Site 151 registered temperatures near 10 to 11° C in Zone P2 (Table 9-3). At both sites, temperatures from 11 to 12° C characterize the remainder of the Paleocene. An exception is the Zone P3 peak near 13° C registered at Site 152.

*Rio Grande Rise*   Late Maastrichtian bottom temperature measurements at a paleodepth about 1500 meters at Site 357 show a slight temperature increase at the bottom just below the Cretaceous-Tertiary boundary. Zone P1a bottom temperature lies near 12° C and remains at that level through P1a, b. Bottom temperature values for Sites 357 and 21 remain near 11 to 12° C through the remainder of the Paleocene (Table 9-3). The low near 10° C at Site 357 in late Zone P3 correlates with late Zone P3 lows at Sites 384, 152, and 356 and suggests a slight drop in bottom temperature late in Zone P3.

*Agulhas Plateau*   Late Maastrichtian oxygen isotope values for the Agulhas Plateau give temperature estimates near 9° C; in the late *Globigerina eugubina* Zone, bottom temperature approaches 10° C (Figure 9-1). Late in Zone P3 to early Zone P4, bottom temperatures are close to a low of 7° C (Table 9-3). All these analyses are based on unispecific samples of *Gavelinella beccariformis*.

*Falkland Plateau*   Two Late Paleocene samples from Zone P4 and P5 give estimated bottom temperatures near 12° C at a paleodepth of about 1000 meters (Table 9-3).

SURFACE CARBON ISOTOPE VALUES

Surface carbon isotope ratios for each species that records the warmest temperature in each sample from the five areas of the Atlantic are shown in Figures 9-2 to 9-4.

There is a problem in interpreting the carbon isotope values in the *Globigerina eugubina* Zone and early Zone P1. Throughout this interval, the planktonic foraminiferal species that register the warmest temperatures consistently record more negative carbon values than cooler species. In the Recent and through most of the Tertiary (Boersma, in perparation), many of the warmest species record the most positive carbon isotope values; deeper (cooler) dwelling species record more negative carbon isotope rations. The observed carbon isotope difference between Recent

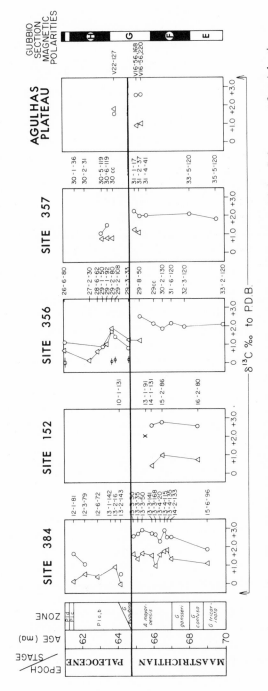

Figure 9–4. Carbon isotope record of benthic and planktonic foraminifera through the Maastrichtian-Paleocene at five Atlantic sites. Symbols are the same as those used in Figure 9–1.

*Globigerinoides sacculifer* and average benthonic foraminifera in the Recent is approximately equivalent to the observed gradient in the water column (Broecker, 1971). Explanations for the anomalous carbon isotope profiles during the Early Paleocene are moot (Boersma et al., in press); for the present discussion, the highest carbon value in a sample during this time is used.

As the surface carbon isotope records at all sites are remarkably similar, they are discussed together. Surface carbon values from the latest Campanian through the Early and Middle Maastrichtian cluster near +2.00 per mil; values increase through the late Maastrichtian to near +2.30 to +2.60 per mil in the *A. mayaroensis* Zone just below the Cretaceous-Tertiary boundary. The lowest values occur on the Agulhas and Falkland plateaus, and the Rio Grande Rise.

There is a significant excursion in the planktonic carbon isotope ratios across the Cretaceous-Tertiary boundary at all sites; carbon isotope ratios become significantly more negative by 1 to 1.5 per mil in the earliest *Globigerina eugubina* Zone. Measurements of bulk carbonate values at Sites 384 and 152 showed a similar drop across the boundary.

Following the temperature maximum in Zone P1d, the modern pattern of most positive carbon at the surface that decreased with depth became re-established. Surface carbon values gradually increased to near +2.2 to +2.7 per mil by Zone P3 and continued to rise to a maximum late in Zone P4. At this time, values close to +4.3 per mil are recorded at Sites 144 (Table 9–3) and 384, so that the difference between surface and bottom carbon isotope value near +3.3 per mil. In Zone P5, carbon isotope values decrease slightly and range near +3.0 to +3.7 per mil in samples analyzed so far.

Although each site demonstrates this same pattern, there are notable discrepancies in timing between Site 356 and the other areas. While other sites register values near +2.2 per mil in Zone P2 and +2.7 per mil early in Zone P3, the carbon values at Site 356 remain at the lower values (near +1.9 per mil) achieved in P1d. Not until late in Zone P3 do the carbon values at Site 356 resemble those of the other sites and continue to do so through the remainder of the Paleocene analyzed there.

## SURFACE TO BOTTOM CARBON ISOTOPE DIFFERENCES

It has been noted that bottom temperature estimates are less reliable when based on samples with varying species composition. This problem is exacerbated in bottom carbon isotope measurements, due to the very large carbon isotope variations between species. (Differences from 0.7 to 1.0 per mil between species are not uncommon.)

| Site 384 | Site 152 151 | Site 356 | Site 357/ 21 | Falkland Plateau | Agulhas Plateau | Zone/Stage |
|---|---|---|---|---|---|---|
| 2.0 |  | 2.8 |  |  |  | P5 |
| 3.2 | 3.3 | 2.5 | 2.9 |  | 2.8 | P4 |
| 2.4 | 3.0 | 2.8 | 2.5 |  | 2.6 | P3 |
| 1.8 | 1.9 | 1.0 | 1.9 |  |  | P2 |
|  |  | 2.1 |  |  |  | P1 |
| .8* |  | (-.5)* | .7* |  |  |  |
| 1.0* | (-.6)* | (-.4)* | .5* |  | .4* | G. eugubina |
| 1.0 | 2.2 | 2.0 | 1.1 |  | 2.3 | Maastrichtian |

*Surface value from highest value, not warmest species; values in %.

Table 9-4.  Near Surface—Bottom Carbon Gradients

This uncertainty must be borne in mind in interpreting estimates of the carbon isotopic gradient through the water column. Representative surface to bottom carbon isotope differences are shown in Table 9–4. Through the Maastrichtian the average difference was around 1.2 per mil; surface values range from +2.0 per mil in the Early Maastrichtian to nearer +2.5 per mil in the Late Maastrichtian.

A major excursion in the carbon isotope gradient occurred at the Cretaceous-Tertiary boundary, and resulted in a negligible gradient in the G. eugubina Zone and in early Zone P1; the highest value for a planktonic foraminifer at this time is equal to, or less than, the carbon values at the bottom. Despite the uncertainty in bottom values mentioned above, the very reduced gradient is still almost certainly real. It is at this time that the warmest (shallowest) foraminifera record very negative carbon values at the ocean surface. These anomalous values must reflect a major alteration in the carbon isotope distribution through the water column at the Cretaceous-Tertiary boundary.

The carbon isotope gradients through the Early Paleocene gradually increase, as the surface-dwelling species indicate gradually more positive 13C values. Deep-dwelling planktonics and the benthics do not show this same increase (Boersma and Shackleton, in preparation). Late in Zone P4, however, both the deep planktonics and the benthic foraminifera become nearly 1.0 per mil more positive, which results in the Late Paleocene carbon excursion noted by other workers (Kroopnick et al., 1977). Surface carbon values drop off slightly in Zone P5, while intermediate and deep values remain at the higher levels achieved during Zone P4; the drop at the surface accounts for the slightly reduced carbon gradients at this time.

## THERMAL GRADIENTS

Figures 9–1 to 9–3 give representative estimates of near surface to bottom paleotemperature. In the Maastrichtian, gradients are less in higher latitude sites, but higher than in the Paleocene at the same site. In the Paleocene, paleotemperature gradients through the water column are lower at mid-latitude Site 384 and on the Agulhas Plateau. They are substantially greater at the warmer Sites 356 and 152, and also during the temperature maxima in the very earliest Paleocene and in basal Zone P3.

I have plotted (Figure 9–5) some temperature estimates from the Late Maastrichtian across the Cretaceous-Tertiary boundary into the *G. eugubina* Zone to show the general trends. In the southern hemisphere the meridional temperature gradients increase between the latest Maastrichtian and the *G. eugubina* Zone, thus they reflect the decrease in temperatures at higher latitudes; Figure 9–6 shows that temperatures at the surface in high southern latitudes are two degrees cooler than in the Late

### UPPER MAASTRICHTIAN BOTTOM TEMPERATURES

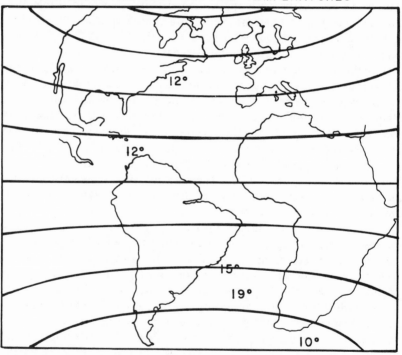

Figure 9–5. Bottom temperature estimates in the Atlantic Ocean during the late Maastrichtian *Abathomphalus mayaroensis* Zone.

Maastrichtian. By contrast in early Zone P3 (Figure 9–7), the low latitudes and the equatorially derived current systems indicate significant warming; these warmer waters are carried to nearly 40 °N and 35 °S paleolatitude.


SUMMARY

The data presented to date indicate the following trends:

1) The major paleotemperature change of the Maastrichtian occurred between the Late Campanian and the Early Maastrichtian *Guembelitria tricarinata* Zone, not at the Cretaceous-Tertiary boundary.

2) A major carbon isotope event occurred precisely at the Cretaceous-Tertiary boundary. A negative excursion in the carbon isotope values is registered in the very earliest Paleocene *Globigerina eugubina* Zone. This

### G. EUGUBINA ZONE NEAR SURFACE TEMPERATURES

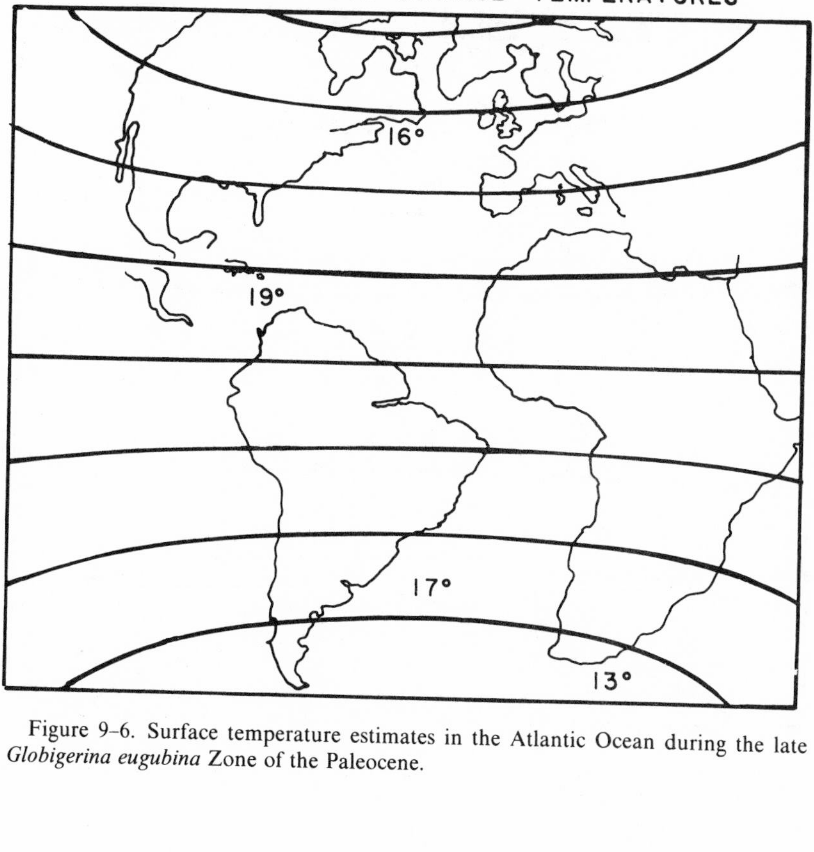

Figure 9–6. Surface temperature estimates in the Atlantic Ocean during the late *Globigerina eugubina* Zone of the Paleocene.

excursion is not registered in the Cretaceous sediments below, but occurs *at* the Cretaceous-Tertiary boundary.

3) Temperature at the surface in the South Atlantic rose in the very Early Paleocene; bottom temperatures decreased, resulting in strong vertical temperature gradients through the water column.

4) A noteworthy increase in surface $\delta$ 13C values occurred during the Paleocene; by the later Paleocene the bottom and intermediate waters also registered increased carbon isotope values that resulted in the Late Paleocene carbon excursion into Zone P4. At this time, both the absolute $\delta$ 13C values and the range of values within the ocean were exceptionally large.

## SPECULATIONS

The events documented by many other authors as well as by our data and ideas are presented in Table 9-5. Refinement of the sequence of events

LOWER ZONE P3 NEAR SURFACE TEMPERATURES

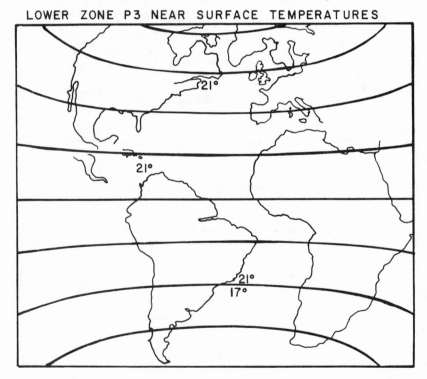

Figure 9-7. Surface temperature estimates in the Atlantic Ocean during the early part of Zone P3, Middle Paleocene.

just prior to the boundary is now possible, due to published analyses of deep marine sequences at Zumaya, Spain (Percival and Fischer, 1977), Baranco del Gredero, Spain (Smit, 1977), the Mendez Shale, Tampico, Mexico (Gamper, 1977) and DSDP Site 384 in the North Atlantic.

*Late Maastrichtian*   The late Maastrichtian scenario is by now familiar; vast transgressions of epicontinental seas spread over low-lying continents (Dunbar, 1960; Hallan, 1968). The increasing bottom and decreasing surface temperatures resulted in decreased thermal gradients through the water column, and possibly caused a reduced thermocline that implies reduced barriers to vertical mixing and reduced niche differentiation. Meridional temperature gradients also decreased because of lower surface water temperatures in low latitudes.

The rising surface carbon isotope values through the Maastrichtian imply increased nutrients and productivity through this time period. Surface carbon values and gradients were high and uniform during the *A. mayaroensis* Zone, which suggests productivity booms and heightened siliceous productivity at this time.

While there is uncertainty in the timing of the change in ridge spreading rate, an hypothesis that postulates that it caused sea-level changes requires that the Late Cretaceous change in spreading rate be closely correlated in time to the regression. These two events must have occurred immediately prior to the boundary, if the regression is considered responsible for the changing faunas in the boundary clays and shales in the land sections discussed below.

According to Tucholke (in press), the distribution of carbonate sediments at Leg 43 sites indicates a drop and then a rise in the CCD of the North Atlantic at the end of the Maastrichtian. A rise in the CCD in the Late Maastrichtian is indicated at Site 358 in the South Atlantic (Boersma, 1977).

The faunal events just prior to the boundary are encompassed in thin, 5 to 10-centimeter bands of marine clay or shale in land sections. Gamper (1977) has reported a zone of small globigerinids above the extinction of *A. mayaroensis* and before the first appearance of *G. eugubina* in the Mendez Shale. Within the boundary shale at Zumaya, Percival and Fischer (1977) have described a 10-centimeter zone just below the boundary where faunal extinctions were already numerous among both the nannoplankton and the globotruncanids. Smit (1977) has reported a thin clay boundary layer between the *A. mayaroensis* Zone and the *G. eugubina* Zone in the Loma de Solana unit at Barranco del Gredero. Within this zone, most globotruncanids, *A. mayaroensis* as well as the rugoglobigerinids, disappear; however, tiny biserial heterohelicids, *Globigerinel-*

| | |
|---|---|
| decreasing surface temperatures[1]<br>decreasing bottom temperatures[2] [3]<br>decreasing vertical gradients<br>reduced thermocline<br>decreasing meridional gradients | THERMAL<br>EVENTS |
| low lying continents[4]<br>transgression[6]<br>extensive epicontinental seas[5] | TECTONIC/<br>STRATIGRAPHIC EVENTS |
| increased carbon isotope gradient<br>increased nutrients through water column<br>increased oxygenations of water column<br>high CCD[7] | CHEMICAL<br>EVENTS |
| increasing productivity[11]<br>changing zonal and floral distribution patterns[10]<br>changing floral-faunal diversity: dominance patterns[8] [9] | BIOTIC EVENTS |

IMMEDIATELY PRIOR TO BOUNDARY

| | |
|---|---|
| increased bottom temperatures<br>decreased thermal gradients | THERMAL |
| decreased spreading rate on MAR[12]<br>regression[6] | TECTONIC/STRATIGRAPHIC |
| CCD drop then rise, North Atlantic[13]<br>increased sediment dissolution at shallowing depths[14] | CHEMICAL |
| decreased diversity invertebrates[8]<br>appearance thoracospheres and *Braarudosphaera*[9]<br>rising extinctions nannofossils[9]<br>beginning extinctions globotruncanids[15] [16]<br>extinction *A. mayaroensis* from land sections [16]<br>increased dominance deep-dwelling planktonic foraminifera,<br>    particularly from O-minimum levels<br>increasing dominance megalospheric heterohelicids | BIOTIC |

BOUNDARY CRISIS

| | |
|---|---|
| | THERMAL |
| hiatuses[17]<br>hardgrounds[18] | STRATIGRAPHIC |
| pyritization of deep marine sediments<br>decreased carbon isotope gradients through water column | CHEMICAL |
| extinctions of numerous siliceous plankton<br>               planktonic foraminifera<br>               calcareous nannofossils<br>               inter alia | BIOTIC |

GLOBIGERINA EUGUBINA ZONE

| | |
|---|---|
| increased surface temperature<br>decreased bottom termperature<br>increased vertical thermal gradients | THERMAL |
| regression[18] | TECTONIC/STRATIGRAPHIC |
| CCD drop, South Atlantic[7]<br>low oxygenation of water column<br>decreased nutrients in water | CHEMICAL |
| decreased productivity<br>decreased size planktonic foraminifera<br>decreased size benthic foraminifera<br>low diversity, high dominance plankton faunas[9]<br>scarcity siliceous plankton<br>abundant resistant, brackish water nannofossils<br>abundant 0-minimum zone and shelf types of<br>  planktonic foraminifera | BIOTIC |

Table 9-5. Documented events and suppositions of events during the Late Maastrichtian, just prior to the Cretaceous/Tertiary boundary, and in the Early Paleocene in the Atlantic Ocean.
References cited: [1] Savin, 1975; [2] Saito and Van Donk, 1974; [3] Douglas and Savin, 1975; [4] Hallam, 1968; [5] Dunbar, 1960; [6] Vail et al., 1977; [7] Boersma, 1977; [8] Kaufman, 1975; [9] Percival and Fischer, 1977; [10] Tschudy, 1971; [11] Tappan, 1968; [12] Pitman and Hays, 1973; [13] Tucholke, in press.; [14] Worsley, 1974; [15] Smit, 1977; [16] Gamper, 1977; [17] Newell, 1963; and [18] Vail, et al., 1977.

*loides* sp., *Guembelitria cretacea* and *Hedbergella monmouthensis,* become relatively more numerous.

At site 384, mixing is intense through the approximately 30-centimeter boundary zone. In general the globotruncanids become less diverse and less abundant through the boundary interval, while the heterohelicids become more abundant and produce more meglospheric individuals. Twisted forms, called *Woodringina,* are more numerous through the boundary zone. Mixing makes it very difficult to sort out further faunal changes just below the boundary at 13-3-30 centimeters. However, an unusual latest Cretaceous fauna occurs in burrow fillings just at the boundary level; this fauna contains small specimens, primarily of heterohelicids, a few corroded globotruncanids, *Globerigerinelloides* spp., *Hedbergella monmouthensis,* and some corroded rugoglobigerinids. This fauna, its color and preservation, is unlike any preceding or following in the *Globigerina eugubina* Zone. While the faunal composition and size may be a result of selective feeding by the burrower, it is strikingly similar to the fauna that has been recorded by Smit (1977).

Together these reports suggest that prior to the boundary event there was an interval of rapidly changing composition in marine microfaunas, which is seen both in land sections and in the deep-sea sediments. The faunas of this interval are small, lower in diversity, and contain small-sized individuals of *Heterohelix, Woodringina, Guembelitria, Hedbergella,* and *Globigerinelloides.* (Such faunas are typical of shallow shelf samples in the late Cretaceous, but the outer shelf would contain small rugoglobigerinids). The nannofossils during this interval are low in diversity and dominated by resistant forms, some of which may be in an encysted state (Percival and Fischer, 1977).

*The Cretaceous-Tertiary boundary*   The Cretaceous-Tertiary boundary is expressed by hiatuses, hard grounds, phosphorites, and pyritization of open marine sediments (Rosenkrantz and Brotzen, 1960; Percival and Fischer, 1977). There is no fluctuation reported in the sulfur isotope record across this interval (Holser, 1977).

The carbon isotope change took place during the interval of time encompassed by the boundary. The latest Cretaceous faunas at 13-3-30 centimeters at Site 384 still record the high Late Maastrichtian values, while the faunas immediately above have significantly lower carbon isotope ratios. Thus, a chemical change in the ocean that is reflected in the carbon isotope rations, occurred at the Cretaceous-Tertiary boundary in the Atlantic.

The final extinctions of Cretaceous plankton took place within this interval, as did whatever evolution was needed to produce the *Globigerina eugubina* fauna.

*Paleocene*   In the early *Globigerina eugubina* Zone, significant changes in temperature are recorded by the foraminifera. An increase in surface and a decrease in bottom temperatures resulted in stronger thermal gradients through the water column in the South Atlantic.

The regression continued along with low carbon isotope ratios and gradients. The isotopes suggest low nutrient levels, low productivity, and low oxygen levels through the water column. In such an environment, small, thin-walled planktonic and benthonic foraminifera, and low diversity foraminiferal and nannofossil faunas are not surprising. It is noteworthy that the planktonic foraminifera that survived the boundary, for example, *Guembelitria cretacea, Woodringina, Hedbergella monmouthensis,* and the smooth biserial heterohelicids were, in the Cretaceous, adapted to either the surface waters over shelves and shelf seas, or to the deeper O-minimum levels of the water column (Boersma, in preparation). The scarcity of siliceous plankton at this time may reflect the very low concentrations of other nutrients.

The eventual rejuvenation of planktonic communities occurred during the Paleocene, but culminated in the Late Paleocene. Increased planktonic diversification closely parallels the rising surface carbon isotope values and thus the increased productivity of the surface waters. From the fluctuating temperatures at the surface and at the bottom, it appears that the increased temperature gradients, both vertical and meridional, produced the ecologic conditions requisite for faunal diversification. Increased nutrients and niches together then resulted in the increased diversification and development of food chains during the later Paleocene.

## ACKNOWLEDGMENTS

The author wishes to thank Nicholas J. Shackleton for freely providing data for this study and for his willingness to accept responsibility for their accuracy. Thanks are also due to Mike Hall and Quentin Given for their expert assistance in the oxygen isotope lab. Drafting was provided by Eric Trachtenberg and Mary Perry. Jim Hays and Lloyd Burckle reviewed the manuscript and I am very grateful for their suggestions and help.

This research is sponsored by NSF contract Oce-76-83000.

## LITERATURE CITED

Berggren, W. A., 1972. A Cenozoic time-scale: implications for regional geology and paleobiogeography. *Lethaia,* 5:195–215.

Boersma, A., 1977. Cenozoic planktonic foraminifera—DSDP Leg 39 (South Atlantic). *Initial Reports f the Deep Sea Drilling Project* (Washington, D.C.: U.S. Govt. Printing Office), vol. 39, pp. 567–588.

Boersma, A., and N. J. Shackleton, 1977a. Tertiary oxygen and carbon isotope stratigraphy, Site 357 (mid-latitude South Atlantic). *Initial Reports of the Deep Sea Drilling Project* (Washington, D.C.: U.S. Govt. Printing Office), vol. 39, pp. 911–924.

———, 1977b. Oxygen and carbon isotope record through the Oligocene at DSDP Site 366 (equatorial Atlantic). *Initial Reports of the Deep Sea Drilling Project* (Washington, D. C.: U. S. Govt. Printing Office), vol. 61, pp. 957–962.

Boersma, A., N. J. Shackleton, M. Hall, and Q. Given, 1979. Carbon and oxygen isotope records at DSDP Site 384 (North Atlantic) and some Paleocene paleotemperature and carbon isotope variations in the Atlantic Ocean. *Initial Reports of the Deep Sea Drilling Project* (Washington, D. C.: U. S. Govt. Printing Office), vol. 43, pp. 695–717.

Broecker, W. S., 1971. A kinetic model for the chemical composition of sea water. *Quat. Res.,* 1:188–207.

Dilley D., 1973. Cretaceous larger foraminifera. In: A. Hallam (ed.), *Atlas of Paleobiogeography* (London: Elsevier), pp. 400–420.

Douglas, R., and S. Savin, 1975. Oxygen and carbon isotope analysis of Tertiary and Cretaceous microfossils from Shatsky Rise and other sites in the North Pacific Ocean. *Initial Reports of the Deep Sea Drilling Project* (Washington, D. C.: U. S. Govt. Printing Office), vol. 31, pp. 550–590.

Dunbar, C., 1960. *Historical Geology* (New York: Wiley & Son), 500 pp.

Flessa, K., 1975. "Evolutionary Pulsations: Evidence from Phanerozoic Diversity Patterns." Brown Univ., Ph. D. thesis. 79 pp.

Gamper, M. A., 1977. Acerca del limite Cretácico-Terciaria en México. *Univ. Nat. Auton. México, Inst. Geol., Rev.,* 1:23–27.

Gignoux, M., 1950. *Geologie Stratigraphique* (Paris: Masson), 709 pp.

Hallam, A., 1968. Major epeirogenic and eustatic changes since the Cretaceous, and their possible relationship to crustal structure. *Amer. Jour. Sci.,* 261:397–423.

Harker, S., and W. Sarjeant, 1975. The stratigraphic distribution of organic-walled dinoflagellate cysts in the Cretaceous and Tertiary. *Rev. Paleobot. Palynol.,* 20:217–235.

Holser, W., 1977. Catastrophic chemical events in the history of the ocean. *Nature,* 267:403–407.

Kauffman, E. G., 1973. Cretaceous Bivalvia. In: A. Hallam (ed.), *Atlas of Paleobiogeography* (London: Elsevier), pp. 353–385.

Kroopnick, P. M., S. Margolis, and C. Wong, 1977. $\delta$ $13_c$ variations in marine carbonate sediments as indicators of the $CO_2$ balance between the atmosphere and oceans. In: *Fate of Fossil Fuel $CO_2$* (New York: Plenum), pp. 295–321.

Lowenstam, H., and S. Epstein, 1954. Paleotemperatures of post-Aptian Cretaceous as determined by the oxygen isotope method. *Jour. Geol.,* 62:207–248.

Luterbacher, H., and I. Premoli-Silva, 1962. Note preliminaire sur une révision du Profil de Gubbio, Italie. *Riv. Italiana Pal. Stratigr.,* 68:253.

Newell, N., 1963. Crises in the history of life. *Sci. Amer.,* 208:76–92.

Percival, S., and A. Fischer, 1977. Changes in calcareous nannoplankton in the Cretaceous/Tertiary biotic crisis as Zumaya, Spain. *Evol. Theory,* 2:1–37.

Pitman, W., and J. Hays, 1973. Upper Cretaceous spreading rates and the great transgression. *Geol. Soc. Amer. Abstr.,* 5:768.

Premoli Silva, I., and H. Bolli, 1973. Late Cretaceous to Eocene planktonic foraminifera and stratigraphy of Leg 15 sites in the Caribbean. *Initial Reports of the Deep Sea Drilling Project* (Washington, D.C.: U.S. Govt. Printing Office), vol. 15, pp. 499–542.

Reid, G., 1977. Stratospheric aeronomy and the Cretaceous/Tertiary boundary extinctions. In: Cretaceous/Tertiary extinctions and possible terrestrial and extraterrestrial causes. *Canadian Nat. Mus. Syllogeus,* 12:75–88.

Russell, D., 1975. L'extinction des sauropsides à la fin de L'ére secondaire une hypothèse. *Cent. Natl. Res. Sci., Colloq. Int.,* 218:513–518.

Saito, T., and J. Van Donk, 1974. Oxygen and carbon isotope measurements of Late Cretaceous and Early Tertiary foraminifera. *Micropaleontology,* 20:152–177.

Sloan, R., 1967. Cretaceous and Paleocene terrestrial communities of western North America. *Proceedings of the North American Paleontological Convention, Part E.,* pp. 427–354.

Savin, S., 1977. The history of the earth's surface temperature during the past 100 million years. *Ann. Rev. Earth Planet. Sci.,* vol. 5, pp. 319–355.

Smit, J., 1977. Discovery of a planktonic foraminiferal association between the *Abathomphalus mayaroensis* Zone and the *Globigerina eugubina* Zone at the Cretaceous/Tertiary boundary in the Barranco del Gredero (Caravaca, Spain): a preliminary note. *K. Ned. Akad. Wet., Proc. Ser. B,* 80:280–301.

Supko, P., et al., 1977. (eds.) *Initial Reports of the Deep Sea Drilling Project* (Washington, D.C.: U.S. Govt. Printing Office), vol. 39.

Tappan, H., 1968. Primary productivity, isotopes, extinctions, and the atmosphere. *Palaeogeogr. Palaeoclimatol. Palaeoecol.,* vol. 4, no. 3, pp. 187–210.

Tappan, H., and A. Loeblich, 1971. Geobiologic implications of fossil phytoplankton evolution and time-space distribution. *Geol. Soc. Amer. Spec. Paper,* 227:247–340.

Termier, H., and G. Termier, 1969. Global paleogeography and earth expansions. In: *Applications of Modern Physics to Earth and Planetary Interiors* (New York: Wiley & Son), pp. 87–101.

Tschudy, R., 1971. Palynology of the Cretaceous/Tertiary boundary in the northern Rocky Mountain and Mississippi embayment regions. *Geol. Soc. Amer. Paper,* 127-65-111.

Tucholke, B. E., and P. R. Vogt, 1979. *Initial Reports of the Deep Sea Drilling Project* (Washington, D. C.: U. S. Govt. Printing Office), vol. 43.

Uffen, P., 1963. Influence of the earth's core on the origin and evolution of life. *Nature,* 198:143–144.

Vail, P. R., R. Mitchum, and S. Thompson, 1977. Seismic stratigraphy and global changes in sea level, part 3. Relative changes of sea level from coastal onlap. *Amer. Assoc. Petrol. Geol. Mem.,* 26:83–99.

Whitehead, D., 1969. Wind pollination in the angiosperms: evolutionary and environmental considerations. *Evolution,* 23:28–35.

Worsley, T., 1974. The Cretaceous/Tertiary boundary event in the ocean. *Soc. Econ. Pal. Min. Spec. Pub.,* 20:94–125.

Chapter 10

# CHANGES IN THE ANGIOSPERM FLORA ACROSS THE CRETACEOUS-TERTIARY BOUNDARY

LEO J. HICKEY

Division of Paleobotany, Smithsonian Institution

## INTRODUCTION

Recent hypotheses have proposed that a universal biotic catastrophe caused by an asteroid impact (Alvarez, W. et al., 1979; Alvarez, L. W. et al., 1980a, b; Smit and Hertogen, 1980; Gagnapathy, 1980), a cometary impact (Hsü, 1980), or a supernova (Russell and Tucker, 1971) terminated the Cretaceous period. Paleontology is severely limited in its ability to specify the causes of the Cretaceous extinctions that so dramatically affected the dinosaurs and numerous groups of marine organisms because it can only infer the operative forces from their effects on the biota. If the observed effects are drastic and cut across taxonomic and regional lines, their causes are likely to be drastic and widespread. If, on the other hand, these effects are limited to specific groups of animals or plants, or to particular regions, then less drastic and more localized causes must be sought. Land plants should form a central element in any comprehensive inquiry into the possible causes of extinctions at the Cretaceous-Tertiary (K/T) boundary because they are a conspicuous and exposed part of the biota and make up the base of the terrestrial food chain. Catastrophic physical changes in the environment should have caused major changes in the plant record, while localized physical events or competitive interactions at higher trophic levels should have had correspondingly less effect.

This review will focus on changes in the record of the flowering plants from Cretaceous to Paleocene time. Angiosperms are an excellent subject for such an investigation because they were in the midst of a vigorous evolutionary expansion and had attained dominance in the land flora only a quarter of a period before. Thus, any reversal or interruption in their pattern of diversification would be evidence for powerful and pervasive environmental changes.

279

## DIFFICULTIES IN WORKING WITH THE
## FOSSIL PLANT RECORD

This examination will include data from both megafloral and microfloral assemblages. Unfortunately, because of the wide variation in objectives and techniques of the authors who generated the studies reviewed here there is a danger that the numerical comparisons will be interpreted too strictly, rather than in general and approximate terms. It would, of course, be ideal for this survey if relatively complete suites of reliably and uniformly identified micro- and megafossil plants from a large number of sites closely bracketing the Cretaceous-Paleocene boundary and distributed over the whole world were available. In fact, almost all these data are defective in one or more ways.

It is surprising how few of the present number of paleofloral studies actually span the Cretaceous-Paleocene boundary. As for the megafossil records, of the 33 reports listed in Table 10–2, only four deal with megafossils. This scarcity may result from widespread marine regressions at the close of the Cretaceous, which caused sediment transport through former depositional areas by lowering the base levels of the streams.

In some cases, the contrast between fossil angiosperm assemblages of Cretaceous and of Paleocene age is exaggerated by widely spaced samples. An example is Stanley's (1965) study (Table 10–2) which shows a striking difference between the pollen content of uppermost Cretaceous and lower Paleocene strata, with only 18% of the gymnosperm and angiosperm species persisting across the boundary. The shortest distance however, between sample sites in the Hell Creek Formation and the Ludlow Member of the Fort Union Formation is 48 kilometers, which is approximately parallel to the dip, which gives a minimum stratigraphic separation of 270 meters, if we assume level topography. Similarly, studies which average assemblages from all or large parts of the Maastrichtian against those for long intervals of the Paleocene tend to falsely heighten the real differences. It is unfortunate that reports that deal with restricted intervals across the boundary are in the minority.

Another factor which elevates the extinction figures that are reported here arises from treating localities separately rather than including them in regions. A form may become locally extinct for a variety of reasons but survive elsewhere across the boundary.

As Tschudy notes in the present volume, palynological studies often place a strong emphasis on labeling a stratigraphic interval as Cretaceous or Paleocene by the use of selected guide fossils rather than by the evaluation of whole floras. Thus, forms that span the erathem boundary tend to be ignored in favor of those that are clearly restricted to a single period.

In addition, changes in sedimentary facies are seldom mentioned in the studies reviewed here. Yet such shifts, as from the terrestrial to the marine sedimentary regime, can drastically alter the pollen spectrum (see Jarzen, 1978).

Further difficulties arise as a result of the extensive misidentification of Cretaceous and Paleocene megaflora, largely by assignment to extant genera on superficial criteria (Pacltová, 1961; Wolfe, 1972; Hickey, 1973; Dilcher, 1974; Hughes, 1976). The process of restudying this material has only begun and it will be a long time before the actual course of evolution during the Late Cretaceous is worked out in detail. Earlier authors also had a tendency to greatly overspeciate their megafloras. Despite these difficulties, I will attempt a preliminary overview of possible angiosperm trends through the Late Cretaceous and Paleocene by dealing with broad morphological groupings of leaves (Table 10–1, Figure 10–4).

Another problem in attempting to integrate these floral investigations is caused by the great variation in the size of the samples studied (from 438 to 10 species in Table 10–2). Most of the megafloras are large; except for the Western Canada figures that Vakhrameev and Achmetev (1977) gave, which were based on only 14 species, they range from 35 (Shoemaker, 1966) to 81 species (Dorf, 1940). Artificial variability also arises when data from pollen studies, where "species" correspond mainly to genera or to groups of actual species within genera, are summarized with those from megafloras where "species" may represent only one of a number of variant leaf forms occurring within a true species. In several studies artificial enhancement of the differences between the time periods occurred when changes in the Cretaceous or in the Paleocene assemblages were analyzed across the boundary without considering the total floral composition of both time periods.

Mistakes in correlation and differences in dating are inevitable whenever paleontological data from a large number of sources are summarized. For example, Tschudy (1970) and Shoemaker (1966) have accepted the so-called "z" coal as the base of the Paleocene Fort Union strata in the Hell Creek area of Montana, whereas the highest dinosaur remains occur approximately 10 meters lower. In the Powder River Basin of Wyoming, Leffingwell (1970) has found that the most significant palynological break lay 2 meters above the highest dinosaur fossils and about 10 meters below the lithologic contact between the Lance and Fort Union Formations. I have accepted Leffingwell's break as the K/T boundary for this paper because of the lack of any absolute dates that would allow more precise correlation in this section. In general, incorrect correlations are believed to have had only a small effect on the results given here. Furthermore, the absence of precisely synchronous biotic

| LEAF MORPHOLOGICAL TYPES / MONOCOTS | Cenomanian | | | | | Turon. | Coniac. | Santon. | Campan. | | | Maastr. | Paleocene | | |
|---|---|---|---|---|---|---|---|---|---|---|---|---|---|---|---|
| **Age →** | Dakota Group | Woodbine Fm | Melozi Fm | Woodbridge Mbr of Raritan Fm | Peruc Fm | Long Island Raritan Fm | Eutaw Fm | Magothy Fm (Severn R) | Middendorf Mbr Black Creek Fm | Magothy Fm (C & D Canal) | Judith River Fm | Ripley Fm | Lance/Medicine Bow Fms | Lower Fort Union Fm | Upper Fort Union Fm |
| 1b Linear; longitudinal parallel ven. | | | | | 3 | 5 | 7 | 9 | 2 | | | 2 | 3 | 2 | 2 |
| 1c Linear; "Pinnate" parallel ven. | | | | | | | | | | | | | 3 | 2 | 3 |
| 2 Leaves sagittate | | | | | | | | | | | | 2 | 1 | | |
| 3a Costapalmate palms | | | | | | | | 4 | 2 | 2 | | 1 | 4 | 3 | 2 |
| 3b Pinnate palms | | | | | | | | 1 | | | | 1 | | | |
| 3c Pure palmate palms | | | | | | | | | | | | 1 | | | |
| **DICOTS** | | | | | | | | | | | | | | | |
| 4 Cordate; "palmate" | 2 | 6 | 13 | 7 | 4 | 4 | 5 | | 2 | | 8 | 10 | 9 | 14 | 12 |
| 5 Peltate; actinodromous-Nymphaephylls | | | | 5 | | | 2 | | 4 | | | 1 | 6 | 4 | 2 |
| 6 Simple, entire; festooned broch. | 4 | 6 | | 7 | 3 | 7 | | 7 | 9 | 4 | | 7 | 4 | 1 | 2 |
| 7 Simple, entire; pinnate broch. | 37 | 37 | 11 | 34 | 31 | 37 | 35 | 45 | 35 | 54 | 8 | 24 | 30 | 23 | 25 |
| 8 Simple, pinnate, crasp., toothed | 6 | 10 | | 1 | 8 | 2 | | 7 | 6 | 2 | 42 | 14 | 3 | 22 | 21 |
| 9 Celastrophylls | 0.5 | 2 | | 13 | 9 | 4 | 5 | 7 | 9 | 4 | | 6 | | 2 | 4 |
| 10 Myrtophyllum type | 0.5 | 4 | 2 | 1 | 3 | 4 | 5 | 7 | 9 | 4 | | 4 | 5 | 2 | |
| 11 Pinnately compound | 6 | | 9 | 8 | 8 | 10 | 10 | | 13 | | | 15 | 1 | 11 | 11 |
| 12 Fontanea type | | | | | 3 | 5 | | 5 | 4 | | | 2 | | | |
| 13 "Bauhinia" type | 0.5 | 2 | | 4 | 1 | | | 4 | 4 | | | | 1 | | |
| 14 Dewalquia-Debeya type | | | | | 1 | 3 | | | 2 | 2 | | | 1 | | |
| 15 Citrophyllum type | 0.5 | | | | | | | | | 2 | | | | | |
| 16 Cornophylls | 0.5 | 2 | | 1 | | 1 | | | 2 | | | | 1 | 1 | 3 |
| 17 Rhamnophylls | 2 | 4 | | 1 | 6 | 20 | | 2 | | | | 8 | 1 | 4 | 2 |
| 18 Palinactinodromus-Platanophylls | 15 | 10 | 32 | 11 | 16 | 14 | 5 | | | | | 5 | 9 | 4 | 5 |
| 19 Betulites-Populites - "Viburnum" | 20 | 6 | 25 | | 4 | 2 | | 4 | | 13 | 33 | 1 | 12 | 3 | 8 |
| 20 Liriodendropsis type | 5 | 8 | 2 | 6 | 4 | | | 4 | | | | | | | |
| 21 Other forms | 0.5 | 2 | | 1 | 2 | 1 | 5 | | 2 | | | 4 | 5 | 1 | 2 |
| Number of Species | 246 | 52 | 44 | 71 | 75 | 113 | 20 | 28 | 47 | 46 | 12 | 83 | 80 | 90 | 62 |
| Number of Types | 15 | 14 | 8 | 14 | 14 | 15 | 10 | 11 | 12 | 13 | 5 | 18 | 17 | 19 | 14 |

Table 10-1. Changes in the relative abundance (expressed as percentages) through the Late Cretaceous and Paleocene of the twenty leaf morphological categories discussed in the text. These values are also graphed in Figure 10-4. The relative sequence of localities is of more significance than their assignment to a particular stage, which is only approximate in some cases and done to facilitate comparisons. In the list of localities below, the source of the floral data is cited immediately following the name of the flora, then the source of the correlation or comments on it are given following the dash: Dakota Gp., Lesquereux (1892)—Cobban and Reeside (1952); E. Kauffman (written comm., 1979); Woodbine Fm., MacNeal (1958)—(same): Melozi Fm., Hollick (1930)—the date of early middle Albian given this unit by Imlay and Reeside (1954) is too early considering the strongly Dakota aspect of its flora and the time of earliest angiosperm appearance in Alaska (Brenner, 1976; Hickey and Doyle, 1977; Smiley, 1966, 1969a, 1969b). The flora is at least late Albian and probably Cenomanian and thus Martin's (in Hollick, 1930) date is retained. Woodbridge Clay, Berry (1911)—Wolfe and Pakiser (1971); Peruč Fm., Velenovsky (1882)—Pacltová (1971); Raritan Fm. at Kreischerville, Long Island, Hollick (1906)—Doyle (1969); Eutaw Fm., Berry (1919)—Tschudy (1975); Magothy Fm. at the Severn River, Maryland, Berry (1916)—Wolfe

changes at the end of the Cretaceous argues indirectly against a catastrophic cause.

Errors introduced from all the sources discussed above, as well as the possibility of reworking, probably account for part of the variation seen in the summary tables and diagrams in this paper. The general accordance of neighboring localities, however, and the coherence of the worldwide pattern implies that the basic trends are a valid reflection of the changes that occur in the angiosperm flora as a whole across the K/T boundary.

## MATERIALS AND METHODS

An important component of this review is a survey I compiled of literature that described either pollen or megafossil assemblages that spanned the K/T boundary. Only the pollen studies were well enough distributed to give at least cursory coverage of the world. Usable megafossil data were restricted to the western United States but yielded important additional climatic and systematic information. I tallied all forms of pollen alike whether some authors treated them as genera or groups of genera or others as "species." I listed them as occurring in the Maastrichtian, in the Paleocene, or in both. The data were then used to plot the persistence of Cretaceous forms across the boundary and to estimate the change in diversity from Maastrichtian into Paleocene strata. Only angiosperm pollen was used, except in a few cases where gymnosperm pollen or even spores were not separated from the totals (such as in the report by Tschudy, 1970).

The procedure for fossil angiosperm leaves was much the same, except that I analyzed data on leaf margin percentage changes across the boundary and made revisions of some of the previously described forms if an examination of specimens from the same or closely correlative formations indicated that this was warranted.

The range and abundance plots (Table 10-1, Figure 10-4) for the various angiosperm leaf morphological groupings through the Late Cretaceous and Paleocene I assembled from a survey of literature and selected

---

(1976); Middendorf equivalent of the Tuscaloosa Fm., Berry (1914)—floral comparison for this study, including the presence of palms; Magothy Fm. at the Chesapeake and Delaware Canal, Berry (1916)—Wolfe (1976); Judith River Fm. Knowlton (1905)—Cobban and Reeside (1952); E. Kauffman (written comm., 1979); Ripley Fm., Berry (1925)—Tschudy (1975); Lance and Medicine Bow Fms., Dorf (1938, 1942)—Cobban and Reeside (1952); E. Kauffman (written comm., 1979); Fort Union Fm. lower and upper, Brown (1962)—Brown (1962).

| No. | Area/Formation | %K Forms Persisting | Diversity | Reference/Remarks |
|---|---|---|---|---|

AQUILLAPOLLENITES PROVINCE

| No. | Area/Formation | %K Forms Persisting | Diversity | Reference/Remarks |
|---|---|---|---|---|
| 1 | Moreno Fm., California | 90 | 29 | Drugg, 1967 |
| 2 | Fox Hills & lower Medicine Bow Fms., Wyoming | 44 | 16 | update & revision of Dorf, 1938 |
| 3 | + Lance Fm., Wyoming | 50 | 16 | update & revision of Dorf, 1940, 1942 |
| 4 | Lance & Ft. Union Fms. | 83 | 37 | Leffingwell, 1970 |
| 5 | N.W. South Dakota | 18 | 5 | Stanley, 1965 |
| 6 | Glendive, Montana | 13 | -68 | Tschudy, 1970 (all pollen & spores) |
| 7 | Hell Creek Fm-Tullock Mbr, Montana | 68 | -4 | Shoemaker, 1970 |
| 8 | Type Hell Creek Fm, Montana | 10 | -70 | Norton & Hall, 1967 |
| 9 | Seven Blackfoot Creek, Montana | 33 | -46 | Tschudy, 1970 (all pollen & spores) |
| 10 | + White Mud Fm., Saskatchewan | 33 | -44 | Vachrameev & Achmetev, 1977 |
| 11 | S.W. Siberian Lowland | 51 | -8 | Zaklinskaya, 1970 |
| 11A | Amuro-Zeyskoy Plain | 58 | -19 | Mamontova, 1977 |
| 12 | Yenesi River-Ulema Lowland | 30 | -72 | Zaklinskaya, 1970 |
| 13 | N.W. Siberia | 17 | -63 | Zaklinskaya, 1970 |

NORMAPOLLES PROVINCE

| No. | Area/Formation | %K Forms Persisting | Diversity | Reference/Remarks |
|---|---|---|---|---|
| 14 | Mississippi Embayment | 17 | -44 | Tschudy, 1970 (all pollen & spores) |
| 15 | Mississippi Embayment | 33 | -33 | Tschudy, 1975 (Normapolles gen. only) |
| 16 | Mississippi Embayment | 42 | 7 | Tschudy, 1975 (Normapolles species only |
| 17 | Atlantic Coastal Plain | 91 | 14 | Zaklinskaya, 1977 |
| 18 | Middle Europe | 93 | 25 | Krutzsch, 1957 |
| 19 | Europe | 69 | — | Tschudy, 1975 (Normapolles genera only) |
| 20 | N. & Central Europe | 62 | 6 | Zaklinskaya, 1970 |
| 21 | N. & Central Europe | 67 | 22 | Zaklinskaya, 1970 |

OTHER AREAS

| No. | Area/Formation | %K Forms Persisting | Diversity | Reference/Remarks |
|---|---|---|---|---|
| 22 | Near Belem, Brazil | 23 | -38 | Boer, van der Hammen, & Wymstra, 1965 |
| 23 | Caribbean Venezuela | 75 | -75 | Germeraad, Hopping, & Muller, 1968 |
| 24 | British Guiana | 100 | 50 | Leidelmeyer, 1956 |
| 25 | N.W. Brazil | 77 | -11 | H. Mueller, 1966 |
| 26 | Egypt | 50 | -43 | Kedeves, 1971 |
| 27 | Nigeria | 63 | -12 | Germeraad, Hopping, & Muller, 1968 |
| 28 | Nigeria | 92 | — | Hoeken-Klinkenberg, 1965 |
| 29 | Sarawak, Borneo | 94 | -6 | J. Muller, 1968 |
| 30 | New Zealand | 59 | -3 | Couper, 1960 |
| 31 | Gippsland Basin, Australia | 88 | 0 | Stover & Evans, 1973 |
| 32 | Gippsland Basin, Australia | 93 | 13 | Stover & Partridge, 1973 |

Table 10–2. Persistence and diversity changes in Cretaceous plant forms into the Paleocene.

Persistence data for the localities on this table have been plotted on Figures 10–5 and 10–7 and diversity data on Figure 10–6. The number beside the points on these figures identifies the entry on this table. Within each of the major regions the localities are arranged by increasing paleolatitude.

specimens in the National Museum of Natural History. With the exception of the Cenomanian Peruč flora from Czechoslovakia, time allowed only the inclusion of North American floras. Comparative abundance during each age was determined by averaging the number of species in each category from the floras of a particular age. The basis for the age assignments of the floras treated in Table 10–1 is given in the notes accompanying them and should be considered tentative. The relatively small sample sizes of some of the floras allow a large margin for error in the sample percentages, especially for types with low representation. The overall picture however, or ranges and variations in abundances for these leaf types agrees with estimates derived from general experience with Late Cretaceous and Paleocene floras. While it is true that earlier authors tended to maximize the number of species in their assemblages, for the purpose of this analysis, I assumed that the proportion of this overspeciation would be the same for each category of leaf in the table and figure.

## CRETACEOUS ANGIOSPERM HISTORY

At present, the earliest valid records of the flowering plants are from strata of Barremian age (the fourth of six ages of the Lower Cretaceous) in England, the eastern United States, Siberia, and equatorial Africa-South America (Hughes, 1976, 1977; Doyle and Hickey, 1976; Hickey and Doyle, 1977; Vakhrameev and Kotova, 1977; and Doyle et al., 1977). These occurrences consist either of pollen of monosulcate form with a columellate exine (*Clavatipollenites*) or a coarsely reticulate tectum (*Retimonocolpites*) and a nonlaminated foot layer or nexine (Van Campo and Lugardon, 1973; Doyle et al., 1975; Hughes, 1976) or small, simple leaves with disorganized, pinnate venation patterns (Doyle and Hickey, 1972; Hickey and Doyle, 1972; Doyle and Hickey, 1976; Hickey, 1978). A number of the morphological features of these early pollen and leaves can be related to the present-day angiosperm subclasses Monocotyledonae and Dicotyledonae (Doyle, 1973; Wolfe, Doyle and Page, 1975; Doyle and Hickey, 1976).

Through the remainder of the Early Cretaceous, angiosperms gradually diversify in numbers of species and variety of morphological types (Hickey and Doyle, 1977; Niklas, et al., 1980). Despite the superficial similarity of some of the leaves to those of modern angiosperm taxa, closer examination indicates that they belong to archaic forms not comparable in the majority of their details to the leaves of living genera or families. This archaic angiosperm radiation continued vigorously into the Late Cretaceous, with flowering plants becoming dominant in the Northern Hemisphere by the Turonian (Penny, 1969; Hughes, 1976). Data from the Southern Hemisphere are more difficult to evaluate, but overall dominance there may have been delayed until as late as the Maastrichtian (Penny, 1969).

The Late Cretaceous pollen record of the flowering plants has been well summarized by Penny (1969), Doyle (1969), and Muller (1970). One of the most important trends during this interval is the gradual breakdown of a relatively homogeneous pollen flora after the Cenomanian (see Tschudy, this volume). By the mid- and late Senonian, two well-defined pollen provinces were in existence at the mid- and higher latitudes of the

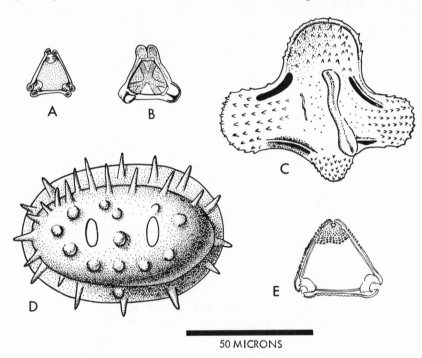

50 MICRONS

Figure 10-1. Some characteristic agiosperm pollen grains of the Late Cretaceous.

1A) *Complexiopollis* from the Cenomanian Peruč Formation of Bohemia (redrawn from Pacltová, 1977).

1B) *Oculopollis concentus* Pflug, the type species of the genus, from the middle Senonian of Germany.

1C) *Aquilapollenites quadrilobus* Rouse, type species of the genus, from the Late Cretaceous Brazeau Formation of Alberta.

1D) *Wodehousia spinnata* Stanley, from the Upper Maastrichtian of Alaska (redrawn from Wiggins, 1976).

1E) *Proteacidites formosus* Samoilovich, type species of the genus, from Danian (Early Paleocene) sediments of the West Siberian Lowland. (Figures 10-1B, 1C, and 1E redrawn from Jansonius and Hills, 1976).

Northern Hemisphere. These provinces appear to owe their discreteness to physical barriers provided by the Mid-Continent Seaway in North America and to the Straits by Turgai east of the Urals (Doyle, 1969). The area from eastern North America across Europe lay in the Normapolles Province. Grains of this complex are characterized by protruding pores with a highly complex, vestibulate structure (Figure 10–1B). The group appears to have arisen from the triangular triporate form known as *Complexiopollis* of Cenomanian age (Figure 10–1A). Normapolles reach the height of their diversity in the late Campanian and gradually decline to extinction through the Paleocene and Eocene (Figure 10–2).

The second pollen province became strongly differentiated during Campanian and Maastrichtian times and extended from western North America nearly across Siberia. This province was dominated by a bizarre complex of forms known as the Aquilapollenites or Triprojectacites group (Mchedleshvili, 1961: Stanley, 1970; Tschudy, this volume) whose pollen is distinguished by having three, or sometimes four, arms projecting from an elongate central body (Figure 10–1C). With these, occur forms such as *Wodehousia* and *Azonia* of the so-called Oculata type (Figure 10–1D; Wiggins, 1976) and reputedly proteaceous pollen of the genus *Proteacidites* (Figure 10–1E). The *Aquilapollenites* group arose in the latest Turonian (Stanley, 1970) and reached the peak of its abundance in the late Maastrichtian. After this it underwent a drastic decline to extinction in the Early Eocene (Figure 10–2). The patterns for both of the major pollen types shown in Figure 10–2 are based on analyses by Zalkinskaya (1977), and they contrast with the drastic reduction in both groups at the end of the Cretaceous seen in the review by Muller (1970, Figure 3). Neither of these dominant pollen types has yet been convincingly related to modern families. Normapolles have been compared with certain of the amentiferous orders such as Juglandales, Betulales, Fagales, and Urticales, all of which have an angular rotund outline and simpler pore structure (Muller, 1970; for other references see Penny, 1969), while Santalalean affinities have been proposed for *Aquilapollenites* (Jarzen, 1977a).

After the middle of the Late Cretaceous, pollen with clearly demonstrable morphological similarities to that of modern families and even genera became progressively more abundant. Palm pollen appeared in the early Senonian (an age comprising the final four ages of the Cretaceous Period) and became a very important component in the diverse tricolpate and tricolporate pollen flora that then characterized the lower latitudes (Muller, 1970). Also in the early Senonian, pollen of *Nothofagus,* or the Southern Hemisphere beech, appeared (Penny, 1969) and, together with proteaceous types, dominated what Muller calls the Australian-Antarctic Province.

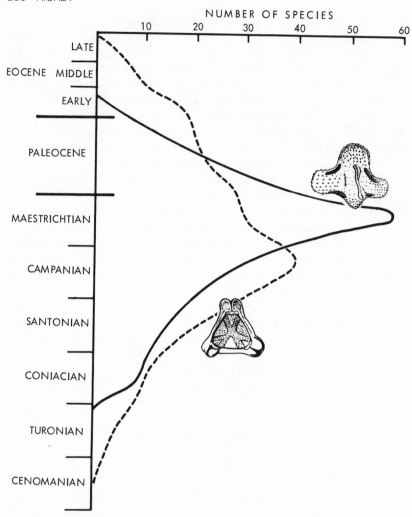

Figure 10-2. Plot of the abundance of species belonging to the Normapolles complex (dashed line) and to the Aquilapollenites complex (solid).
The Normapolles summary is world-wide while the Aquilapollenites curve is based on Eurasia only. Redrawn from Zaklinskaya, 1977. Pollen symbols redrawn from Jansonius and Hills, 1976.

The graph of the stratigraphic increase in the number of modern genera in Figure 10-3 shows a modest acceleration during the Late Maastrichtian with only a slight increase across the K/T boundary. Actual major periods of modernization occurred in the Middle Oligocene and the Early Miocene and apparently correlated with major climatic changes

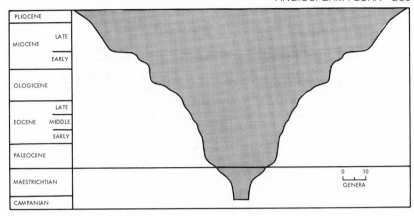

| PLIOCENE | |
| MIOCENE | LATE |
| | EARLY |
| OLOGICENE | |
| EOCENE | LATE |
| | MIDDLE |
| | EARLY |
| PALEOCENE | |
| MAESTRICHTIAN | |
| CAMPANIAN | |

Figure 10–3. Increase in the number of modern genera of Angiosperms through time. (Mainly after J. Muller, 1970.)

and the development of grassland and semi-xeric areas (Wolfe and Hopkins, 1967).

I emphasize the megafloral record in this summary of Late Cretaceous vegetational trends because the previous picture of a relatively modern angiosperm flora of this time, which resulted from the substantial misidentification of their leaf remains, has now been essentially discredited by palynological studies, and there is need of a reinterpretation based on more modern methods. Inferences concerning the affinities and relationships of these megafossils, however, are now in such a chaotic state due to the superficial and conflicting identifications of leaf remains with modern forms that the only way, for the present, to understand basic trends in the archaic radiation of the flowering plants is by attempting to analyze the assemblage in terms of character-morphological complexes as has been advocated by Doyle (1969), Wolfe (1972), Doyle and Hickey (1976), Hughes (1976), Hickey and Doyle (1977), and Krassilov (1977). I have used these methods to break down the Late Cretaceous leaf record into groups based on morphological similarities. In these categories I include related forms as well as those forms whose features represent evolutionary convergences. However, because certain combinations of leaf architectural characters are strongly correlated with particular higher taxa (Hickey and Wolfe, 1975), knowledge of the ranges and diversity of these morphological groups can yield limited information on angiosperm history without an understanding of the detailed systematic relationships. The ranges and abundance of the most important angiosperm leaf architectural complexes through the Late Cretaceous and Paleocene are summarized in Table 10–1 and Figure 10–4. In the discussion below, the number in parentheses following the first mention of each leaf type iden-

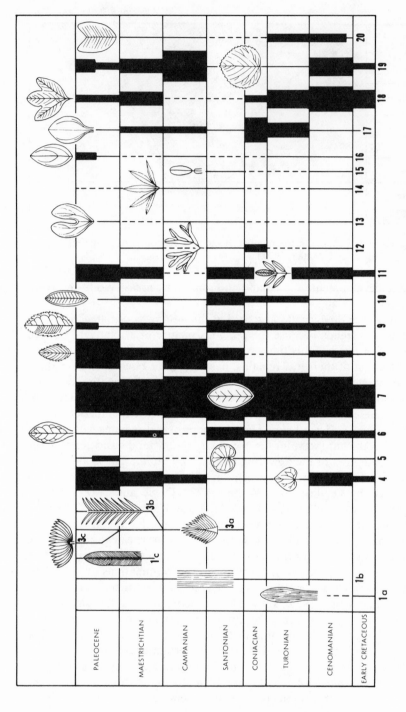

Figure 10-4. Relative abundance of various leaf architectural groups through the Upper Cretaceous and Paleocene based on the fifteen floras analyzed in Table 10–1. The number under each column refers to the description of the leaf given in the table.

tifies the group on this table and figure. Figure 10-4 also includes a figure of each leaf type illustrating its salient features in outline form.

Elongate leaves with cross-veins and at least two orders of rigidly parallel venation (1b in Table 10-1 and Figure 10-4)—characters found principally among monocotyledonous leaves—appeared in the Cenomanian (see Teicheira, 1948; 1950). Pollen with monocotyledonous features (*Liliacidites,* Doyle, 1973), however, and *putatively monocotyledonous* leaves of spatulate and saggitate form (2) as well as an elongate leaf type with irregularly parallelodromous venation (1a) are known from Lower Cretaceous strata (Doyle, 1973; Doyle and Hickey, 1976; Hickey and Doyle, 1977). Monocotyledonous leaves with a parallel vein-set emerging pinnately from a mid-costa (1c), such as characterize the modern Ginger Order (Zingiberales), are not encountered until the Maastrichtian. Pilicate leaves with parallel venation in several orders and cross-veins traversing several of the parallel set (palmophylls of Krassilov, 1977) apparently belong to the palms, and they appeared during the early Senonian synchronously with the first record of palm pollen. At first, these leaves are of the costapalmate type (3a), that is, with a hastula or projection of the petiole usually on the upper surface of the leaf, a feature which characterizes the more primitive subfamilies of palms such as the Coryphideae (Moore and Uhl, 1973). Pinnate palm leaves (3b) are first seen in Maastrichtian sediments[1] and completely palmate forms (3c) first, in low numbers, during the Paleocene and more abundantly in the Eocene.

Among leaves of the dicotyledons, a number of important architectural types had become established in mid- and late Albian times continued a vigorous radiation through the Late Cretaceous. In addition to these were a number of other types, quantitatively less important but occasionally of rather bizarre form (Figure 10-4), that gave a distinctive aspect to the angiosperm flora of that time.

Simple leaves with entire margins and pinnate venation (7) remained the single most important element of the angiosperm flora throughout the Late Cretaceous. Features such as leaf shape and the spacing, angle of origin, and marginal behavior of the secondary veins allow subdivision of this generalized group, and certain combinations of these features tend to be associated with particular higher systematic categories of dicots today (Hickey and Wolfe, 1975). In this review, however, only one type, consisting of simple leaves with disorganized (that is, "low rank"; see Hickey, 1971, 1977) pinnate venation and "festooned brochidodromous" (Mou-

---

[1] This constitutes a range extension from that which I previously reported in Doyle, 1973.

ton, 1970) secondary veins (6) is separated, because this morphology represents the most primitive leaf form seen in the angiosperm record (Doyle and Hickey, 1972; Hickey and Doyle, 1972; Wolfe, 1972; Wolfe, Doyle and Page, 1975; Doyle and Hickey, 1976; Hickey and Doyle, 1977) and is found today only in dicots regarded as primitive on the basis of other features (Hickey, 1971; Doyle and Hickey, 1976; Hickey, 1977). This architectural type persisted through the Late Cretaceous virtually unchanged in numbers, then declined in the Paleocene. It is inferred to represent the survival of a generalized stock or primitive dicots, many probably vesselless, of the lower magnoliid grade of evolution.

Cordate leaves with actinodromous or acrodromous primary venation (4,5) undergo a minor radiation in the Cenomanian and a second period of diversification in the latest Cretaceous and Paleocene after a diminution in the medial Late Cretaceous. It is improbable that this group represents a single systematic entity and uncertain even if the Cenomanian and latest Cretaceous forms are closely related. Many of the latest Cretaceous and Paleocene leaves and associated fruiting structures, however, have their closest morphological counterparts among the present-day hamamelid orders Trochodendrales and Cercidiphyllales.

Palmately lobed leaves (18) of the platanoid or platanophyll type (Krassilov, 1977) are diverse and abundant in the angiosperm record of the Late Cretaceous. An important characteristic of the group is the general retention of palinactinodromous primary venation. Other diagnostic features are rigidly organized, transverse tertiary veins with subparallel quaternaries. Two distinct maxima are apparent in the Late Cretaceous and Paleocene range of the platanophylls (Figure 10–4). The earlier of these was already vigorously underway in the late Albian before its peak in the Cenomanian and gradual decline into the early Senonian. This radiation was characterized by a number of unusual lobate and even peltate genera described as *Protophyllum, Aspidiophyllum,* and *Pseudoprotophyllum,* in addition to more conservative, lobate forms such as *Credneria.* The second radiation is relatively minor and may represent only the continuation of certain members of the previous one. It is at its climax in the Maastrichtian Stage and gradually tapers off into the Paleocene. This secondary climax consists of forms whose morphology agrees more closely with that of the modern genus *Platanus,* including the genus *Credneria,* leaves of *Platanus* itself (Krassilov, 1973a), and a palmately compound form improperly identified as *Cissus.*

In addition to the platanophylls, a group of simple, pinnately veined leaves inferred to have been derived from them by suppression of their lateral primary veins and lobes (19) also makes up a significant portion of the angiosperm megaflora of the Late Cretaceous. The inference of derivation is supported by crowding of the basal secondaries, the presence of outer branches on these, and details of the higher order venation and marginal teeth (see Hickey and Doyle, 1977). Cenomanian genera such as *Betulites* and *Populites* are probably members of this line, as are many of the Maastrichtian and Paleocene forms referred to as *Viburnum.* The history of this group parallels that of the Platanophylls in apparently having two diversity peaks in the Late Cretaceous, except that the later peak is the more important.

Simple, pinnately veined leaves with toothed margins and craspedodromous secondary veins (8) undergo a marked fluctuation during the latter half of the Late Cretaceous. Since the time of Bailey and Sinnott (1915) a gain in the proportion of non-entire leaves in a fossil flora has been interpreted as evidence of climatic cooling. Although the proportions of such leaves are higher in the latest Cretaceous and increases from the Maastrichtian into the Paleocene, the validity and significance of these trends cannot be determined from the sample analyzed.

Simple leaves with closely spaced or semi-craspedodromous secondaries and closely spaced marginal serrations, here termed the celastrophyll type (9), appeared in the late Albian (see Vakhrameev, 1952; Doyle and Hickey, 1976) and persisted as a minor element through the Late Cretaceous. If at least part of this type is represented in the Pinnate Dilleniidae (Hickey and Wolfe, 1975), as its morphology suggests, then the climatic requirements of their present analogues, as well as the virtual absence of the group from higher latitude floras in Table 10–3 suggest that their decline into the lower Paleocene may have been due to climatic cooling. These forms began to increase again in the Late Paleocene and Eocene, a shift correlated with warming climates (Wolfe and Hopkins, 1967). The percentage of pinnately compound leaves (11) in these floras, although probably lower than in actuality because of the tendency of individual leaflets to become detached before fossilization, remains remarkably constant through the Late Cretaceous and Paleocene. Peltate, actinodromously veined leaves (5) of probable affinity to the modern orders Nymphaeales and Nelumbonales occurred in low numbers

| | LATE CRETACEOUS | EARLY PALEOCENE | % CHANGE INTO PALEOCENE |
|---|---|---|---|
| FORMATION | Lance-Fox Hills-Medicine Bow | Lower Fort Union | |
| SPECIES | 82 | 94 | +15 |
| FORMATION | Hell Creek | Tullock | |
| SPECIES | 25 | 24 | −4 |

Table 10–3. Change in megafossil diversity from Late Cretaceous into Early Paleocene strata in Wyoming and Montana.

through the Late Cretaceous and seem to show an affinity for higher latitude areas and inferred intervals of cooler climate.

Minor leaf types of no numerical importance but especially characteristic of the Late Cretaceous show two patterns of distribution. The first
group consists of genera apparently related to the bizarrely lobate *Fon-
tanea* (12), the secondarily simple genus *Citrophyllum* (15) and the *Lirio-
dendropsis* line (20) and died out by the end of the Cretaceous Period.
Members of the second group persisted into the Paleocene or even later
before disappearing. Among these is a line, of which *Myrtophyllum* (10) is
a characteristic genus, having simple, entire leaves with well-developed
intramarginal veins. At least some of this complex probably belong to
either the Ochnalean line of the Dilleniidae (Hickey and Wolfe, 1975) or
to the Myrtales. The so-called genus *Bauhinia* (13) with its unusual,
deeply emarginate shape has been identified in Early Paleocene sediments (Brown, 1962) although its relationship to the Cenomanian to
Santonian *"Bauhinia"* line is not definitely established. Another distinctive group that persists into the Paleocene with a diversity peak in the
Turonian and Coniacian (?) is recognized by leaves with acrodromous
primary veins that radiate from the leaf base but remain separate far
down into the petiole (17). Today such morphology is found only in the
primitive Rhamnaceae and in the Coriariaceae. A fourth type consists of
unusual palinactinodromously compound leaves called by the names *De-
beya* and *Dewalquia* (14), which range from the Cenomanian to the late
Oligocene. Early members of the line average five leaflets per leaf, while
later members generally have three. The simple-leaved genus *Dryophyl-
lum* of latest Cretaceous and Early Tertiary age and the modern order
Fagales may be derivatives of this line (Rüffle, 1978). A final type that
consists of simple leaves with pinnate secondary veins arranged in ascending concentric eucamptodromous arches, might be termed corno-
phylls (16) after the typical expression of this architecture in the modern
genus *Cornus*. This line persists in low numbers through the Late Cretaceous and Paleocene.

As yet, my analysis of megafossil distributions is not complete enough
to permit provincial differences or latitudinal gradients to be recognized
in leaves as in pollen. The few critical comparisons of Late Cretaceous
angiosperm leaves which have been made, however, indicate that forms
with the architecture of modern families, and rarely genera, began to appear gradually during the Campanian and Maastrichtian within these
leaf morphological groups. For example, in the Cenomanian, the platanophyll complex contained a large number of genera and had a far greater
variety of architectural features than are currently found in the Hama-
melididae or the basal Rosidae (Saxifragales) where its closest morphological affinities lie. In the latter half of the Late Cretaceous, however,

leaves (Krassilov, 1973a) and associated pollen (Pacltová, 1978) that are closely comparable to those of the modern genus *Platanus* begin to appear. In all the lines which were to persist, a similar gradual modernization would continue into the Eocene.

## PATTERNS OF CHANGE ACROSS THE CRETACEOUS-TERTIARY BOUNDARY

Data from 32 floral assemblages were used in the quantitative summary of floral change across the K/T boundary tabulated in Table 10–2 and shown in Figures 10–5, 10–6, and 10–7. The changes are expressed as the percentage of Cretaceous forms persisting into the Paleocene and as the

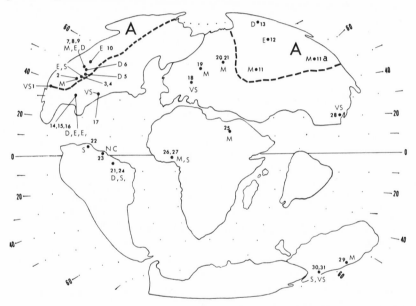

Figure 10–5. Severity of plant extinction (mostly among angiosperms) across the Cretaceous-Paleocene boundary.

The numbers beside the localities are keyed to Table 10–2. The dashed line represents the approximate limits of the Aquilapollenites Province (designated by the letter A) covering northwest North America and northeast Asia. The letters indicate the range of extinction, as follows:

NC no change;
VS very slight, 90–99% survival of Cretaceous forms;
S slight, 75–90% survival;
M moderate, 50–75% survival;
E extensive, 25–50% survival;
D drastic, less than 25% survival.

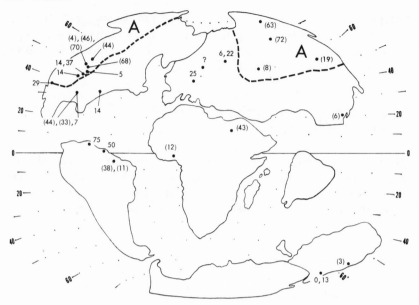

Figure 10-6. Changes in diversity across the Cretaceous-Paleocene boundary.
The localities and designation of the Aquilapollenites Province are the same as those in Figure 10-5. In this case a number standing alone indicates the percentage of increase occurring at the locality; while a number in parenthesis indicates a percentage decrease.

percentage increase or decrease in floral diversity from that of the Cretaceous. Localities are grouped by area, as follows: western North America and Siberia, coinciding roughly with the Aquilapollenites Pollen Province of the Late Cretaceous; eastern North America and Europe, coinciding with the Normapolles Province; and then South America; Africa; the East Indies; and Australia-New Zealand. Within each group, localities are arranged in order of increasing paleolatitude, using figures derived from plotting their positions on the Paleocene continental drift position maps given in Smith and Briden (1977).

Even though the persistence and diversity comparisons across the boundary are subject to the limitations discussed above (see also Tschudy, this volume) some patterns are apparent from them. Most important is that the highest level of extinction generally occurs within the Aquilapollenites Province. Elsewhere, the rate of extinction is moderate to slight except for drastic levels at one locality in the Mississippi embayment (Tschudy, 1970) and one in Brazil (Boer, van der Hammen, and Wymstra, 1965). Jarzen (1978) has suggested that a change in facies from terrestrial to marine is the cause of the anomalously high level of extinc-

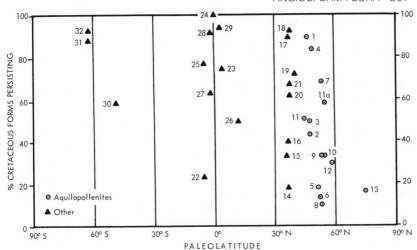

Figure 10-7. Survival of Cretaceous land-plant species (mainly angiosperms) across the K/T boundary as a function of paleolatitude.

Localities falling in the Aquilapollenites Province have been distinguished from those in other regions. Reference numbers beside the points are keyed to the entries in Table 10-2. Note the generally high levels of floral survival from high southern to mid-northern paleolatitudes. (Paleolatitudes from Smith and Briden, 1977.)

tion at the Mississippi embayment locality. The extensive amount of extinction in the Normapolles group there, which Tschudy (1975) has also reported, probably reflects facies bias in samples as well as the general decline of this archaic line taking place during Maastrichtian and Paleocene times. The Brazilian figures are from a small flora where the emphasis appears to have been on the distribution of guide fossils rather than on the total pollen spectrum; a later study in this area by Müller (1966) reported that more than three-quarters of the Cretaceous forms survived. Nevertheless, the Mississippi Embayment and northeastern Brazilian localities were included for the sake of objectivity.

Within the Aquilapollenites Province, there seems to be an imperfect pattern of increase in the severity of angiosperm extinction toward the north. This trend suggests that a climatic deterioration accompanied by a steepened temperature gradient from south to north may have occurred at the end of the Cretaceous. The admittedly weak trend shown by the survival data in the Aquilapollenites Privince in Figure 10-7 is reinforced by a number of paleobotanical reports from the region. Srivastava (1970) has proposed a climatic deterioration accompanied by the onset of more seasonal regimes during the Maastrichtian of central Alberta, which he ascribed to the withdrawal of the Mid-Continental Seaway. His supposition, however, was partially based on the ecology of modern forms inferred to have relationship with the fossils, a procedure which may be suspect for such a remote time. Smiley (1972) has inferred a shift to cooler and moister climates in northern Alaska during the Maastrichtian as indicated by a decrease in dicot pollen from 50 to 20% of the flora with increases in fern spores from 4 to 20% and in conifer pollen from 20 to 60%.

Although authors such as Hughes (1976) and McLean (1978) have hypothesized a drastic short-term temperature increase as a possible cause of the latest Cretaceous extinctions, the preponderance of evidence from both biologic and isotopic sources favors a temperature decline at or near the K/T boundary (Dorf, 1959; 1963; 1969; Voight, 1964; Krassilov, 1973b; 1978; Savin et al., 1975; Van Valen and Sloan, 1977; Boersma and Shackleton, 1978; and Jarzen, 1978). A short-term temperature increase at the boundary (Boersma et al., 1979; Boersma and Shackleton, 1979) could not be observed at the time scale of the floral studies reviewed here and does not seem to have had a drastic effect on plant extinction levels, in any case.

Throughout the rest of the world, floras show relatively little disturbance across the K/T boundary except in the cases noted above (Figure 10-7). Only a very slight change occurs in the angiosperm pollen assemblages of the Atlantic Coastal Plain of eastern North America; slight to moderate changes in Europe, over most of South America, Africa, Borneo, and in Australia and New Zealand (Table 10-2). Diversity figures tend to fluctuate with less regularity (Table 10-2 and Figure 10-6), but the largest and most consistent pattern of decrease across the boundary occurs within the confines of the Aquilapollenites Province. It should be noted that these are "worst case" figures including components added by continuous extinction, facies change, and local extinction.

Since the studies by Dorf (1940), the megafloral record of the Western Interior United States has been interpreted as indicating a drastic change across the K/T boundary. Table 10-4 shows that Dorf found only 16% of his Lance species surviving into Fort Union strata of Paleocene age. An update of these findings based on species revisions since that time and on a more complete knowledge of forms occurring in the lower Fort Union Formation (especially from Brown, 1962; and my own data) shows that while the change across the boundary is substantial, it is not merely so great as originally thought. Thus, 46% of Lance species persist into the Paleocene as do the same proportion of the combined Lance-Fox Hills-lower Medicine Bow species. A recent study by Shoemaker (1966), showing 68% persisting, is in line with these figures as well. A comparison of the megafloral diversity of the Lance Medicine Bow and Hell Creek For-

ANGIOSPERM SPECIES

| FORMATION | TOTAL | K only | K-Paleo | % Persisting |
|---|---|---|---|---|
| Type Lance (Dorf, 1940, fig. 2) | 55 | 48 | 7 | 16 |
| Reanalysis of Lance | 52 | 28 | 24 | 46 |
| Fox-Hills-Lower Medicine Bow | 50 | 23 | 27 | 54 |
| Combined Lance-Fox-Hills-Medicine Bow | 82 | 44 | 38 | 46 |
| Hell Creek (Shoemaker, 1966) | 25 | 8 | 17 | 68 |

Table 10–4. Persistence of Cretaceous megafossil angiosperm species from the Lance, Fox Hills, lower Medicine Bow, and Hell Creek Formations into the Paleocene.

Under "Angiosperm species" are tabulated the total number of species used in each flora, the number restricted to the Cretaceous (under "K only"), the number persisting into the Paleocene (under "K-Paleo") as well as the percentage of the flora made up of those persisting. The reanalysis was based on the species described by Dorf, 1942 and 1938. A detailed tabular summary of the species in these floras, their ranges, and current taxonomic status is available from the author upon request.

mations with that of the lower Fort Union Formation and its basal Tullock Member shows small and probably insignificant changes into the Paleocene (Table 10–3). An analysis of leaf margin percentages in these floras indicates the same general trend toward increase in non-entire forms as in the overall analysis of the Late Cretaceous and Paleocene records (Table 10–5). As stated above, this trend is compatible with inferences of climatic cooling at that time.

Where stratigraphic sections containing both the last dinosaurs and plants have been carefully observed—as it turns out only in western North America—the Cretaceous flora persists above the level of the highest unreworked dinosaur bone. In Alberta, the floral change occurs 6 meters higher (Lerbekmo et al., 1979). In east-central Montana (Clemens and Archibald, 1980), in northwestern Wyoming (my observations as well as those of P. Gingerich, pers. comm.) and in southwestern Wyo-

| FORMATION | No. of Sp. | Non-Entire | % Non-Entire |
|---|---|---|---|
| Lance-Fox Hills & lower Medicine Bow | 65 | 22 | 34 |
| Hell Creek | 16 | 11 | 68 |
| Tullock | 14 | 10 | 71 |
| Lower Fort Union | 81 | 48 | 59 |

Table 10–5. Change in leaf margin percentages from the Late Cretaceous Lance, Medicine Bow, and Hell Creek Formations into the Early Paleocene strata of the Lower Fort Union Formation and its Tullock Member.

ming (Leffingwell, 1970), the change occurs a minimum of 2 to 3 meters higher. In Colorado the change lies an unspecified distance above the last dinosaur (Newman, 1979). Based on a sedimentation rate of 65 meters/myr for the Edmonton Group (Van Valen and Sloan, 1977), the stratigraphic interval between the last dinosaur and the floral change would represent approximately 50 to 90 thousand years. In addition, several authors noted an attenuation in dinosaur abundance and diversity toward the boundary (Clemens and Archibald, 1980; Van Valen and Sloan, 1977), a trend that I also found in my field work in northwestern

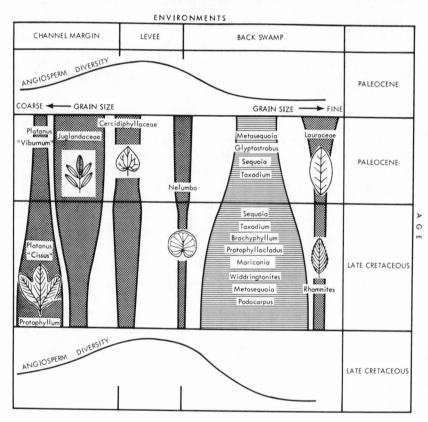

Figure 10–8. Plant community relationships across the Cretaceous-Paleocene boundary.

Inferred environments are listed at the top, with the average grain size of their sediments below. The relative diversity of angiosperm from near-channel to back-swamp environments is indicated by the curves. Only the major leaf architectural types found in each environment are shown. The position of each type remains stable across the boundary with conifers gradually becoming less diverse and more modern in aspect.

Wyoming. Such a pattern argues for considering Late Cretaceous extinctions to be a continuous process rather than the result of a single episode at the boundary.

Megafloras are more sensitive indicators of community ecology and distribution than are pollen and spores (Hickey, 1976; Hickey and Doyle, 1977). A striking characteristic of the megafloral assemblages across the K/T boundary is the unbroken continuity in the association of particular leaf types with certain lithofacies, a phenomenon believed to be the result of the absence of severe disturbance of the plant communities into the Paleocene epoch. A pattern of high angiosperm diversity, including platanophylls and compound leaves, in levee and near-channel deposits and of conifer dominance in backswamp sediments was established in lowland fluvial deposits as early as the Albian Stage before the achievement of overall angiosperm dominance (Doyle and Hickey, 1976; Hickey and Doyle, 1977). This pattern was still apparent at the close of the Late Cretaceous with platanophylls dominating the near-channel stream deposits, a diverse mixture of compound- and simple-leafed angiosperms of archaic and more rarely modern affinities on inferred levee and crevasse splay deposits, and a mixture of archaic and modern conifer genera such as *Moriconia, Protophyllocladus, Brachyphyllum, Sequoia, Metasequoia, Rhamnites* dominant in the backswamp (Parker, 1975). This same pattern persists into and through the Paleocene, with further enrichment and modernization of the angiosperm component, extinction of the archaic conifers, and the intrusion of some additional flowering plants into the backswamp (Figure 10-8).

## THE SIGNIFICANCE OF THE OBSERVED CHANGES

From the data summarized above, it appears that the main trends in angiosperms after the Cenomanian were increasing provinciality and the gradual introduction of forms of modern familial and even generic affinity. The Maastrichtian marks a time of gradual acceleration in the trend toward floral modernization and, at least in western North America, of decrease in total angiosperm diversity and a shift toward forms of cooler and more seasonal climates (Srivastava, 1970; Smiley, 1972).

Floral change occurring at the K/T boundary was relatively minor and geographically variable on a world-wide scale (Table 10-2, Figures 10-5, 10-6, 10-7; Penny, 1969; Jarzen, 1977b). The most noteworthy changes were the decimation of the Aquilapollenites Province and its elimination as a distinct floral region. The highest levels of extinction seem to have occurred within this province with a suggestion of increasing severity toward the north (Figure 10–7). Throughout the rest of the world, the magnitude of floral change was much less, seldom exceeding 50% of Cretaceous forms. In addition, equatorial and Southern Hemisphere floras were relatively little affected (Figure 10–7) in contrast to the situation in the marine realm where Tethyan communities showed the highest levels of extinction (Kauffman, 1979; this volume).

The plant extinction figures reported here include extinctions that occurred at the boundary as well as those that occurred continuously during the latest Cretaceous. By way of comparison, the same method of analysis was applied to extinction across the Paleocene-Eocene boundary, where no catastrophe has ever been invoked. In North Dakota one-third of the palynoflora (Bebout, 1977) and one-half of the megaflora (Hickey, 1977) became extinct in the final 10 meters of the Paleocene section. Similar levels of extinction are seen in comparing the total Clarkforkian (latest Paleocene) with the total Wasatchian (Early Eocene) megaflora of the Bighorn Basin in Wyoming (Wing, in preparation).

These findings make it unnecessary for the paleobotanist to invoke a catastrophic or unusual mechanism to explain the observed patterns of localized extinctions and apparent increase in seasonal-climate forms seen in angiosperms at mid- and higher latitudes in the Northern Hemisphere. Conversely, these data require that any drastic mechanism hypothesized to explain the extinctions of the dinosaurs and various groups of marine organisms be compatible with the relatively low levels of extinction in the plant record.

Nevertheless, a number of relatively spectacular scenarios have been proposed over the past 15 years, that identify a cosmic or extraterrestrial catastrophe as the cause of biotic extinction at the end of the Cretaceous. Most recently Alvarez et al. (1979; 1980a, 1980b) have hypothesized that the impact of an asteroid, of some 6 to 14 kilometers in diameter, threw up a dust cloud that effectively cut off sunlight and stopped photosynthesis for several years, thereby causing the collapse of the earth's food chain. As this dust settled and was incorporated into the sedimentary record it bore with it an anomalously high content of the platinum-group metal iridium. Alvarez et al. also have speculated that frictional heating of the atmosphere may have been a cause of extinction.

A second hypothesis by Hsü (1980) invokes a cometary collision with the earth as the terminal Cretaceous extinction mechanism.. Hsü has speculated on the possibility of the poisoning of the earth's ecosystem by material such as cyanide in the head of the comet as well as frictional heating of the atmosphere to as much as 190° C.

Until the advent of the asteroid and comet hypotheses, the most active area of speculation on catastrophic mechanisms involved hypotheses of a supernova or a solar flare that subjected the earth to a geologically brief period of elevated radiation (Terry and Tucker, 1968; Russell and Tucker; 1971; Roy, 1977; Tucker, 1977). Although the probabilities of such occurrences as well as the levels and duration of the resultant radiation received by the earth are all open to serious question (Laster, 1968; Tucker and Terry, 1968; Feldman, 1977), only the postulated terrestrial and biological effects are examined in this summary. Three basic scenarios have been proposed for the operation of such bursts of radiation on the earth:

1) long-term radiation—gamma radiation lasting for months to tens of years at levels of several hundred to several thousand roentgens per year, followed by levels of a hundred roentgens per year persisting for thousands of years (Tucker, 1977);

2) short-term radiation—a burst or flash of gamma radiation lasting only hours or days and carrying levels from 15 roentgens in 3 hours (Tucker, 1977) to thousands of roentgens a day (Terry and Tucker, 1968);

3) very short-term climatic deteriorations caused by stripping of the earth's ozone layer due to chemical reactions caused by a greatly increased x-ray or ultraviolet flux (Russell and Tucker, 1971; Ruderman, 1974).

In the case of all three hypotheses, the relative continuity of the angiosperm record across the K/T boundary; the generally moderate and geographically variable levels of land-plant extinction; and the lack of synchroneity in plant-dinosaur extinctions make it highly unlikely that a universal biotic catastrophe occurred. Blockage of solar radiation, massive injections of poison, or intense heating should have had a drastic and universal effect on vegetation that simply is not seen in plant survival figures, biased as they are by representing total rather than instantaneous extinction.

In similar fashion, these figures effectively rule out long-term exposure to radiation as a cause of terminal Cretaceous extinction because the susceptibility of forest plants and of woody and late successional vegetation, in general, over periods of months to years is approximately the same as for modern larger vertebrates—and, presumably, for dinosaurs (Woodwell, 1963; 1967; Armentano and Woodwell, 1976; Bond et al., 1965; Cronkite and Bond, 1960). Indeed, it is the occurrence of a substantial number of localities where the flora survived virtually intact (Figures 10–5, 10–7) that most effectively falsifies these hypotheses.

However, the most ingenious of these proposals have sought to decouple plants from animals in their responses to catastrophe and thus to explain the paradox of land plant survival during the wave of animal extinction. Thus some authors (Terry and Tucker, 1968; Russell and Tucker, 1971; Alvarez et al., 1980) called upon survival from seeds, dormant buds, and subterranean parts to carry plants over the relatively brief period of lethal conditions from which most animals have no equivalent refuge. Terry and Tucker (1968) have correctly pointed out that plants are much less susceptible to mortality from short-term radiation—by a factor of 10 over animal susceptibility or over their own chronic (long-term) susceptibility—and seeds are even more resistant (Woodwell, 1963).

Although validly based, these arguments fail to account for the pattern of floristic change seen at the close of the Cretaceous. Despite a lower thermal gradient at that time (Gartner, 1979; Saito and Van Donk, 1974) due to lack of polar ice caps and a generally higher average earth temperature, Northern Hemisphere floras show an increase of cooler temperature forms with inreasing latitude (Savin, 1977), though more gradually than at present. This suggests that, by analogy with today's trends, the means by which plants dealt with intervals of environmental stress, including physiological and morphological dormancy mechanisms, protective coverings, cryptic placement of early developmental phases, tissue desiccation, chromosomal redundancy (polyploidy), would have been relatively poorly developed in the tropics and increasingly important at higher latitudes (Raunkiaer, 1934; Cain, 1950; Woodwell, 1963). These are the same mechanisms that confer elevated radiation resistance on plants (Woodwell, 1963). Given such a latitudinal increase in the efficacy of such dormancy and carry-over mechanisms, the pattern of relatively elevated levels of extinction at higher northern altitudes and relatively low levels of damage in tropical latitude floras is exactly the opposite of what would be expected from the various catastrophic models (Figures 10–5, 10–6, 10–7).

A modification of the hypothesis of Alvarez et al., (1980) involving an attenuated dust cloud, rather than a completely opaque one, might be developed to fit a part of the pattern seen in the plant data. This modifiction would have caused relatively less light to reach high latitudes due to the greater screening power of dust at the more oblique angles of insolation that occur there. The result would have been a steepened gradient of declining temperatures toward higher latitudes and an attendant increase in plant extinctions. However, the effect of such a dust cloud would have been of very short duration. In addition, this modified version of the Alvarez hypothesis fails to account for the lack of appreciable extinction or decline in diversity at high southern latitudes (Figures 10–5, 10–6, 10–7) or the irregular occurrence of such effects elsewhere.

Although the angiosperm data reviewed here set limits on the agents which can be invoked for biotic extinction at the end of the Cretaceous, they provide no clear insights into the true causes. On the other hand, the pattern of change in land plants, the increasingly cooler affinities of latest-Cretaceous to Early Paleocene palynofloras (Savin, 1977; Samilovich, 1967; Smiley, 1972; Srivastava, 1970; and Krassilov, 1978), and the increase in the percentage of toothed leaves in floras from the western United States suggest that a climatic deterioration was underway at the end of the Cretaceous. A paleotemperature curve for Montana and Wyoming suggested by leaf margin data (Hickey, 1980) and numerous marine paleotemperature curves (Savin, 1977) also point to climatic cooling across the K/T boundary with gradual recovery late in the Paleocene. The decline of the dinosaurs and the stratigraphic separation of plant and dinosaur extinctions are consistent with a climatic-cooling model.

In addition, the withdrawal of eperic seaways that was under way at the end of the Cretaceous would tend to accentuate seasonality as well as to lower overall taxonomic diversity by combining formerly separate floral provinces. During the Late Cretaceous, the gradual rise to dominance of the flowering plants caused fundamental changes in the type and availability of plant food resources, in successional patterns, and in the whole fabric of coevolutionary relationships of plants and animals. Coupled with the added stresses of climatic deterioration and marine withdrawal, these changes may have been enough to trigger a rapid collapse of the dominant land vertebrates.

In conclusion, although a catastrophic cause for the widespread extinctions at the end of the Cretaceous appears unlikely, the problem of the extinctions remains. Before a general theory explaining these extinctions can be framed, additional data must be gathered and evaluated on the timing and magnitude of extinctions as well as on changes in the physical environment across the K/T boundary.

## LITERATURE CITED

Alvarez, L. W., W. Alvarez, F. Asaro, and H. V. Michael, 1979. Extraterrestrial cause for the Cretaceous-Tertiary extinction: experiment and theory. *Calif. Univ., Lawrence Berkeley Laboratory Publ.*, 9666, 84 pp.
———, 1980. Extraterrestrial cause for the Cretaceous-Tertiary extinction. *Science*, 208:1095–1108.
Alvarez, W., L. W. Alvarez, F. Asaro, and H. V. Michael, 1979. Anomalous iridium levels at the Cretaceous/Tertiary boundary at Gubbio, Italy: negative results of a test for a supernova origin. In: W. K. Christensen and

T. Birkelund (eds.), *Cretaceous-Tertiary Boundary Events. Copenhagen Univ., Proc. Sympos. "Cretaceous-Tertiary Boundary Events"*, vol. 2, pp. 50–53.

Armentano, T. V., and G. M. Woodwell, 1976. The production and standing crop of litter and humus in a forest exposed to chronic gamma irradiation for twelve years. *Ecology,* 57:360–366.

Bailey, I. W., and E. W. Sinnott, 1915. A botanical index of Cretaceous and Tertiary climates. *Science,* 41:832–833.

Bebout, J. W., 1977. *Palynology of the Paleocene-Eocene Golden Valley Formation of Western North Dakota.* Penn. State Univ. Dept. of Geosciences, unpubl. Ph.D thesis, 391 pp.

Berry, E. W., 1911. The flora of the Raritan Formation. *New Jersey Geol. Surv. Bull.,* 3, 233 pp.

————, 1914. The Upper Cretaceous and Eocene floras of South Carolina and Georgia. *U. S. Geol. Surv. Prof. Paper,* 84, 200 pp.

————, 1916. Angiospermophyta. In: *Maryland Geol. Surv.,* pp. 806–901; 952–987.

————, 1919. Upper Cretaceous floras in the eastern Gulf region in Tennessee, Mississippi, Alabama, and Georgia. *U. S. Geol. Surv. Prof. Paper,* 112, 177 pp.

————, 1925. The flora of the Ripley Formation. *U. S. Geol. Surv. Prof. Paper,* 136, 94 pp.

Boer, N. P. de, T. van der Hammem, and T. A. Wymstra, 1965. A palynological study on the age of some borehole samples from the Amazonas Delta area, N. W. Brazil. *Geologie en Mijnbouw,* 44:254–258.

Boersma, A., and N. Shackleton, 1978. Some oxygen and carbon isotope variations from late Campanian across the Cretaceous/Tertiary Boundary at five Atlantic Ocean DSDP sites. *Geol. Soc. Amer. Abstr. Progr.,* 10:368.

Boersma, A., N. Shackleton, M. Hall, and Q. Given, 1979. Carbon and oxygen isotope records at DSDP site 384 (North Atlantic) and some Paleocene paleotemperatures and carbon isotope variations in the Atlantic Ocean. *Initial Reports of the Deep Sea Drilling Project* (Washington, D. C.: U. S. Govt. Printing Office), vol. 43, pp. 695–718.

Bond, V. P., T. M. Fliedner, and J. O. Archambeau, 1965. *Mammalian Radiation Lethality* (New York: Academic Press), 340 pp.

Brenner, G. J., 1976. Middle Cretaceous floral provinces and early migrations of angiosperms. In: C. B. Beck (ed.), *Origin and Early Evolution of Angiosperms* (New York: Columbia Univ. Press), pp. 23–47.

Brown, R. W., 1962. Paleocene flora of the Rocky Mountains and Great Plains. *U. S. Geol. Surv. Prof. Paper,* 375, 119 pp.

Cain, S. A., 1950. Life-forms and phytoclimate. *Bot. Rev.,* 16:1–32.

Clemens, W. A., and J. D. Archibald, 1980. Evolution of terrestrial faunas during the Cretaceous-Tertiary transition. *Soc. Géol. France Mém.,* 139:67–74.

Cobban, W. A., and J. B. Reeside, Jr., 1952. Correlation of the Cretaceous formations of the Western Interior of the United States. *Geol. Soc. Amer. Bull.,* 63:1011–1043.

Couper, R. A., 1960. New Zealand Mesozoic and Cainozoic plant microfossils. *New Zealand Geol. Surv. Pal. Bull.*, 32, 87 pp.

Cronkite, E. P., and V. P. Bond, 1960. *Radiation Injury in Man* (Springfield, Ill.: Thomas), 200 pp.

Dilcher, D., 1974. Approaches to the identification of angiosperm leaf remains. *Bot. Rev.*, 40:1–157.

Dorf, E., 1938. Upper Cretaceous floras of the Rocky Mountain region. 1: Stratigraphy and palaeontology of the Fox Hills and lower Medicine Bow Formations of southern Wyoming and northwestern Colorado. *Carnegie Inst. Wash. Publ.*, 508:1–78.

———, 1940. Relationship between floras of type Lance and Fort Union Formations. *Geol. Soc. Amer. Bull.*, 51:213–236.

———, 1942. Upper Cretaceous floras of the Rocky Mountain region. 2: Flora of the Lance Formation at its type locality, Niobrara County, Wyoming. *Carnegie Inst. Wash. Publ.*, 508:79–159.

———, 1959. Climatic changes of the past and present. *Michigan Univ. Mus. Pal. Contrib.*, 13:181–210.

———, 1963. The use of fossil plants in palaeoclimatic interpretations. In: A. E. M. Nairn, *Problems in Palaeoclimatology* (London: Interscience), pp. 13–48.

———, 1969. Paleobotanical evidence of Mesozoic and Cenozoic climatic changes. *Paleoclimatology, N. Amer. Paleontol. Conv. Proc., Part D*, pp. 323–346.

Doyle, J. A., 1969. Cretaceous angiosperm pollen of the Atlantic Coastal Plain and its evolutionary significance. *Jour. Arnold Arb.*, 50:1–35.

———, 1973. Fossil evidence on early evolutions of the monocotyledons. *Quart. Rev. Biol.*, 48:399–413.

Doyle, J. A., P. Biens, A. Doerenkamp, and S. Jardine, 1977. Angiosperm pollen from the pre-Albian Lower Cretaceous of Equitorial Africa. *Soc. Natl. Petrol. Aquitaine, Cent. Rech. Pau, Bull.*, 1:451–473.

Doyle, J. A. and L. J. Hickey, 1972. Coordinated evolution in Potomac Group angiosperm pollen and leaves. *Amer. Jour. Bot.*, 59:660 (abstract.)

———, 1976. Pollen and leaves from the mid-Cretaceous Potomac Group and their bearing on early angiosperm evolution. In: C. B. Beck (ed.), *Origin and Early Evolution of Angiosperms* (New York: Columbia Univ. Press), pp. 139–206.

Doyle, J. A., M. Van Campo, and B. Lugardon, 1975. Observations on exine structure of *Eucommiidites* and Lower Cretaceous angiosperm pollen. *Pollen et Spores*, 17:429–486.

Drugg, W. S., 1967. Palynology of the upper Moreno Formation (Late Cretaceous-Paleocene) Escarpado Canyon, California. *Palaeontographica*, 120 (B):1–71.

Feldman, P. A., 1977. Astronomical evidence bearing on the supernova hypothesis for the mass extinctions at the end of the Cretaceous. *Syllogeus*, 12:125–135.

308 HICKEY

Germeraad, J. H., C. A. Hopping, and J. Muller, 1968. Palynology of Tertiary sediments from tropical areas. *Rev. Paleobot. Palynol.,* 6:189–348.

Hickey, L. J., 1971. Evolutionary significance of leaf architectural features in the woody dicots. *Amer. Jour. Bot.,* 58:469 (abstract).

———, 1973. Classifications of the architecture of dicotyledonous leaves. *Amer. Jour. Bot.,* 60:17–33.

———, 1976. Relationship of lithofacies to Cretaceous and Tertiary megafloral assemblages. *Bot. Soc. Amer. Abstr. 1976,* p. 26.

———, 1977. Stratigraphy and paleobotany of the Golden Valley Formation (Early Tertiary) of western North Dakota. *Geol. Soc. Amer. Mem.,* 150, 183 pp.

———, 1978. Origin of the major features of angiosperms leaf architecture in the fossil record. *Forsch.-Inst. Senckenberg, Cour.,* 30:27–34.

———, 1980. Paleocene stratigraphy and flora of the Clarks Fork Basin. *Michigan Univ. Mus. Pal. Paper,* 24:33–49.

Hickey, L. J., and J. A. Doyle, 1972. Fossil evidence on evolution of angiosperm leaf venation. *Amer. Jour. Bot.,* 59:661 (abstract).

———, 1977. Early Cretaceous fossil evidence for angiosperm evolution. *Bot. Rev.,* 43:3–104.

Hickey, L. J., and J. A. Wolfe, 1975. The bases of angiosperm phylogeny: vegetative morphology. *Missouri Botan. Garden, Ann. Rept.,* 62:538–589.

Hoeken-Klinkenberg, P. M. J. van, 1966. Maastrichtian, Paleocene and Eocene pollen and spores from Nigeria. *Leidse Geol. Meded.,* 38:37–48.

Hollick, A., 1906. The Cretaceous flora of southern New York and New England. *U. S. Geol. Surv. Monogr.,* 50, 219 pp.

———, 1930. The Upper Cretaceous floras of Alaska. *U. S. Geol. Surv. Prof. Paper,* 159, 123 pp.

Hsü, K. J., 1980. Terrestrial catastrophe caused by cometary impact at the end of the Cretaceous. *Nature,* 285:201–203.

Hughes, N. F., 1976. *Palaeobiology of Angiosperm Origins* (Cambridge: Cambridge Univ. Press), 242 pp.

———, 1977. Paleo-succession of earliest angiosperm evolution. *Bot. Rev.,* 43:105–127.

Imlay, R. W., and J. B. Reeside, Jr., 1954. Correlation of the Cretaceous formations of Greenland and Alaska. *Geol. Soc. Amer. Bull.,* 65:223–246.

Jansonius, J., and L. V. Hills, 1976. Genera file of fossil spores and pollen. *Univ. Calgary Dept. Geol. Spec. Pub.*

Jarzen, D. M., 1977a. Aquillapollenites and some Santalalean genera. A botanical comparison. *Grana,* 16:29–39.

———, 1977b. Angiosperm pollen as indicators of Cretaceous-Tertiary environments. *Syllogeus,* 12:39–49.

———, 1978. The terrestrial palynoflora from the Cretaceous-Tertiary transition, Alabama, U. S. A. *Pollen et Spores,* 20:535–553.

Kauffman, E. G., 1979. The ecology and biogeography of the Cretaceous-Tertiary extinction event. In: W. K. Christensen and T. Birkelund (eds.),

*Cretaceous-Tertiary Boundary Events. Copenhagen Univ., Proc. Sympos.* "*Cretaceous-Tertiary Boundary Events*", vol. 2, pp. 29–37.

Kedeves, M., 1971. Présence de types sporomorphes importants dans les sédiments prequarternaires Egyptiens. *Act. Bot. Sci. Hung.*, 17:371–378.

Knowlton, F. H., 1905. Fossil plants of the Judith River Beds. In: T. W. Stanton and J. B. Hatcher (eds.), Geology and paleontology of the Judith River Beds. *U. S. Geol. Surv. Bull.*, 257, pp. 129–174.

Krassilov, V. A., 1973a. Cuticular structure of Cretaceous angiosperms from the Far East of the USSR. *Palaeontographica*, 142 (B):105–115.

———, 1973b. Novye dannye po flori i fitostratigraphii verchnego Mela Sachalina. In: V. A. Krassilov (ed.), Iskopaemye flori i fitostratigrafiya Dal'nego Vostoka. *Vladivostok, Acad. Nauk, Far East Geol. Inst.*

———, 1977. The origin of angiosperm. *Bot. Rev.*, 43:143–176.

———, 1978. Late Cretaceous angiosperms from Sakhalin and the terminal Cretaceous event. *Paleontology*, 21:893–905.

Krutsch, W., 1957. Sporen- und Pollengruppen aus der Oberkreide und dem Tertiär Mitteleuropas und ihre stratigraphische Verteilung. *Zeits. angewandte Geol.*, 3:509–548.

Laster, H., 1968. Cosmic rays from nearby supernovas: biological effects. *Science*, 160:1138.

Leffingwell, H. A., 1970. Palynology of the Lance (Late Cretaceous) and Fort Union (Paleocene) Formations of the type Lance area, Wyoming. In: R. M. Kosanke and A. T. Cross (eds.), Symposium on palynology of the Late Cretaceous and Early Tertiary. *Geol. Soc. Amer. Spec. Paper*, 127, pp. 1–64.

Leidelmeyer, P., 1966. The Paleocene and Lower Eocene pollen flora of Guyana. *Leidse Geol. Meded.*, 38:49–70.

Lerbekmo, J. F., M. E. Evans, and H. Baadsgaard, 1979. Megnetostratigraphy, biostratigraphy, and geochronology of Cretaceous-Tertiary boundary sediments, Red Deer Valley. *Nature*, 279:26–30.

Lesquereux, L., 1982. The flora of the Dakota Group. *U. S. Geol. Surv. Monogr.*, 17, 400 pp.

MacNeal, D. L., 1958. The flora of the Upper Cretaceous Woodbine Sand in Denton County, Texas. *Philadelphia Acad. Nat. Sci. Monogr.*, 10, 152 pp.

Mamontova, I. V., 1977. Palinoflora perochodnych sloev verchnego Mela i Paleogena Amuro-Zeiskoi depressi. In: V. A. Krassilov (ed.), Paleobotanika nq Dal Vostoke. *Vladivostok, Acad. Nauk, Inst. Biol. Pedol.*, pp. 32–37.

Mchedlishvili, N. D., 1961. *Triprojectacites.* In: S. R. Samilovitch (ed.), Pyl'tsa i Neft spory zapadnoi Sibiri, Yura-Paleotsen. *Leningrad Vses Nauchwo-Issled. Geol.-Razevd. Inst. (VNIGRI), Trudy,* 177:203–229.

McLean, D. M., 1978. A terminal Mesozoic "greenhouse": lessons from the past. *Science*, 201:401–406.

Moore, H. E., Jr., and N. W. Uhl, 1973. Palms and the origin and evolution of monocotyledons. *Quart. Rev. Biol.*, 48:414–436.

Mouton, J. A., 1970. Architecture de la nervation foliare. *92$^e$ Congr. Soc. Savantes, Compt. Rend.*, 3:165–176.

Müller, H., 1966. Palynological investigation of Cretaceous sediments in northeast Brazil. In: J. E. van Hinte (ed.), *Proceedings of the Second West African Micropaleontological Colloquium*, pp. 123–136.

Muller, J., 1968. Palynology of the Pedawan and Plateau Sandstone Formations (Cretaceous-Eocene) in Sarawak, Malaysia. *Micropaleontology*, 14:1–37.

———, 1970. Palynological evidence on early differentiation of angiosperms. *Biol. Rev.*, 45:417–450.

Newman, K. R., 1979. Cretaceous/Paleocene boundary in the Denver Formation at Golden, Colorado, U. S. A. In: W. K. Christensen and T. Birkelund (eds.), *Cretaceous-Tertiary Boundary Events. Copenhagen Univ., Proc. Sympos. "Cretaceous-Tertiary Boundary Events"*, vol. 2, pp. 246–248.

Niklas, K. J., B. H. Tiffney, and A. H. Knoll, 1980. Apparent changes in the diversity of fossil plants. In: M. K. Hecht, W. C. Steere and B. Wallace (eds.), *Evolutionary Biology*, 12:1–89.

———, 1978. Evolutionary trends of platanaceoid pollen in Europe during the Cenophytic. *Forsch.-Inst. Senckenberg, Cour.*, 30:70–76.

Norton, N. J., and J. W. Hall, 1969. Palynology of the Upper Cretaceous and Lower Tertiary in the type locality of the Hell Creek Formation, Montana, U. S. A. *Palaeontographica*, 125 (B):1–92.

Pacltová, B., 1961. Zur Frage der Gattung *Eucalyptus* in der bohmischen Kreideformation. *Preslia*, 33:113–129.

———, 1971. Palynological study of angiosperms from the Peruč Formation (? Albian-Lower Cenomanian) of Bohemia. *Bratislava, Ústředni Ústav Geol., Sborn. Geol. Věd. Pal., Řáda P*, 13:105–141.

———, 1977. Cretaceous angiosperms of Bohemia-Central Europe. *Bot. Rev.*, 43:128–142.

———, 1978. Evolutionary trends of platanaceoid pollen in Europe during the Cenophytic. *Forsch.-Inst. Senckenberg, Cour.*, 30:70–76.

Parker, L. R., 1975. The paleoecology of the fluvial coal-forming swamps and associated floodplain environments in the Blackhawk Formation (Upper Cretaceous) of central Utah. *Brigham Young Univ. Geol. Studies*, 22:99–116.

Penny, J. S., 1969. Late Cretaceous and Early Tertiary palynology. In: R. H. Tschudy and R. A. Scott (éds.), *Aspects of Palynology* (New York: Wiley-Interscience), pp. 331–376.

Raunkiaer, C., 1934. *The Life Forms of Plants and Statistical Plant Geography; being the Collected Papers of C. Raunkiaer* (Oxford: Clarendon Press), 632 pp.

Roy, J. R., 1977. Variations of the luminosity of the sun and "super" solar flares: possible causes of extinctions. *Syllogeus*, 12:89–110.

Ruderman, M. A., 1974. Possible consequences of nearby supernova explosions for atmospheric ozone and terrestrial life. *Science*, 184:1079–1081.

Rüffle, L., 1978. Evolutionary and ecological trends in Cretaceous floras particularly in some Fagaceae. *Forsch.-Inst. Senckenberg, Cour.*, 30:77–83.

Russell, D., and. W. H. Tucker, 1971. Supernovae and the extinction of the dinosaurs. *Nature*, 229:553–554.

Saito, T., and J. Van Donk, 1974. Oxygen and carbon isotope measurements of

Late Cretaceous and Early Tertiary foraminifera. *Micropaleontology*, 20:152–177.

Samilovich, S. R., 1967. Tentative botanico-geographical subdivision of northern Asia in Late Cretaceous times. *Rev. Paleobot. Palynol.*, 2:127–139.

Savin, S. M., 1977. The history of the earth's surface temperature during the past 100 million years. *Ann. Rev. Earth Planet Sci., 1977*, 5:319–355.

Savin, S. M., R. C. Douglas, and F. G. Stehli, 1975. Tertiary marine paleotemperatures. *Geol. Soc. Amer. Bull.*, 86:1499–1510.

Shoemaker, R. E., 1966. Fossil leaves of the Hell Creek and Tullock Formations of Eastern Montana. *Palaeontographica*, 119 (B):54–75.

Smiley, C. J., 1966. Cretaceous floras of the Kuk River Area, Alaska: stratigraphic and climatic interpretations. *Geol. Soc. Amer. Bull.*, 77:1–14.

————, 1969a. Cretaceous floras of the Chandler-Colville region, Alaska: stratigraphic and preliminary floristics. *Amer. Assoc. Petrol. Geol. Bull.*, 53:482–502.

————, 1969b. Floral zones and correlations of Cretaceous Kukpowrak and Corwin Formations, northwestern Alaska. *Amer. Assoc. Petrol. Geol. Bull.*, 53:2049–2093.

————, 1972. Plant megafossil sequences, North Slope Cretaceous. *Geoscience and Man*, 4:91–99.

Smit, J., and J. Hertogen, 1980. An extraterrestrial event at the Cretaceous-Tertiary boundary. *Nature*, 285:198–200.

Smith, A. G., and J. C. Briden, 1977. *Mesozoic and Cenozoic Paleocontinental Maps.* (Cambridge: Cambridge Univ. Press), 63 pp.

Srivastava, S. K., 1970. Pollen biostratigraphy and paleoecology of the Edmonton Formation (Maastrichtian), Alberta, Canada. *Palaeogeogr. Palaeoclimatol. Palaeoecol.*, 7:221–276.

Stanley, E. A., 1965. Upper Cretaceous and Paleocene plant microfossils and Paleocene dinoflagellates and hystrichosphaerids from northwestern South Dakota. *Amer. Pal. Bull.*, 49:179–383.

————, 1970. The stratigraphical, biogeographical, paleoautecological and evolutionary significance of the fossil pollen group Triprojectacites. *Georgia Acad. Sci. Bull.*, 28:1–44.

Stover, L. E., and P. R. Evans, 1973. Upper Cretaceous-Eocene spore-pollen zonation, offshore Gippsland Basin, Australia. *Aust. Geol. Soc. Spec. Publs.*, 4:55–72.

Stover, L. E., and A. D. Partridge, 1973. Tertiary and Late Cretaceous spores and pollen from the Gippsland Basin, southeastern Australia. *Roy. Soc. Victoria Proc.*, 85:237–286.

Teixeira, C., 1948. Flora Mesozoica Portuguesa. Part I. (*Lisbon: Serv. Geol. Portugal*), n.p..

————, 1950. Flora Mesozoica Portuguesa. Part II. (*Lisbon: Serv. Geol. Portugal*)

Terry, K. D., and W. H. Tucker, 1968. Biologic effects of supernovae. *Science*, 159:421–423.

Tschudy, R. H., 1970. Palynology of the Cretaceous-Tertiary boundary in the northern Rocky Mountain and Mississippi embayment regions. In: R. M.

Kosanke and A. T. Cross (eds.), Symposium on palynology of the Late Cretaceous and Early Tertiary. *Geol. Soc. Amer. Spec. Paper,* 127:65–111.

———, 1975. Normapolles pollen from the Mississippi embayment. *U. S. Geol. Surv. Prof. Paper,* 865.

Tucker, W. H., 1977. The effect of a nearby supernova explosion on the Cretaceous-Tertiary environment. *Syllogeus,* 12:111–121.

Tucker, W. H., and K. D. Terry, 1968. Cosmic rays from nearby supernovae: biological effects (reply). *Science,* 160:1138–1139.

Vakhrameev, V. A., 1952. Stratigrafiya i iskopaemaya flora melovykh otlozhenii zapadnogo Kazakhstana. Regionalnaya stratigrafiya SSSR, 1. (Moscow: Akad. Nauk.)

Vakhrameev, V. A., and M. A. Archmetev, 1977. Vysshie rasteniya potom dannym izucheniya listev. In: V. A. Vachrameev (ed.), Razvitie flor na granitse Mezozoya i Kainozoya. *Nauka,* pp. 39–65.

Vakrameev, V. A., and I. Z. Kotova, 1977. Drevnie pokrytosemennye u sopulstvuiushe im rasteniya iz niznemelobych otlozhenii Zabaikalia. *Pal. Zhur.,* 4:101–109.

Van Campo, M., and B. Lugardon, 1973. Structure grenue infratectale de l'ectixine des pollens de quelques Gymnospermes et Angiospermes. *Pollen et Spores,* 15:171–187.

Van Valen, L., and R. E. Sloan, 1977. Ecology and the extinction of the dinosaurs. *Evol. Theory,* 2:37–64.

Valenovsky, J., 1882. *Die Flora der Bonmischen Kreideformation* (Vienna: Holder), 75 pp.

Voight, E., 1964. Zur Temperatur-Kurve der obern Kreide in Europa. *Geol. Rundsch,* 54:270.

Wiggins, V. D., 1976. Fossil oculate pollen from Alaska. *Geoscience and Man,* 15:51–64.

Wolfe, J. A., 1972. Significance of comparative foliar morphology to paleobotany and neobotany. *Amer. Jour. Bot.,* 59:664 (abstract).

———, 1976. Stratigraphic distribution of some pollen types from the Campanian and lower Maastrichtian rocks (Upper Cretaceous) of the Middle Atlantic States. *U. S. Geol. Surv. Prof. Paper,* 977, 18 pp.

Wolfe, J. A., J. A. Doyle, and V. M. Page, 1975. The bases of angiosperm phylogeny: paleobotany. *Missouri Botan. Garden, Ann. Rept.,* 62:801–824.

Wolfe, J. A., and D. M. Hopkins, 1967. Climatic changes recorded by Tertiary land floras in northwestern North America. In: K. Hatai (ed.), Tertiary correlations and climatic changes in the Pacific. *11th Pacific Sci. Congr. Tokyo, 1966,* pp. 67–76.

Wolfe, J. A., and H. M. Pakiser, 1971. Stratigraphic interpretations of some Cretaceous microfossil floras of the Middle Atlantic States. *U. S. Geol. Surv. Prof. Paper,* 750 (B):B35–B47.

Woodwell, G. M., 1963. The ecological effects of radiation. *Sci. Amer.* 208 (6):40–49.

———, 1967. Radiation and the patterns of nature. *Science,* 156:461–470.

Zaklinskaya, E. D., 1970. Pozdnemelobye i Rannepaleogenovye flori (po Palino-logiche skim dannym). In: V. A. Vakhrameev, I. A. Dobruskina, E. D. Zal-kinskaya, and S. V. Meyen (eds.), *Paleozoiskie i Mesozozoiskie flori Evrazii i fitogeografiya etogo vremeni.* Nauka, Moscow, pp. 302–331.

———, 1977. Pokrytosemennye po palinologicheskim dannym. In: V. A. Vackrameev (ed.), *Razvitie flor na granitse Mezozoya i Kainozoya.* Nauka, Moscow, pp. 66–119.

Chapter 11

# PALYNOLOGICAL EVIDENCE FOR CHANGE IN CONTINENTAL FLORAS AT THE CRETACEOUS TERTIARY BOUNDARY

ROBERT H. TSCHUDY

U.S. Geological Survey

## INTRODUCTION

The greater part of the evidence supporting a drastic and widespread terminal extinction at the end of the Cretaceous is from the marine realm. If indeed a catastrophic change did occur, the record from continental deposits would have coincided in severity with that from the marine record. Pollen and spores are widely disseminated by wind and water, are extremely abundant and diverse (often thousands of grains per gram of rock) and well-preserved in many continental rocks. Thus they can provide the most abundant and perhaps the most reliable source of evidence for change in the land plant biota at the Cretaceous-Tertiary boundary. In this paper, I will review pollen and spore data for evidence of a catastrophic change in land plants at the end of Cretaceous time.

In the evaluation of data concerning drastic extinctions at the Cretaceous-Tertiary boundary one should be mindful of the following:

1) Mass extinctions such as the episode at the Cretaceous-Tertiary boundary were not unique in the history of life. Indeed, period boundaries were originally proposed primarily according to prominent discontinuities in the fossil record. Some of these discontinuities were due to mass extinctions of some biotic lineages at the end of the Late Devonian, the Late Permian and the Late Triassic.

2) A rapidly expanding paleontological data base has shown that extinction was not an abrupt event among all the groups involved in the terminal Cretaceous mass extinction. For example, the extinction of the ammonites was a gradual event; many genera had recently become extinct by Maastrichtian time. "Extinction ... must be explained in general

terms as part of the widespread Late Cretaceous extinctions. Some causal factor linking the Late Cretaceous regression and the decline of the ammonites appears most plausible" (Kennedy and Cobban, 1976, p. 81). Sloan (1976, p. 136) stated, "Dinosaur extinction was hardly instantaneous . . . it is known that at least in North America the process took place over about 12 million years. During this time there was a steady and progressive decline in the number of taxa of dinosaurs."

3) The fossil record exhibits a continuum of extinctions and evolutionary innovations from the Precambrian to the present. Extinctions are not confined to specific time intervals; they are only more pronounced at certain times than at others. The same generalization applies to innovations. The introduction of new forms did occur, at times, in bursts, as did diminution in the rates of evolutionary diversification.

With these ideas in mind one can ask, is there evidence for catastrophic extinctions of whole floras, or of specific groups or lineages of plants at the Cretaceous-Tertiary boundary? Can observed local changes be attributable to changes in ecological conditions, facies changes, effects of tectonism or of hiatuses in the record? Were the changes of sufficient magnitude to be considered as unusual or catastrophic, and were they world-wide in scope?

FLORAL CHANGES DURING THE CRETACEOUS-TERTIARY TRANSITION

Angiosperms were introduced into the world land flora during late-Early Cretaceous (Albian) time. The Late Cretaceous saw the explosive diversification of angiosperms, until they became the dominant element in the land flora. Palynologically, monosulcate angiosperm pollen, and simple tricolpate pollen grains were introduced in the Albian, followed successively by the introduction of tricolporate, triporate, and morphologically more complex pollen in the Cenomanian (Doyle, 1969; Muller, 1970). Cenomanian time saw the gradual introduction of new angiosperm taxa and the extinction of some of the older taxa, accompanied by a decline in the relative representation and diversification of gymnosperms, ferns, and their allies. These latter plant groups remained dominant only in specific ecological habitats.

By the end of the Cenomanian the angiosperms became the dominant element in the land flora. From Turonian time through the remainder of the Late Cretaceous and early Tertiary, angiosperms continued their ascendency, and the other elements of the whole flora continued to be progressively more poorly represented, except in isolated instances. The more primitive, less adaptive plants were continuously replaced by more advanced and specialized taxa until the flora attained a somewhat modern aspect by Late Eocene or early Oligocene time. A significant number of the plant taxa that appeared during Late Cretaceous time also became extinct, and the relationship of most of these extinct plants to modern genera or even families is still not known.

## BIAS IN PALYNOLOGICAL DATA

In assessing the data presented in the palynological literature, it is surprising to find that there has been no detailed evaluation of the whole flora across the Cretaceous-Tertiary boundary. Indeed, remarkably few reports deal with pollen and spore changes across the boundary, and these emphasize the readily recognizable changes, often limited to only a few taxa, rather than to the changes in whole floras. This lack of data directly applicable to the present survey is, in part, due to the direction or emphasis given to the original reports.

Much palynological work has been directed toward correlations or age determinations, rather than to evolutionary changes in the palynomorph biota. The extinction (or disappearance due to other factors such as change in facies) of one or several taxa at a specific stratigraphic level serves as a useful correlation criterion. The appearance of a distinct new taxon that persists for a significant interval in rocks above a given stratigraphic level also serves as a useful correlation criterion. These "guide" fossils facilitate correlation, and as a consequence, the remainder of the palynomorph flora is often ignored or given but fleeting recognition. "Tropical pollen floras are very rich in species and the average type collection may easily contain 800–1,000 different species. For stratigraphic purposes generally less than 200 are of importance per area. For a comprehensive review, such as this, a further reduction is desirable and only 49 species are discussed" (Germeraad et al., 1968, p. 191).

Many reports dealing with correlations or fossil zonation record only those taxa that are characteristic of the horizon or zone and omit those accessory species that are members of whole floras. These incomplete suites are of limited value in the present review.

On a world-wide scale, palynology has suffered from the absence of adequate and accurate stratigraphic control. Most fossil pollen and spore assemblages are derived from continental rocks that too often have not yielded fossils of other types that could provide biostratigraphic age baselines. This lack of accurate basic data in some areas of the world has led to the practice of referring plant microfossils, in the absence of more definitive information, to broad time intervals such as the Senonian (Coniacian through Maastrichtian Ages) or to the Paleogene (Paleocene through Oligocene Epochs). Such designations obviously make attempts to trace palynological changes at the Cretaceous-Tertiary boundary very difficult.

The areal palynological coverage leaves many gaps, and the quantity and quality of the data are not uniform. The stratigraphic coverage is also erratic. Primary palynological emphasis from any one continent has been at least partially influenced by the exploration for economic deposits (chiefly oil). From Venezuela, for example, there are as yet no palynological reports on rocks older than Cretaceous because all petroleum and bauxite deposits are in Cretaceous and Tertiary rock sequences.

Preservational bias is not confined to palynological work; limitations such as accidents of preservation (or complete lack of preservation) affect all fossil groups. A notable example involving land plants is the almost complete absence of plant megafossils (leaves) from the Paleocene of eastern North America (Berry, 1916), yet from the same area Paleocene pollen and spores that represent an extensive and varied land flora are found in abundance (Tschudy, 1971).

Some deficiency is present in each of the examples cited in the following discussions, and these will be noted in the appropriate place. Nevertheless, the palynological data provide a consistent view of the floral changes that occurred at the Cretaceous-Tertiary transition, so that valid conclusions emerge.

PROVINCIALISM

After the Cretaceous land flora became established it remained more or less homogeneous through the Cenomanian. Subsequently, broad floral provinces began to become recognizable. After its inception, provincialism continued to become increasingly prominent up to the present. Present-day floral provinces are readily recognizable even on individual continents. For example, Gleason and Cronquist (1964) recognize 10 floral provinces in present-day North America. It is more difficult to recognize some ancient floral provinces from the palynological samples provided from widely scattered localities.

At the end of Cretaceous time several somewhat distinct floral provinces were already in existence—the Atlantic-European, the east Siberian-North Pacific, the Malesian, the central Atlantic, and the Australian-Antarctic (Muller, 1970). The Malesian-Central Atlantic or Pantropical Province occupied the land areas adjacent to the ancestral Tethys (Germeraad et al., 1968; Muller, 1970) and the Australia-Antarctic Province occupied the south circumpolar area (Muller, 1970).

An examination of palynological data from North America and Eurasia shows that two, almost mutually exclusive, suites of palynomorphs are present. One suite is from the western North American-Siberian-Pacific Cretaceous Floral province (W. Krutzsch in Góczán et al., 1967, p. 519), or in floral terms—the *Aquilapollenites*-Triprojectacites province (Zaklinskaia, 1967; Stanley, 1970). A second suite derived from rocks of the western or European part of Russia, Europe, and eastern North America consists of fossil pollen grains generally foreign to the *Aquilapollenites* area or province. The flora from this area is characterized by pollen that has been placed in the Normapolles group of pollen genera (Pflug, 1953); hence this floral province has been referred to as the Atlantic-European or Normapolles province. The eastern and western parts of North America were isolated from each other during most of Cretaceous time by a mid-continental epeiric sea, which inhibited floral mixing between the western *Aquilapollenites* flora and the eastern Normapolles flora. The European and Siberian areas were likewise isolated from each other by an arm of the Tethys Sea known as the Turgai Strait, which was located at approximately the present position of the Ural Mountains. The eastern and western parts of the Normapolles province (Europe and eastern North America) were united by a north Atlantic land connection; likewise, the eastern and western parts of the *Aquilapollenites* province were united by a land connection from western North America across Beringia to eastern Asia.

## THE POLLEN AND SPORE EVIDENCE

As might be anticipated, the greatest number of papers dealing with the Cretaceous-Tertiary floral change originated in North America, western Europe and eastern Europe, a lesser number originated from southeast Asia and Australia and comparatively few from South America and Africa.

The greatest body of evidence for a prominent change in the land flora across the Cretaceous-Tertiary boundary is from the western North American *Aquilapollenites* Province. These data are from the same area as has provided the greatest body of information concerning dinosaur extinction—the Hell Creek, Lance Creek, Tullock and Fort Union Formations that were at one time the subject of the classic Laramie controversy (Dorf, 1942). Leaf floral data derived from the same area do not show so drastic a change as once was thought (Hickey, this volume).

A significant body of data is also available from the Normapolles Province. Data have been assembled from the Pantropical Province and from the Austral Province. Palynological evidence will be presented from the several world floral provinces, along with some inferences concerning general trends and world-wide patterns.

## THE *AQUILAPOLLENITES* PROVINCE

A few specimens of the nominate genus for this province, *Aquilapollenites,* are present in almost every productive Maastrichtian sample from western North America. Specimens of this genus, however, are not the dominant element of the flora. One sample may yield taxodiaceous pollen, as the dominant, yet succeeding samples may yield specimens of the genus *Kurtzipites* or *Gunnera* as the dominant taxon. Accompanying these taxa commonly are specimens of other taxa characteristic of the Late Cretaceous such as *Proteacidites* and *Wodehouseia.* The implication is that successive samples partially represent the regional flora and partially the florule adjacent to the site of deposition. Other factors such as facies, or lithotype may also influence palynomorph variation from sample to sample.

In western North America a marked change in pollen and spores accompanied by the elimination of many taxa characteristic of the Late Cretaceous such as those mentioned above, coincides closely with the stratigraphically highest occurrence of dinosaurian bones. A few genera and species of palynormorphs persist until the end of the Cretaceous before becoming extinct (Stanley, 1965, 1970; Srivastava, 1967; Norton and Hall, 1969; Snead, 1969; Leffingwell, 1971). However, for each genus or species that became extinct at this horizon, several others persisted unchanged across the Cretaceous-Tertiary boundary (*Alnipollenites, Ulmipollenites, Azolla, Equisetosporites,* among others). In fact, at least one species each of *Aquilapollenites* and *Wodehouseia* occur in lower Tertiary rocks from western North America.

A sample-to-sample change through the Hell Creek Formation (Maastrichtian) into the basal Tullock Formatin (Paleocene) is shown on Figure 11-1. The number of taxa per sample differs significantly, which reflects lithological, preservational or other parameters. The most significant aspect of this figure, is the apparent somewhat constant extinction of taxa throughout the section, culminating in more pronounced extinctions at the end of the Cretaceous. Simultaneously, new taxa were appearing, the greatest number of innovations in the basal Paleocene sample.

Data presented by Srivastava (1970) is from the upper part of the Edmonton Formation (Maastrichtian) of Canada (Figure 11-2). These data cannot be directly compared with data from the Hell Creek Formation because Srivastava was trying to zone this part of the section, and terminated his evaluation at the base of the Tertiary. His data were pre-

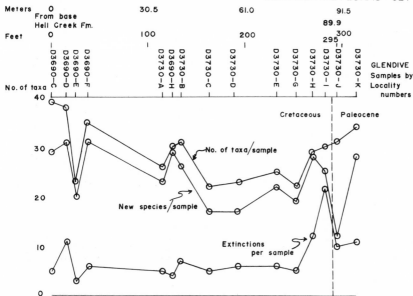

Figure 11-1. Sample-to-sample changes in Hell Creek and Tullock Formations, Glendive, Montana.

Samples were first counted (200 counts) to determine the principal components. The remainder of each slide was then scanned in a search for additional taxa that did not appear in the count. The number of extinctions and number of new species represented in sample D3690-C (the basal Hell Creek sample) was obtained from the differences between this sample and the underlying sample from the Colgate Member (not shown) of the Fox Hills Sandstone.

sented as zonal changes rather than sample-to-sample changes. Nevertheless, the data presented shows a continuum of innovations and extinctions throughout the uppermost part of the Cretaceous. Further numerical data relating to the Cretaceous-Tertiary palynoflora from the Hell Creek and contiguous Formations are presented in Table 11-1. The numerical data demonstrates some of the difficulties in interpreting material from different sources. When two workers present data, each from different samples or sample localities, and have not treated the samples in the same manner it is difficult to reconcile discrepancies. My data are from a measured section of the Hell Creek Formation from Seven Blackfoot Creek, Montana, and Norton and Hall (1969) have presented data from the type Hell Creek Formation some 20 miles to the east. My data show that of the Cretaceous Colgate and Hell Creek palynofloras only 13% of the species persist into the Paleocene. If one considers only angiosperm species, however, 32% persist into the Paleocene. Similarly, 37% of the Colgate species persist into the Hell Creek Formation, indicating that a floral change was

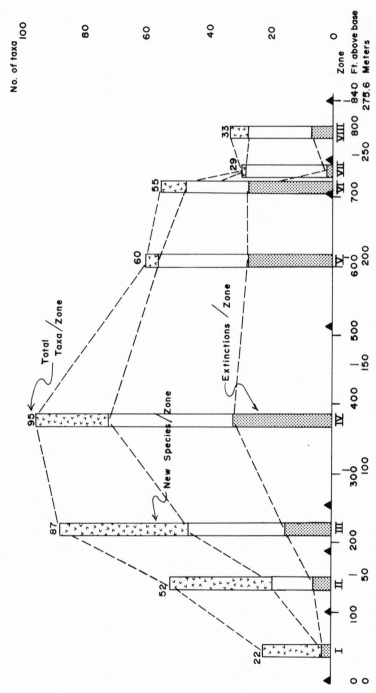

Figure 11–2. Angiosperm distribution by zone in the Late Cretaceous (Maastrichtian) Edmonton Formation, Alberta, Canada. Data extracted from Srivastava (1970).

| Reference | Locality | Remarks | Formation | No. of taxa | No. taxa persisting into basal Paleocene | % |
|---|---|---|---|---|---|---|
| Tschudy, unpub. | Seven Black-foot Creek Montana | Whole flora | Fox Hills (Colgate Mbr.) and Hell Creek | 132 | 22 | 13 |
| | | Angiosperms | Do. | 61 | 20 | 32 |
| | | Whole flora | Fox Hills (Colgate Mbr.) | 67 | [25 persist to Hell Cr.] | [37] |
| | | Whole flora | Hell Creek | 99 | 15 | 15 |
| | | Angiosperms | Do. | 75 | 8 | 10 |
| Norton & Hall 1969 | Hell Creek (type) | Whole flora | Hell Creek | 88 | 34 | 38 |
| | | Angiosperms | Do. | 56 | 26 | 46 |
| Tschudy, unpub. | Mississippi Embayment | Whole flora | McNairy and Owl Creek | 298 | 64 | 21 |
| | | Angiosperms | Do. | 191 | 40 | 21 |
| | | Whole flora | McNairy | 194 | [67 persist to Owl Cr.] | [34] |

Table 11-1. Persistence of Cretaceous palynomorph taxa from the Late Cretaceous into the early Paleocene.

The data from the Seven Blackfoot Creek locality were derived from 13 Maastrichtian samples (Colgate Formation, 11 Hell Creek Formation) and 5 Paleocene samples (Basal Tullock Formation).

The data from the Mississippi Embayment were derived from 7 Maastrichtian samples (4 McNairy Formation, 3 Owl Creek Formation) and 4 Paleocene samples (basal Clayton Formation). Note the greater floral diversity in the Maastrichtian of the Mississippi Embayment when compared with the Maastrichtian of Montana, even though fewer samples were employed in obtaining the Mississippi Embayment data.

occurring during the Cretaceous that was more prominent at the Cretaceous-Tertiary boundary.

The data from Norton and Hall further emphasize the danger inherent in applying numerical values to data from different workers even though dealing with the same rock sequences. Norton and Hall have reported only 88 taxa from the Hell Creek Formation and I list 99 from the Hell Creek and 132 from the Colgate and Hell Creek combined. Trends represented by the figures are valid, but little validity can be attributed to the discrepancies in the percentage values from the two reports.

Palynological sequences generally similar to those found in western North America and western Canada can be traced through Alaska into Siberia (Zaklinskaia, 1967; Bratzeva, 1967: Stanley, 1970). However, the extinctions of members of the *Aquilapollenites* flora in Sibera is not as abrupt as it appears to be in western North America. Mchedlishvili (1964) noted that the pollen of *Aquilapollenites* is commonly found in the

Paleocene of Siberia, although in reduced quantities compared to that of the Cretaceous.

A case can be made for an increased rate of extinction of the Late Cretaceous flora of western North America and Siberia at the end of the Cretaceous. However, a massive elimination of taxa is not indicated. In North America, Jarzen (1977, p. 42) arrived at the following conclusions:

1) extinctions of the angiosperms at the Cretaceous-Tertiary boundary are observed at the specific and generic level, but they were by no means extensive or drastic,

2) a gradual, and perceptible transition of angiosperm floristic composition takes place as the evolutionary trends, which developed during the early Upper Cretaceous, continue through the latest Cretaceous and into the Tertiary period,

3) although these floristic changes across the Cretaceous-Tertiary boundary do not suggest sharp climatic changes, the possibility of short-term minor climatic fluctuations cannot be ruled out.

Mchedlishvili (1964) has suggested that the terminal Cretaceous change in plant life in this province was brought about by a decrease in temperature. Norris (1977), who summarized the evidence for periodic oceanic temperature changes, has indicated that global temperatures declined during the latest Cretaceous to a minimuim at the Cretaceous-Tertiary boundary, and then started to rise, thus he presented a picture of gradual rather than catastrophic temperature change.

THE NORMAPOLLES PROVINCE

The name of this province is derived from a group of extinct genera of fossil pollen grains possessing distinct and "bizarre" apertures (Pflug, 1953). Some specimens representing Normapolles genera are commonly present in virtually every Late Cretaceous and early Tertiary sample from this province. However, members of this group of genera are seldom the dominant pollen type. The palynomorph flora from the Normapolles province is, as a whole, distinctly different from the palynormorph flora of the *Aquilapollenites* province. In eastern North America the Normapolles flora occupies the Mississippi Embayment area and the eastern seaboard as far north as New York.

The change in the flora of the Normapolles Province across the Cretaceous-Paleocene boundary is pronounced and easily recognizable, but it is much less abrupt than that found at the boundary in the *Aquilapollenites* Province.

Data from the Mississippi Embayment area provided by Tschudy (unpublished) are shown on Table 11-1. These data show that the Late Cretaceous palynomorph flora of the Mississippi Embayment region is much more diverse than that found in western North America. The data presented in Table 11-1 is from four samples of the McNairy Formation type section, and three from the overlying Owl Creek Formation, yet 298 distinct taxa were identified in this limited Cretaceous sequence. More Cretaceous species were found as additional McNairy samples were studied. Consequently the Cretaceous flora enumerated is not the complete Late Cretaceous (Maastrichtian) flora of the region. When data from the McNairy and Owl Creek Formations were considered together as Maastrichtian Formations, it was found that 21% of the taxa persisted into the Paleocene. In comparing data from the McNairy and Owl Creek Formations, only 34% of the McNairy taxa persisted into the Owl Creek Formation. Thus, the same picture emerges—extinctions were occurring throughout the Late Cretaceous, with a slightly greater incidence of extinctions at the Cretaceous-Tertiary boundary.

All members of the Normapolles group are now extinct. However, the extinction was gradual, as is shown by the data presented by Góczán et al. (1967). These data have been re-drafted in Figure 11-3 to include data from the Mississippi Embayment. The group as a whole had its inception in Europe and North America during the Cenomanian, reached its acme in Late Cretaceous time, and declined to total extinction in early Oligocene time. The slopes of the curves are similar, as the European curve reflected a decline that began in Campanian time, while, from the Mississippi embayment, the decline began in Maastrichtian time. No pronounced extinction of Normapolles taxa is evident at the Cretaceous-Tertiary boundary.

Two fossil gymnosperm genera, *Pristinuspollenites* and *Rugubivesiculites,* are present in Mississippi Embayment rocks, and become extinct at the end of the Cretaceous in that region. These two genera are also present in the *Aquilapollenites* province but became extinct there during the Campanian time rather than at the end of the Cretaceous. Thus some other argument rather than a terminal Cretaceous catastrophic event must be adduced to explain these extinctions. Climatic deterioration that occurred earlier in the Rocky Mountain region than in the Mississippi Embayment region is a probable cause.

THE PANTROPICAL PROVINCE

Palynological work in South America, Africa, and southeast Asia was initiated by oil companies whose primary interest was in the correlation and zonation of strata. Only recently have such problems as the evolution of fossil floras and internationally defined paleontological boundaries

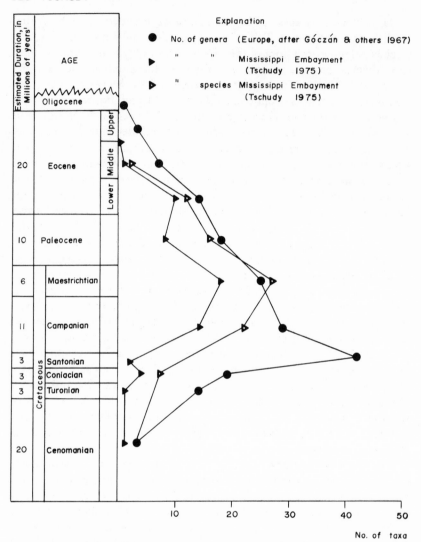

Figure 11-3. Normapolles distribution.
Owing to differences in the presentation of data, control points have been plotted at the approximate mid-points of the respective ages represented. Modified from Kulp (1961), Holmes (1965) and Gill and Cobban (1966).

begun to receive some attention. Almost all the palynological reports are deficient, therefore, in one or more aspects relating to the compilation of reliable records of changes of whole floras across the Cretaceous-Tertiary boundary. However, in general substance the reports show a marked similarity in that no catastrophic change is detectible in the palynomorph flora at the end of Cretaceous time. The geographical areas of northern South America, northern Africa, and southeast Asia will be discussed separately.

*Northern South America*  Hammen (1954) has published the first palynologic report on a complete stratigraphic section traversing the Cretaceous-Tertiary boundary in South America in which he showed ranges of 165 taxa from Columbia. He examined about 50 samples from several sections and claimed that the sections studied represented continuous deposition from the Maastrichtian into the Tertiary. The Cretaceous-Tertiary boundary was located on the basis of occurrences of appropriate taxa of pelagic foraminifera. On the basis of palynological results, Hammen proposed several zones and subzones as shown in Table 11–2. This work was a pioneering effort, as is indicated by the statement that subzones D-1 and D-2 yielded many species not yet identified and not included in the floral diagram. A gradual floral change is evident during the Maastrichtian. Of the 161 Cretaceous species listed all but 16 had become extinct by the end of the Cretaceous. Gradual extinction is shown by the number of extinctions recorded in each Maastrichtian subzone. Extinction of taxa was more evident within the Maastrichtian than at the Creta-

| Zone | Extinctions |
|------|-------------|
| D-2 | |
| D-1 | 15 |
| Maestrichtian / Tertiary | |
| C-2 | 22 |
| C-1 | 28 |
| B-3 | 17 |
| B-2 | 11 |
| B-1 | 10 |
| A-3 | 7 |
| A-2 | 39 |
| A-1 | 10 |

Table 11–2. Zonation of the Late Cretaceous-early Tertiary (Hammen, 1954) and extinctions observed at the end of each zone.

ceous-Tertiary boundary. Later, Sanroma (1970) concluded that the major change in pollen flora in Columbia did not take place at the Cretaceous-Paleocene boundary but rather occurred within the Maastrichtian, and was out of phase with the faunal change delienated by ammonites and foraminifers.

Correlation horizons were extended across northern South America to the Guianas and south to Brazil (Hammen, 1957; Hammen et al., 1961; Wijmstra and Hammen, 1964; Boer et al., 1965; Leidelmeyer, 1966; Müller, 1966; Doubinger, 1972; Solé de Porta, 1972; Herngreen, 1975). Virtually the same palynomorph floras in the Maastrichtian and Paleocene were found as those reported by Hammen in 1954, although most of this subsequent work dealt with partial rather than complete stratigraphic sections across the Cretaceous-Tertiary boundary. Furthermore, all the above-mentioned workers were primarily concerned with correlation, and they listed only those few species that previously had been found useful in the recognition of specific segments of the stratigraphic column. The complete fossil flora, with the exception of records of some new taxa, was largely omitted from consideration.

Hammen and Wijmstra (1964) noted great similarity of the pollen change in British Guiana rocks to the pollen changes previously recorded from Colombia. They found a transitional flora at the Cretaceous-Tertiary boundary rather than a catastrophic elimination of taxa. The boundary was marked primarily by a change in percentage representation of several taxa rather than by a prominent change brought about by floral extinctions.

In northern South America many new taxa were introduced into the Paleocene. This trend continued until a more or less modern flora emerged during the Miocene.

Hammen (1954, 1957) has attributed the observed palynological changes to climate fluctuations (mainly temperature effects) brought about in part by the withdrawal of the widespread epeiric sea during the Late Cretaceous. He noted evidence of provincialism of late Paleocene and Eocene floras and suggested that migration of taxa may account for the enrichment of the Eocene South American tropical palynoflora.

*Northern Africa.* The palynormorph transition at the Cretaceous-Tertiary boundary in Africa is more obscure than it is in South America, perhaps because detailed work in rocks of the transition interval is sparse. Boltenhagen (1965) charted the percentage occurrence of groups of taxa and recognized the transition from the Cretaceous to the Tertiary by the last occurrence of pollen of the Proteales and the first appearance of pollen of *Rhizophora* and Nymphaeaceae. The basal Paleocene was not represented in Boltenhagen's samples and the exact age of samples from the Paleocene-Eocene sequence is in doubt. The stratigraphic distribution of 34 palynomorph taxa from Nigeria was presented by Hoeken-Klinkenberg (1966), who studied the interval from the Late Cretaceous to the middle Eocene. Of the group of taxa reported, only two were limited to the Maastrichtian, eleven were restricted to the Maastrichtian and Paleocene, and six taxa extended from the upper Paleocene into the Eocene. Although a great many taxa almost certainly must have been ignored, this study likewise does not document notable floral extinction at the end of the Cretaceous. The palynomorph change (introduction of new taxa) appeared to be more prominent in the Eocene than at the Cretaceous-Paleocene boundary.

Several workers, notably Jardiné and Magloire (1963), Belsky et al. (1965), Boltenhagen (1967), and Kieser (1967) have examined African Upper Cretaceous rocks palynologically, but their discussions did not encompass the Cretaceous-Tertiary transition. The most complete record of palynomorph change across this transition interval in this area was provided by Kedves (1971) from an examination of Egyptian rocks. Here again, one finds data that is deficient as it relates to the present review. The palynomorph recovery was somewhat sparse and clearly incomplete. Of the 56 Maastrichtian taxa mentioned, 30 persist into the Danian, 19 into the early Eocene and 32 into the upper Eocene. The Paleocene was not represented. No unusual diminution of taxa is observable at the Cretaceous-Tertiary transition.

*Southeast Asia*   A landmark biostratigraphic paper by Germeraad et al. (1968) traced pollen zones from South America to Africa and then to Borneo. Although the greatest emphasis was on pollen changes during the Tertiary, they did discuss the Cretaceous-Tertiary transition. This zonal boundary was based upon the top occurrences of one or two taxa common to northern South America and Nigeria. Germeraad et al. (1968) placed great emphasis on the influence of local facies factors on time-stratigraphic pollen distribution data, and he claimed (p. 205) that in the tropical realm at least, "time stratigraphic correlation lines are not easily detected."

The general outline of plant succession presented by Germeraad et al. (1968) is reinforced for the Malaysian region by Muller (1968). The interval from the Senonian to the Eocene showed no prominent floral change that would suggest mass extinctions or indicate the Cretaceous-Tertiary boundary. The Late Cretaceous was identified by a pollen spectrum similar to that found in the *Proxapertites* (Maastrichtian) pollen zone of Colombia. The upper part of the *Proxapertites* Zone in Malaysia yielded Paleocene foraminifera, which suggests that the Cretaceous-Tertiary boundary was somewhere within the *Proxapertites* Zone. Here is certainly no evidence of notable extinction of taxa, but rather a gradual elimination and replacement of taxa during the Late Cretaceous and the early Tertiary. The more easily recognizable floral changes are marked by the comparatively abrupt increase or decrease in percentage representation of several taxa.

*The Austral Province*   Until very recently the palynological data applicable to the Cretaceous-Tertiary boundary in this region has been scanty and somewhat unreliable. Early work by Couper (1953) in New Zealand was hampered by lack of information from other fossil groups so that accurate dating of palynological assemblages was next to impossible. In 1960, Couper presented more refined comparisons with fossils from other areas, but he still was limited by poor biostratigraphic data from the Paleocene. At that time (1960), a severe restriction of flora at the Cretaceous-Tertiary boundary was not supported by palynological data. Indeed, on the basis of palynological data without the aid of other fossils, it would have been difficult to pick a horizon representing the boundary—a horizon representing a greater than usual extinction of taxa.

Early palynological work in southeastern Australia (summarized in Stover and Partridge, 1973) was hindered by the absence of information from other fossils; therefore, exact placement of rocks in a time framework was difficult. The first attempt to use palynomorphs in correlation of surface and subsurface sections was based on samples from the Birregurra-1 bore (Cookson, 1953). Assemblages of three distinct ages were recognized: Cretaceous, Paleocene to early Eocene, and Eocene. Palynomorph assemblages from surface sections were correlated with the well assemblages. Subsequently, the important papers by Stover and Evans (1973) and Stover and Partridge (1973) provided more detailed, more complete, and reliable floral data than had been available up to that time. Ten palynomorph zones were recognized in the Late Cretaceous to Miocene section. "The zonation expresses an essentially uninterrupted though gradually changing sequence of palynologic assemblages and is based on the vertical ranges of about 150 spore-pollen species and on the compositional consistency of the palynomorph assemblages from each zone" (Stover and Partridge, 1973, p. 245). Stover and Evans (1973, p. 62) further stated:

> Because of the large number of core samples examined and the density of wells in the Gippsland Basin, the lateral continuity of zones can be clearly demonstrated and the same vertical succession of spore-pollen zones is seen repeatedly in well after well. There is no evidence of a

major palynologic break from the Late Cretaceous to the top of the Eocene section, or to express it in zone terms, from the *Nothofagidites senectus* to the *N. asperus* Zones. The zonal concepts and definitions established for the Gippsland Basin may prove to be equally applicable in the adjacent Bass Basin as well as the Otway Basin.

The species distribution chart presented in Stover and Partridge (1973, p. 246) does not show a single extinction at the Cretaceous-Tertiary boundary. This is not to claim that no extinctions occurred at that time, but rather that extinction, if any, may have been limited to taxa represented by very few specimens. The Cretaceous-Tertiary boundary is not identified clearly enough in this area to preclude the possibility that the boundary may occur at the horizon marked by the extinction of *Nothofagidites senectus*, although Stover and Partridge presented evidence that this taxon persists into the basal Paleocene before becoming extinct.

The data from Australia is more definite and conclusive than that from other areas of the earth. However, the evidence from all continents seems to point in the same direction. There appears to be no convincing palynological data supporting the idea of a world-wide catastrophe resulting in mass extinction of land plants.

## THE PHYSICAL WORLD DURING THE CRETACEOUS-TERTIARY TRANSITION

I can think of no more appropriate introduction to a discussion of world physical geology than is found in the words of Ager (1973, p. 83): "It would be utterly inappropriate in 1972 to try to consider the nature of the stratigraphical record, or indeed of any major aspect of geology, without seeking its relationship with the ideas of sea-floor spreading and plate tectonics. This is not just climbing on to a fashionable band-wagon, it is facing up to the fact that for the first time in the history of our science we are approaching a general theory of the earth." If this statement was appropriate in 1972, how much more applicable it is at present, after the accumulation of many more data that tend to confirm the theory of plate tectonics. Details of the major features of plate tectonics and sea-floor spreading are available from numerous sources and will not be elaborated upon here. An overview may be obtained by consulting Dietz and Holden (1971), Marvin (1973), and Bullard (1975).

It now appears certain that the positions and configurations of the continental land masses as well as the adjacent oceanic basins have changed drastically through time. As new sea floor appeared at mid-ocean ridges and as the continental plates were pushed apart or caused to collide with one another, the world biota could not have escaped some of the consequences. The changes in continental positions, configurations and land-water relationships were gradual, and the effects upon the land flora must also have been gradual. The only other land-based biotic group classically discussed in relation to catastrophic extinctions is the dinosaurs. Even the extinction of this group appears to be losing some of its catastrophic attributes. Axelrod and Bailey (1968) postulated that gradual changes in climatic equability during the Late Cretaceous time may have produced extremes of high and low temperatures that exceeded the tolerance limits of the dinosaurs. And Sloan (1976) also suggests that gradual changes on land caused the eventual extinction of the dinosaurs. "Dinosaur extinction was hardly instantaneous. While few details of distribution of the latest Cretaceous dinosaurs are known for most continents, it is known that at least in North America the process took place over about 12 million years" (Sloan, 1976, p. 136).

All the factors that were operative in restricting the marine biota at or near the Cretaceous-Tertiary boundary were not necessarily operative on the continents. From the Silurian to the present the land flora has been affected by fluctuations in climatic factors caused by the episodic shifting of continents and ocean basins. The restrictions of marine habitats caused by the withdrawal of the epeiric seas necessarily coincided with a gradual expansion of habitats of land plants as the seas retreated. The tempering effect of the warm shallow seas was removed, which resulted in increased continentality and decreased equability of land climates, with greater seasonal temperature extremes (Axelrod and Bailey, 1968; Hallam, 1973). The decrease in mean ocean temperature helped to induce a temperature drop on land. Krasilov (1975) noted a universal temperature drop everywhere at the end of the Cretaceous.

CONCLUSIONS

No prominent world-wide extinction of land plants at the end of Cretaceous time can be postulated from an examination of the pollen and spore record. The land flora during latest Cretaceous and early Tertiary time was changing owing primarily to the explosive evolution and diversification of the angiosperms and the concomitant reduction in proportional representation of gymnosperms, ferns and their allies. The land flora did not stabilize and attain its modern aspects until about late Eocene time. During the Late Cretaceous however, two prominent floral trends emerge.

Provincialism of floras began to be evident after Cenomanian time, and by Maastrichtian time several world floral provinces were well established. The pollen and spore record from the northern hemisphere provides the only evidence for a significant change at the end of the Cretaceous, and this change was most evident in the *Aquilapollenites* province. Many taxa that characterize the Late Cretaceous palynomorph flora became extinct at or near the Cretaceous-Tertiary boundary so that the *Aquilapollenites* province virtually lost its identity. A similar change in the Normapolles province cannot be documented, even though a significant proportion of taxa became extinct at that time (Table 11–1). Normapolles genera reached their acme of diversification during Campanian-Maastrichtian time and gradually declined to total extinction by early Oligocene time. The floral change at the end of the Cretaceous was accompanied in western United States by tectonism resulting in a rise in land surface, and in the Mississippi Embayment area by the final withdrawal of the epeiric sea (Tschudy, 1971). Dorf (1942) has suggested that the early Paleocene floral change was accentuated by the invasion in the Rocky Mountain region of plants heretofore prevented from migrating by the presence of the mid-continental seaway.

The southern hemisphere—South America, Africa and the equatorial belt extending to Indonesia—fails to reveal evidence for a pronounced floral change. In Australia, only extremely minor changes are seen at the Cretaceous-Tertiary boundary.

A second trend is revealed by the palynomorph record. Throughout the Maastrichtian, rocks from all floral provinces yield evidence of a gradual floral change. In some Maastrichtian floral sequences the rate of change is more prominent within the Maastrichtian than at the end of this time interval.

The palynological data, though somewhat incomplete, supports the hypothesis of a gradually changing world flora across the Cretaceous-Tertiary boundary with no more abrupt changes than may be observed at epoch boundaries or even within subdivisions of epochs. It is therefore unnecessary to invoke supranormal or catastrophic mechanisms to account for the observed changes. The causes of these observed changes in the land flora may be sought in the climatic changes brought about by the consequences of continental-plate movements and sea-floor spreading and by the normal operation of the evolutionary process.

## LITERATURE CITED

Ager, D. V., 1973. *The Nature of the Stratigraphical Record* (New York: John Wiley and Sons), 114 pp.

Axelrod, D. I., and H. P. Bailey, 1968. Cretaceous dinosaur extinction. *Evolution*, 22:595–611.

Belsky, C. Y., E. Boltenhagen, and R. Potonié, 1965. Sporae dispersae der Oberenkreide von Gabun, Aguatoriales Afrika. *Pal Zeitschr.*, 39 (1/2):72–83.

Berry, E. W., 1916. The Lower Eocene floras of southeastern North America. *U. S. Geol. Surv. Prof. Paper,* 91:1–481.

Boer, N. P. de, T. van der Hammen, and T. A. Wymstra, 1965. A palynological study on the age of some borehole samples from the Amazonas Delta area, N. W. Brazil. *Geol. Mijnbouw,* 44 (6)254–258.

Boltenhagen, E., 1965. Introduction à la palynologie stratigraphique du bassin sédimentaire de l'Afrique Équatoriale. *Bur. Rech. Geól. Min. Colloq. Intl. Micropal. Mém.,* 32:305–327.

―――, 1967. Spores et pollen du Crétacé supérieur du Gabon. *Pollen et Spores,* 9 (2):335–355.

Bratzeva, G. M., 1967. The problem of the Tsagainsk flora with regard to spore-and-pollen analytical data. *Rev. Palaeobot. Palynol.,* 2:119–126.

Bullard, E. C., 1975. Overview of plate tectonics. In: A. G. Fischer, et al. (eds.) *Petroleum and Global Tectonics* (Princeton, N. J.: Princeton Univ. Press), pp. 5–19.

Cookson, I., 1953. A palynological examination of no. 1 bore, Birregurra, Victoria. *Roy. Soc. Victoria Proc.* n.s., 66:119–128.

―――, 1960. New Zealand Mesozoic and Cainozoic plant microfossils. *New Zealand Geol. Surv., Pal. Bull.,* 32, 87 pp.

Dietz, R. S., and J. C. Holden, 1971. The breakup of Pangaea. *Continents Adrift* (San Francisco: Freeman and Company), pp. 102–113.

Dorf, E., 1942. Upper Cretaceous floras of the Rocky Mountain region: 1. Stratigraphy and paleontology of the Fox Hills and Lower Medicine Bow Formations of southern Wyoming and northwestern Colorado; 2. Flora of the Lance Formation at its type locality, Niobrara County, Wyoming. *Carnegie Inst. Wash. Contr. Pal. Publ. 508,* 168 pp.

Doubinger, J., 1972. Evolution de la flore (pollen et spores) au Chile central (Arauco), du Crétacé supérieur au Miocène. *Soc. Biogégr. Paris, C. R.,* 427:17–25.

Doyle, J. A., 1969. Cretaceous angiosperm pollen of the Atlantic Coastal Plain and its evolutionary significance. *Jour. Arnold Arb.,* 50 (1), 35 pp.

Germeraad, J. H., C. A. Hopping, and J. Muller, 1968. Palynology of Tertiary sediments from tropical areas. *Rev. Paleobot. Palynol.,* 6:189–348.

Gill, J. R., and W. A. Cobban, 1966. The Red Bird section of the Upper Cretaceous Pierre Shale in Wyoming. *U. S. Geol. Surv. Prof. Paper,* 393 (A):A1–A71.

Gleason, H. A., and A. Cronquist, 1964. *The Natural Geography of Plants* (New York and London: Columbia Univ. Press), 420 pp.

Góczán, F., J. J. Groot, W. Krutzsch, and B. Pacltová, 1967. Die Gattungen des "Stemma Hormapolles Pflug 1953b" (Angiospermae)—Neubeschreibungen und revision europäischer Formen (Oberkreide bis Eozän). *Pal. Abh., sec. B,* 2 (3):427–633.

Hallam, A., 1973. Provinciality diversity and extinction of Mesozoic marine invertebrates in relation to plate movements. In: D. H. Tarling and S. K. Runcorn (eds.), *Implications of Continental Drift to the Earth Sciences* (New York: Academic Press), vol. 1, pp. 287–294.

Hammen, T. van der, 1954. El desarrollo de la flora Colombiana en los periodos geologicos. 1. Maestrichtiano hasta Terciario mas inferior. *Bol Geol.*, 2 (1):49–106.

———, 1957. Climatic periodicity and evolution of South American Maastrichtian and Tertiary floras. *Bol. Geol.* 5 (2):49–91.

Hammen, T. van der, and T. A. Wymstra, 1964. A palynological study on the Tertiary and Upper Cretaceous of British Guiana. *Leidse Geol. Meded.*, 30:183–241.

Hammen, T. van der, T. A. Wymstra, and P. Leidelmeyer, 1961. Paleocene sediments in British Guiana and Surinam. *Geol. Mijnbouw,* 40 (6):231–232.

Herngreen, G. F. W., 1975. Palynology of middle and upper Cretaceous strata in Brazil. *Meded. Rijks Geol. Dienst.* n. ser., 26 (3):39–91.

Hoeken-Klinkerberg, R. M. J. van, 1966. Maastrichtian Paleocene and Eocene pollen and spores from Nigeria. *Leidse Geol. Meded.,* 38:37–48.

Holmes, A., 1965. *Principles of Physical Geology, revised ed.* (New York: The Ronald Press Co.), 1288 pp.

Jardine, S., and L. Magloire, 1963. Palynologie et stratigraphie du Cretace des bassins du Senégal et de Cote d'Ivoire. *Bur. Rech. Géol. Min., Colloq. Intl. Micropal., Mem.*, 32:187–245.

Jarzen, D. M., 1977. Angiosperm pollen as indicators of Cretaceous-Tertiary environments. In: Cretaceous-Tertiary extinction and possible terrestrial and extraterrestrial causes. *Syllogeus,* Canadian National Museum Toronto, 12:39–49.

Kedves, M., 1971. Présence de types sporomorphes importants dans les sédiments préquaternaires Égyptiens. *Act. Bot. Sci. Hung.*, 17 (3–4):371–378.

Kennedy, W. J., and W. A. Cobban, 1976. Aspects of ammonite biology, biogeography and biostratigraphy. (London: Pal. Assoc. London) Special Papers in Paleontology, n. 17, 94 pp.

Kieser, G., 1967. Quelques aspects particuliers de la palynologie du Crétacé supérieur du Sénégal. *Rev. Palaeobot. Palynol.*, 5 (1–4):199–210.

Krasilov, V. A., 1975. *Paleoecology of Terrestrial Plants—Basic Principles and Techniques* (New York: Wiley & Son), 283 pp.

Kulp, J. L., 1961. Geologic time scale. *Science,* 133 (3459):1105–1114.

Leffingwell, H. A., 1971. Palynology of the Lance (Late Cretaceous) and Fort Union (Paleocene) Formations of the type Lance area, Wyoming. In: R. M. Kosanke and A. T. Cross (eds.), Symposium on palynology of the Late Cretaceous and Early Tertiary. *Geol. Soc. Amer. Spec. Paper,* 127:1–64.

Leidelmeyer, P., 1966. The Paleocene and Lower Eocene pollen flora of British Guiana. *Leidse Geol. Meded.,* 38:49–70.

Marvin, U. B., 1973. *Continental Drift. The Evolution of a Concept* (Washington, D. C.: Smithsonian Inst. Press), p. 239.

Mchedlishvili, N. D., 1964. The significance of angiospermous plants for stratigraphy of upper Cretaceous deposits. *Vsesoiuznyi Neftanoi Nauchmoissledo-*

*vatelskii Geologo-Razvedochnyi Institut* (*WNIGRI*) *Trudy, 1965,* pp. 5–34. [Translated by the Canadian multilingual services division, 1976, G. S. C. Trans. no. 912.]

Müller, H., 1966. Palynological investigations of Cretaceous sediments in north-eastern Brazil. In: J. E. van Hinte (ed.), *Proceedings of the 2nd West African Micropalaeontological Colloquim, Ibadan,* 1965 (Leiden: Brill), pp. 123–136.

Muller, J., 1968. Palynology of the Pedawan and Plateau Sandstone Formations (Cretaceous-Eocene) in Sarawak, Malaysia. *Micropaleontology,* 14 (1):1–37.

———, 1970. Palynological evidence on early differentiation of angiosperms. *Biol. Rev.,* 45:417–450.

Norris, G., 1977. Phytoplankton changes near the Cretaceous-Tertiary boundary. In: Cretaceous-Tertiary extinctions and possible terrestrial and extraterrestrial causes. *Syllogeus,* Canadian National Museum Toronto, 12:51–57.

Norton, J. J., and J. W. Hall, 1969. Palynology of the upper Cretaceous and lower Tertiary in the type locality of the Hell Creek Formation, Montana, U. S. A. *Palaeontolographica,* sec. B, 125:1–64.

Obradovich, J. D., and W. A. Cobban, 1975. A time scale for the Late Cretaceous of the Western Interior of North America. In: W. G. E. Caldwell (ed.), The Cretaceous System in the Western Interior of North America. *Geol. Soc. Canada Spec. Paper,* 13:31–54.

Pflug, H. D., 1953. Zur Entstehun und Entwicklung des Angiospermiden Pollens in der Erdgeschichte. *Palaetonographica, sec. B,* 95:60–171.

Sanroma, N. S., 1970. Estudio esporo-polínico de la Formación Guaduas (Maastrichtiense-Paleoceno) en la Sabana de Bogotá (Columbia). (Barcelona: Universidad de Barcelona), pp. 1–10.

Sloan, R. E., 1976. The ecology of dinosaur extinction. In: C. S. Churcher (ed.), Athlon, Essays on palaeontology in honour of Loris Shano Russell. *Roy. Ontario Mus., Life Sci. Misc. Publ.,* pp. 134–154.

Snead, R. G., 1969. Microfloral diagnosis of the Cretaceous-Tertiary boundary, central Alberta. *Alberta Res. Counc. Bull.,* 25, 148 pp.

Sole de Porta, N., 1972. Palinologia de la formacion Cimarrona (Maastrichtiense) en el valle medio del Magdalena, Columbia. *Studia Geol.,* 4:103–142.

Srivastava, S. K., 1967. Upper Cretaceous palynology—a review. *Bot. Rev.,* 33 (3):260–288.

———, 1970. Pollen biostratigraphy and paleoecology of the Edmonton Formation (Maastrichtian), Alberta, Canada. *Paleogeogr. Paleoclimatol. Paleoecol.,* 7:221–276.

Stanley, E. A., 1965. Upper Cretaceous and Paleocene plant microfossils and Paleocene dinoflagellates and hystrichosphaerids from northwestern South Dakota. *Amer. Pal. Bull.,* 49 (222):175–384.

———, 1970. The stratigraphical, biogeographical, paleoautecological and evolutionary significance of the fossil pollen group Triprojectacites. *Georgia Acad. Sci. Bull.,* 28:1–44.

Stover, L. E., and P. R. Evans, 1973. Upper Cretaceous-Eocene spore-pollen zonation, offshore Gippsland Basin, Australia. *Geol. Soc. Australia, Sydney, Pub.,* 4:55–72.

Tschudy, R. H., 1971. Palynology of the Cretaceous-Tertiary boundary in the northern Rocky Mountain and Mississippi Embayment regions. In: R. M. Kosanke and A. T. Cross (eds.), Symposium on palynology of the Late Cretaceous and early Tertiary. *Geol. Soc. Amer. Spec. Paper,* 127:65–111.

————, 1975. Normapolles pollen from the Mississippi Embayment. *U.S. Geol. Surv. Prof. Paper,* 865, 42 pp.

Wijmstra, T. A., and T. van der Hammen, 1964. Palynological data on the age of the bauxite in British Guiana and Surinam. *Geol. Mijnbouw,* 43 (4):143.

Zaklinskaia, E. D., 1967. Palynological studies on Late Cretaceous-Palaeogene floral history and stratigraphy. *Rev. Palaeobot. Palynol.,* 2:141–146.

Chapter 12

# MAMMAL EVOLUTION NEAR THE CRETACEOUS-TERTIARY BOUNDARY

J. DAVID ARCHIBALD

Department of Biology and Peabody Museum of Natural History,
Yale University

WILLIAM A. CLEMENS

Department and Museum of Paleontology,
University of California

## INTRODUCTION

Since their discovery and recognition, dinosaurs have been viewed with fascination by laymen and scientists alike. Not the least of this fascination has been the proposed hypotheses for the causes of their extinction. Contemporaries of the dinosaurs, such as mammals, have understandably not been viewed with such wonderment. Explanations are not hard to find. Dinosaurs, mistakenly or not, seem so alien to us when compared with our modern mammalian dominated terrestrial biota, and their supposedly geologically instantaneous disappearance enhances this perception. In contrast, the diminutive mammals did not die out with the passing of the Mesozoic, but rather underwent an explosive radiation during and after this transition. To what degree dinosaurs and mammals interacted is not known; however, studies of mammals during the Cretaceous-Tertiary transition provide data not only concerning their own history, but also that of other terrestrial organisms, including dinosaurs.

Too many of the theories that consider Cretaceous-Tertiary terrestrial extinctions, particularly dinosaur extinction, suggest causation based on generalized data but do not attempt to corroborate theory with documentation from fossiliferous sequences that transgress this boundary. Such documentation will never prove one or another of the various theories, but as we hope to show, it affords a closer examination of some of them. Unfortunately, the aforementioned sequences are extremely rare. A brief review of Late Cretaceous-Early Paleocene mammalian faunas makes the point all too clearly.

Late Cretaceous mammals are known from Asia, Europe, and North and South America. The European record is represented by one therian lower molar (Ledoux et al., 1966). The South American record is a small collection from southern Peru near Lake Titicaca and is considered to be latest Cretaceous (Maastrichtian) based on charophytes (Grambast et al., 1967; Sigé, 1968, 1971, 1972). Although isolated, this occurrence is of interest for several reasons. Didelphid and pediomyid marsupials, which are common in the latest Cretaceous of North America, have been recognized at this site (the latter family only questionably). A possible condylarth has also been reported. Although a common element of Early Tertiary faunas in several parts of the world, latest Cretaceous condylarths are well documented only from eastern Montana. The South American material is tantalizing and further hints at the complexity of latest Cretaceous mammal evolution, particularly in paleobiogeographic patterns, but beyond that is not of great help.

In Asia, Late Cretaceous mammals are best known from collecting areas in the Gobi Desert, Mongolian People's Republic. The fossils include exquisitely presesrved skull and skeletal material of mammals as well as other vertebrates. Gradziński and others (1977) have estimated the range of these fossil-bearing beds to be ?upper Santonian and/or ?lower Campanian and ?lower Maastrichtian. These faunas are distinctly more "upland" than Late Cretaceous faunas of North America and because of the lack of marine ties, stage/age assignments are based on the grade of evolution of the terrestrial vertebrates and invertebrates. The only mammal possibly of latest Cretaceous age in Asia is a eucosmodontid multituberculate (Kielan-Jaworowska and Sochava, 1969; Trofimov, 1975), which Kielan-Jaworowska (1974) believed to be either Maastrichtian or Paleocene in age.

North American Late Cretaceous mammals are not as spectacularly preserved as those from Asia. However, the North American sequence is composed of samples that range in age from Early Campanian through Late Maastrichtian. A complete listing of these occurrences can be found in Clemens and others (1979). For the purposes of this paper, the most important occurrences are those that lie near the Cretaceous-Tertiary boundary. The best known of these are from Alberta (Lillegraven, 1969), Montana (Sloan and Van Valen, 1964; Van Valen and Sloan, 1965; Van Valen, 1978; Archibald, 1980), and Wyoming (Clemens, 1964, 1966, 1973).

It is of further importance to note that well-documented Early Paleocene mammalian faunas are known only from western North America. Various localities of this age are known from Colorado, Montana, New

Mexico, Utah, and Wyoming. If mammal-bearing sequences are sought that bracket the Cretaceous-Tertiary boundary, the list of areas is drastically reduced to one, eastern Montana. However, this regrettably unique area has lately been supplemented by a sequence in Alberta (Paul Johnston, written communication).

The sequence in eastern Montana was first reported from McCone County by Van Valen and Sloan (1965). Complementary fossil-bearing strata have been found approximately 70 kilometers to the west in Garfield County, Montana (Archibald and Clemens, 1977; Archibald, 1979, 1980; Clemens and Archibald, 1980).

We will briefly examine the nature of this record, emphasizing the mammalian portion, and we will suggest how this evidence affects various views of biological events in the terrestrial realm at the Cretaceous-Tertiary boundary. As has been suggested in this brief introduction, the terrestrial record of the Cretaceous-Tertiary transition is not well-known or well understood. The wider applicability therefore, of the conclusions that we will discuss for eastern Montana await testing.

## LITHOGRAPHIC AND TIME-STRATIGRAPHIC BOUNDARIES IN EASTERN MONTANA

In almost every study dealing with paleontologic and stratigraphic data, time, time-stratigraphic, and rock-stratigraphic units must be reconciled. Major controversies tend to arise most often when widely recognized time-stratigraphic/time boundaries are crossed, such as stage/ages. Understandably, but unfortunately, the controversy becomes even more acute when system/period and erathem/era boundaries are crossed. This has been to some extent the case in the terrestrial sequence in eastern Montana. Possibly more regrettable for work in Montana has been the entanglement of concepts based on time (including paleontologic evidence) with those of rock stratigraphy. Beyond the legalistic problem of defining the Cretaceous-Tertiary boundary in eastern Montana, is the problem of more accurately placing the boundary for use in discussing paleontologic events.

By definition and usage in eastern Montana, the Cretaceous-Tertiary boundary has come to be recognized as the contact between the Hell Creek and the Tullock formations. This dual boundary is usually given as the base of the lowest lignite above the last dinosaur. R. Brown (1952, p. 92) popularized this concept in the following "formula":

Search for remains of dinosaurs as high as they can be found. Then look for the first coal zone, no matter how thin. As no authentic, indigenous, dinosaurian bones have ever been found above the base of this zone, the contact is considered to be at that level. It marks not only the inauguration of new environmental conditions but also the beginning of Tertiary time, if the primary assumption that the passing of the last dinosaur signalled the close of the Cretaceous, be accepted.

The earliest paleontologist to realize the significance of the fossils in Garfield and McCone counties, Montana, Barnum Brown, did not propose such a formula, but his observations clearly are in this vein. He named the dinosaur-bearing "Hell Creek beds," and placed the contact with the overlying "lignite beds" (part of the present Tullock Formation) at the lowest lignite (1907, figure 3). In one passage he noted in italics (p. 835), *"It is a most remarkable and significant fact that in no instance has a fragment of dinosaur bones been found in or above the lignite series by any of our party during five years' work in this region."*

During the decades between B. Brown's observations and R. Brown's "formula," considerably differing views were held as to where the Cretaceous-Tertiary boundary and the Hell Creek-Tullock (or Fort Union) contact should be placed. Further adding to this debate was the concern as to whether the Paleocene was "real," and if so, whether it could be recognized in North America. It is beyond the scope of this paper to examine these controversies, but some of their history can be traced in the U.S. Geological Survey coal reports of the region from Calvert (1912) onwards.

Although our views are somewhat different today, especially because more detailed work has been done in the region, there still remains a need to determine the Cretaceous-Tertiary boundary and the Hell Creek-Tullock contact. We will touch on only the major conclusions here before considering in more detail the mammalian sequence in eastern Montana, as a more detailed discussion can be found in Archibald (1980).

To establish the Hell Creek-Tullock contact, unlike the Cretaceous-Teritary boundary, is only a regional problem. Most of our work has concentrated on the type area of the Hell Creek Formation along Hell Creek, Garfield County, but other parts of the county have been included. The Hell Creek Formation has been extended to most of eastern Montana, parts of western North and South Dakota, and parts of northern Wyoming. The name, and sometimes the lithologic attributes of the formation overlying the Hell Creek Formation vary in these different areas, but the overlying contact is taken to be the lowest lignite immediately above the last dinosaur. Such criteria have been extended farther south into east-central Wyoming for the contact between the Lance and Fort Union formations in the type area of the Lance Formation (for example, Clemens, 1964). North, in Alberta, a somewhat analogous series of coal seams are present. There, however, a coal or lignite is not used for both the Cretaceous-Tertiary boundary and a formational contact. The Cretaceous-Tertiary boundary occurs within the Scollard Formation (Russell and Singh, 1978; Lerbekmo et al., 1979). However, as in the more southern regions, a coal, in this case the Nevis or No. 13 (not the Ardley as reported by Russell and Singh, 1978 and D. A. Russell, written communication), occurs near the estimated Cretaceous-Tertiary boundary.

To sum up, coal deposition is clearly associated with the Cretaceous-Tertiary transition in a wide north-south belt, but it definitely commenced *during* the Late Cretaceous in Alberta (Gibson, 1977), whereas (by definition) the inception of major coal deposition occurred at to the Cretaceous-Tertiary boundary in our study area (Archibald, 1980). In Garfield County we continue to place the Hell Creek-Tullock contact at the base of the lowest persistent coal (or lignite). However, the presence or absence of dinosaurs is not used in the redefinition. Dinosaurs (unlike the coals) do not compose a substantial portion of the rock. Their extinction is a paleontologic event (or events) that should not enter into a consideration of a lithologic boundary. This does not of course diminish the problem of correlating lignites from one isolated outcrop to another. Within our study area, laterally traceable lignites change in lithology from coals to mudstones; they pinch, swell, or split into several units; they may be absent in some areas; and they are known to underlie *in situ* dinosaur remains. In this last instance, it has not been determined if the lignite is laterally persistent, and thus, by our redefinition, whether it should be taken as the base of the Tullock Formation. If this is determined to be the case, then the formational contact (that is, the first persistent lignite) in this particular case is latest Cretaceous in Age. This *cannot* be used as proof for the existence of Paleocene dinosaurs.

In our study area we have continued to use dinosaur extinction as the most utilitarian method of establishing the Cretaceous-Tertiary boundary. The highest dinosaur remains occur from about 1.8 meters to 6.1 meters below the formational contact, which we have averaged to an estimate of about 3.0 meters for simplicity. Even after eliminating the use of the formational contact as the Cretaceous-Tertiary boundary, the use of dinosaur extinction to mark this time/time-stratigraphic boundary is not without potential problems. For good reason, using extinctions to mark such boundaries has been a practice avoided by most invertebrate paleontologists (although the Cretaceous-Tertiary boundary is a notable exception).

There is not sufficient data to demonstrate that dinosaur extinction was a globally synchronous event (even within the framework of geologic time). Based on admittedly flimsy evidence, Van Valen and Sloan (1977) have suggested that dinosaurs may have become extinct later in southern regions. Butler and others (1977) have presented magnetostratigraphic evidence that suggested dinosaur extinction occurred later in the San Juan Basin, New Mexico than in the terminal Cretaceous marine extinctions in Italy, although this result has been questioned (for example, Lerbekmo et al., 1979). The synchroneity of dinosaur extinction may have occurred at different times in different environments.

Even if dinosaur extinction can be demonstrated to have been a globally synchronous event, there remains the problem of relating this to the type marine sections in Europe where the Cretaceous-Tertiary boundary was first defined. Because of the incongruity of the terrestrial and marine records, such correlations must be made through direct marine intertonguings, as can be done for the coastal lowland deposits of western North America, or by nonpaleontologic means such as magnetostratigraphy and radiometric dating.

Radiometric dating is the only method available for determining geochronologic dates and a variety of dates have been determined in or near our area of study (for example, Folinsbee et al., 1965, 1970; Shafiqullah et al., 1968; Lerbekmo et al., 1979). The margins of error for these dates are at least a million years. Based on our work in eastern Montana the span of time encompassing the mammalian sequence discussed in the following pages is probably less than a million years. Thus the radiometric dates do not afford fine enough resolution for correlation.

Magnetostratigraphy has been applied to two terrestrial sequences that cross the Cretaceous-Tertiary. One study was conducted in the San Juan Basin, New Mexico (Butler et al., 1977; Lindsay et al., 1978). The second was done in the Red Deer Valley, Alberta (Lerbekmo et al., 1979). As noted, Butler and Lindsay are also conducting magnetostratigraphic studies in eastern Montana, as well as in other areas in the western United States. Various controversies have arisen regarding the published studies (for example, Butler and Linsay, 1980; Lerbekmo et al., 1980), but if and when these are resolved, magnetostratigraphy may offer the most accurate and precise method for correlating paleontologically incongruous regions.

At present, then, correlations within the terrestrial sequences must rely on paleontologic data. Besides dinosaur extinction, palynofloral evidence has most often been used to estimate the Cretaceous-Tertiary boundary in continental sediments of western North America. In eastern Montana palynofloral studies are in general agreement in recognizing a floristic change near the Cretaceous-Tertiary boundary (Hall and Norton, 1967;

Norton and Hall, 1969; Oltz, 1969; Tschudy, 1970). The change appears to occur at the lowest, laterally persistent lignite, which corresponds to the Hell Creek-Tullock formational contact. As noted earlier, dinosaur extinction seems to have occurred at about 3.0 meters below this formational contact. The discrepancy of palynofloral change and dinosaur extinction may be artifical, but a similar relationship was reported in the Red Deer Valley, Alberta (Lerbekmo et al., 1979). Which, if either, of these paleontologic markers most closely approximates the Cretaceous-Tertiary boundary as defined on marine invertebrates in Europe is not certain. One magnetostratigraphic study (Lerbekmo et al., 1979) has argued that dinosaur extinction in Red Deer Valley, Alberta, and the Cretaceous-Tertiary foraminiferal extinctions in the Gubbio section, Italy, are very close in time. This result remains to be tested elsewhere. Our choice of dinosaur extinction to mark the Cretaceous-Tertiary boundary in eastern Montana is, as noted, a utilitarian decision. The data that suggest that this choice coincides with the marine extinctions in Europe are based on as yet poorly tested circumstantial evidence.

GEOLOGIC SETTING

In Garfield and McCone counties, Montana, the relatively flat-lying Hell Creek and Tullock formations are exposed in myriad badlands surrounding the Fort Peck Reservoir on the Missouri River (Figure 12-1a). The Hell Creek Formation overlies the Fox Hills Sandstone and Bearpaw Shale (Figure 12-1b). The last two formations represent, respectively, the final regressive strand line and marine deposits of the last marine incursion that covered eastern Montana during parts of the Campanian and Maastrichtian (Jensen and Varnes, 1964; Waage, 1968; Feldman, 1972; Gill and Cobban, 1973; Obradovich and Cobban, 1975).

The Hell Creek Formation is dominantly a prograding fluviatile deposit that built up a low, broad plain as the sea regressed toward the east. The superjacent Fort Union Group, particularly the Tullock Formation (probably lower and middle Paleocene) and lower parts of the Lebo Formation (middle Paleocene) represent, with slight sedimentological changes, a continuation of fluviatile deposition (Van Valen and Sloan, 1977; Archibald, 1980). Contemporaneously, farther to the east in North and South Dakota, sediments of the Cannonball Formation were being deposited in a greatly reduced epicontinental sea (Fox and Olsson, 1969; Sloan, 1970).

The Hell Creek Formation was named by Barnum Brown (1907, p. 829) for deposits "typically exposed on Hell Creek and nearby tributaries

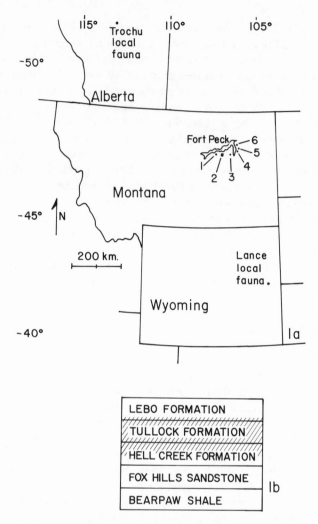

Figure 12-1. Index map showing fossil localities discussed in this paper, 1a, and a generalized stratigraphic section showing Cretaceous and Paleocene formations exposed in Garfield County, 1b.

In 1a:

1) McKeever Ranch localities;
2) Hell's Hollow local fauna, Garbani and Farrand channel deposits;
3) Flat Creek local fauna;
4) Bug Creek Anthills and Bug Creek West local faunas;
5) Purgatory Hill local fauna;
6) Harbicht Hill local fauna.

The diagonal lines in 1b show the interval that has been most thoroughly investigated in Garfield County.

of the Missouri River." He noted (p. 844), "Lithologically the Hell Creek beds of Montana are similar in almost every respect to the Ceratops beds (Lance Formation) of Converse County, Wyoming." The formation varies from 95 meters (B. Brown, 1907) to a reported 168 meters (Thom and Dobbin, 1924) in Garfield County. Although the latter figure may be too great, it is not unusual for the formation to vary in thickness because of its unconformable contact with the underlying Fox Hills Sandstone and general trend of thinning eastwards.

The lower part of the formation is primarily composed of sandstones (B. Brown, 1907; Jensen and Varnes, 1964), while the upper portion (approximately 50 meters) is dominated by light to medium gray siltstones. Some earlier geologists (for example, Collier and Knechtel, 1939) referred to these as the "somber" beds. However, beds in this unit are locally reddish purple, tan to medium brown, or greenish gray. No persistent lignites are present in the formation, but in the upper part localized carbonaceous shales and lignites up to 10 centimeters in thickness do occur infrequently. In some places, mudstones up to 3 meters in thickness contain numerous fresh-water molluscs and gastropods.

Most beds within the upper portion of the formation do not presist laterally for more than 0.4 kilometers in any direction. The siltstones represent small, silty channel-fill and associated floodplain deposits. Large, tan to brown, cross-stratified channel-fill deposits of sandstone occur rarely in the upper 30 meters of the formation (only in the upper 12 to 15 meters or less in Garfield County). These are the first of a series of cyclic deposits of alternating coals and large channels that occur in the overlying Tullock Formation.

As discussed previously, the boundary between the Hell Creek and Tullock formations is traditionally placed at the base of the lowest lignite (or coal) overlying the last dinosaur. We continue to place this boundary at the base of the lignite, but the presence or absence of dinosaurs is excluded as part of the definition. In most areas, the Tullock and Hell Creek formations can be distinguished on lithologic critera. Also the lowest lignite is the first of a series of lignites present throughout the stratigraphic extent of the Tullock Formation. In McCone County, the lignites in the Tullock Formation were designated as the Z (lowermost) through V (uppermost) coals by Collier and Knechtel (1939). This usage has been extended westward into Garfield County. As noted earlier, the lignites vary tremendously from outcrop to outcrop. There is a low probability that the same lignite can be traced over both McCone and Garfield counties. Thus the latter designation refers to a lignite in approximately the same stratigraphic position from one outcrop to the next (Figure 12–2). In many instances a lignite is actually composed of a series of lig-

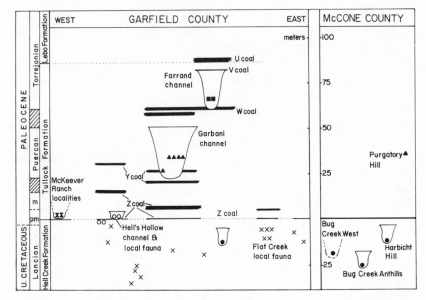

Figure 12-2. Summary of stratigraphic relationships in Garfield County, Montana. *Key to symbols on diagram:*

■ = Torrejonian localities;
▲ = Puercan localities;
ɪ = mantuan localities;
○ = pre-mantuan localities;
● = Bug Creek faunal-facies localities;
× = Hell Creek faunal-facies localities.

*Abbreviations in legend:*
pm = pre-mantuan;
m = mantuan

Localities in McCone County are given for comparison (Sloan and Van Valen, 1965; Sloan, 1976). Geographic placement is not to scale.

nites with various amounts of intervening non-lignitic sediment. In the case of the Z coal, the highest of this series is termed the upper Z coal, while the lowest is the lower Z coal.

The Tullock Formation is dominated by siltstones and fine-grained sandstones which are mostly tan to brown in color. Field observations indicate that this coloration is probably due to a greater amount of iron oxide and carbonaceous material than is seen in the Hell Creek Formation. Within the Tullock Formation, individual beds are usually traceable over several square kilometers. These represent the floodplain deposits associated with the rare, large channel-fill deposits found throughout the stratigraphic extent of the formation.

## FOSSIL LOCALITIES AND THEIR
## STRATIGRAPHIC CONTEXT

In the early 1900s the Hell Creek Formation became well known after Barnum Brown recovered the skeletons of several dinosaurs, most notably those of *Tyrannosaurus rex*, from its strata. Although a few mammals were also discovered by Brown and later studied by Simpson (1927), these were not well known until the 1960s when Sloan and Van Valen described rich microvertebrate sites in McCone County, Montana. These authors were also the first to describe mammals from a locally superjacent Tullock site, a small sample from Purgatory Hill in McCone County. The locality yielded isolated teeth of the early primate *Purgatorius* and such mammals as the multituberculate *Taeniolabis taoensis*, which suggest a late-Early Paleocene age. The current research in Garfield County began in the early 1970s after Harley Garbani discovered a rich mammal site in the Tullock Formation in the headwaters of Hell Creek. Since that time, a joint University of California-Natural History Museum, Los Angeles County, group has greatly expanded the search for fossil mammals in the two formations.

Megavertebrates (especially dinosaurs) are found irregularly throughout the lateral and most of the vertical extent of the Hell Creek Formation, although locally rich concentrations commonly occur in channel-fill deposits. In Garfield County, partial skeletons of dinosaurs, in addition to more numerous isolated teeth and bones, have been discovered up to 3 meters and possibly 1.8 meters below the Hell Creek-Tullock contact. Above this contact, several rich localities within the uppermost 3 meter interval of the Hell Creek Formation have yielded samples of local faunas including a variety of fish, amphibians, reptiles (including turtles, eosuchians, and crocodilians of large size), and mammals, but no dinosaurs. This negative evidence suggests that dinosaurs became extinct before deposition of the Hell Creek Formation ended and it is on this basis that we have approximated the Cretaceous-Tertiary boundary to be about 3 meters below the formational contact.

Two types of microvertebrate assemblages are recognized in much of the Hell Creek Formation, each generally associated with distinctive sediments (Figure 12–2). The two assemblages are the Bug Creek faunal-facies and the Hell Creek faunal-facies. These terms are used in a descriptive sense as compared to the various "communities" recognized by Estes and Berberian (1970) and Van Valen and Sloan (see 1977 and references).

Excluding the uppermost 3 meters, the Hell Creek faunal-facies localities occur stratigraphically throughout the formation, and most commonly are found in small, poorly sorted, silty channel-fill deposits. Although many Hell Creek faunal-facies localities have been discovered in the Hell Creek Formation, most produced only a few specimens, usually isolated teeth. The richest known locality of this faunal-facies is only 5.2 meters below the Hell Creek-Tullock contact. However, there is enough information to suggest that there was little, if any, evolutionary change in the mammals of the Hell Creek faunal-facies during the deposition of the formation.

Among the mammals of the Hell Creek faunal-facies, multituberculates and marsupials predominate. The taxa of this faunal-facies are similar to other Lancian local faunas from the type Lance Formation, Wyoming (Clemens, 1964, 1966, 1973) and the Scollard Formation, Alberta (Lillegraven, 1969). Collectively these are referred to as the "typical latest Cretaceous" mammalian local faunas.

In contrast, the Bug Creek faunal-facies localities preserve samples of Lancian local faunas in which eutherians and a different assemblage of multituberculates predominate. Localities yielding this faunal-facies are generally restricted to relatively sandier, usually larger channel-fill deposits in the upper 24 meters of the Hell Creek Formation. The Bug Creek faunal-facies is best represented in McCone County by three stratigraphically ordered localities—Bug Creek Anthills (lowest), Bug Creek West, and Harbicht Hill (highest)—described by Sloan and Van Valen (1965). This faunal-facies is only questionably known from Garfield County, based on two localities. Unlike the Hell Creek faunal-facies localities, the stratigraphically ordered series of Bug Creek faunal-facies localities (and overlying lowermost Paleocene localities) show rapid faunal turnover, allowing a biostratigraphic check of correlations based on physical stratigraphy (Van Valen and Sloan, 1977; Van Valen, 1978; Archibald, 1980).

Despite the differences in faunal content and environment of deposition, the two faunal-facies are coeval after the appearance of the Bug Creek faunal-facies 24 meters below the Hell Creek-Tullock formational contact. Thus both faunal-facies are referred to the Lancian land mammal age (Late Maastrichtian correlative). Both faunal-facies disappear as units about 3 meters below the formational contact (the Cretaceous-Tertiary boundary). However, as will be seen, the Paleocene mammal assemblages are essentially a continuation of the Bug Creek faunal-facies.

Within the Tullock Formation, four cycles of large (greater than 3 meters), sandy, mammal-producing, channel-fill deposits have been recognized in or near the valley of Hell Creek (Figure 12-2). The deposits are separated by regionally extensive lignites. The mammals suggest that these deposits range in age from earliest to Middle Paleocene.

The lowest of these four cycles, the Hell's Hollow channel deposit, includes two mammal-producing localities. The fossils from these localities constitute Hell's Hollow local fauna. As can be seen in Figure 12–2, the Hell's Hollow channel deposit is capped well below the upper Z coal and its base rests upon, and in some places may slightly breach, the lower Z coal. Earlier preliminary reports (Archibald, 1979; Clemens et al., 1979; Clemens and Archibald, 1980) were incorrect in placing this channel deposit and the included mammals in the uppermost Hell Creek Formation. Similarly, the suggestion of a latest Cretaceous age rather than an earliest Paleocene age for these localities is also incorrect. A reassignment to the earliest Paleocene is based on the lack of dinosaurs (although two water-worn, and probably re-worked hadrosaur teeth were recovered) and to a lesser extent on mammalian biostratigraphy. It is important to emphasize that reassignment of this deposit to the lowermost Tullock Formation is *not* the basis for the suggested earliest Paleocene age for the localities and fossils. Two new localities, one potentially very productive, have recently been discovered just below the lower Z coal within the Hell Creek Formation. The localities are now under study by the second author, and a preliminary analysis suggests that the mammals are similar to those from the Hell's Hollow local fauna. These new localities also lack dinosaurs and are thus in general agreement with our thesis that the uppermost approximately 3 meters of the Hell Creek Formation is earliest Paleocene in age. These two new localities are also noteworthy because they occur in paludal or overbank deposits. This is in marked contrast to the restriction of the Bug Creek faunal-facies localities to larger, sandy, channel-fill deposits. A similar restriction also appears to be the usual case for the mammal-producing localities in the Tullock Formation. As will be discussed below, the mammals from the Hell's Hollow local fauna and the two localities just below the lower Z coal are assigned to the informal premantuan "level" (approximately earliest Paleocene).

The next highest, mammal-producing channel-fill deposit in the Tullock Formation occurs about 15 kilometers northwest of the valley of Hell Creek (Figure 12–1). This deposit cannot be physically traced to the deposits in the valley of Hell Creek, and thus correlations are based on physical stratigraphy and faunal comparisons. The channel base and the two mammal-producing localities, the McKeever Ranch localities, rest immediately upon the lower Z coal. If complications involving the channel deposits are ignored, it can be seen in Figure 12–2 that the McKeever Ranch localities are stratigraphically lower than the Hell's Hollow local fauna localities. However, the small sample of mammals that have been

recovered from the McKeever Ranch localities suggest that these localities are younger than those in the Hell's Hollow channel deposits. This
example indicates the cautions that must be applied when correlating localities in channel-fill deposits. Assuming one cycle of deposition is involved, the channel base represents the maximum age of the channel deposit, while the capping marks represent the minimum age. The younger
age suggested by the mammals from the McKeever Ranch localities thus
predicts that the channel-capping for this deposit should be stratigraphically higher than that for the Hell's Hollow channel deposit. Unfortunately, the capping has not been found for the channel deposit that includes the McKeever Ranch localities, but this deposit appears to be
thicker than the Hell's Hollow channel deposit, and the former may simply have cut deeper into the underlying sediments. Other possible explanations are possible, such as mis-correlations of the lower Z coal, but we
favor the former explanation and accordingly believe the correlation
based on the mammals is more accurate. The mammals from the two
McKeever Ranch localities are informally referred to the mantuan
"level" (approximately earliest Paleocene).

The third highest channel-fill deposit, the Garbani channel deposit,
spans about 27 meters and is capped below the W coal and cuts into the
upper lignites assigned to the Y coal (Figure 12-2). Much of the fauna
from this deposit is currently being studied, but accounts of some of the
mammals have been published (Clemens, 1974; Novacek, 1977; Novacek
and Clemens, 1977). The mammals from the various localities in this deposit appear to be similar to the Purgatory Hill local fauna (Van Valen
and Sloan, 1965; Van Valen, 1978), and like the latter, appear to be
Puercan (early Paleocene) in age.

The fourth, and stratigraphically highest mammal-producing localities
in the Tullock Formation occur in the Farrand channel deposit. This deposit is capped immediately below the V coal and cuts into the top of the
underlying W coal (Figure 12-2). A very preliminary examination suggests a Torrejonian (middle Paleocene) age. The mammals from this deposit and the underlying Garbani channel deposit will not be discussed,
rather, attention will be focused on the mammals preceding and immediately following the Cretaceous-Tertiary boundary.

## MAMMALIAN BIOSTRATIGRAPHIC AND
## PALEOECOLOGIC COMPARISONS

For purposes of discussion, we divide the mammal-bearing localities
somewhat arbitrarily into two groups. The first group take in the Hell

Creek faunal-facies localities. The second group includes the Bug Creek faunal-facies localities and the pre-mantuan and mantuan "levels."

## HELL CREEK FAUNAL-FACIES

Before the discovery of moderately rich Hell Creek faunal-facies localities in Garfield County, large samples of "typical latest Cretaceous" mammals were known only from the Lance local fauna of Wyoming and Trochu local fauna of Alberta (local fauna used *sensu* Tedford, 1970). The localities in Garfield County are geographically intermediate between these other two areas, 530 kilometers northwest of Lance and 680 kilometers southeast of Trochu (Figure 12–1a).

Over 50 Hell Creek faunal-facies localities have been discovered in Garfield County. However, only one locality included in the Flat Creek local fauna has produced a large enough screen-washing sample for comparison with the Trochu and Lance local faunas. This local fauna includes all but five of the 31 mammalian species known to occur in the Hell Creek faunal-facies of eastern Montana. In Table 12–1, these five species are designated with an "0" in the column headed "Flat Creek local fauna." Mammals known from the Lance and Trochu local faunas, as well as Hell Creek faunal-facies mammals occurring at Bug Creek faunal-facies localities are also noted in this table. The first three columns are arranged in geographic sequence, with the more southerly Lance local fauna to the left.

The geographically intermediate position of the Flat Creek local fauna between those of Wyoming and Alberta is reflected in its taxonomic diversity and composition. Thirteen species are present in all three local faunas. This number is raised to 15 when taxa known from other Hell Creek faunal-facies localities are added to those from the Flat Creek local fauna. No species are common to the Lance and Trochu local faunas that are not also found at Hell Creek faunal-facies localities in Montana. However, four species are exclusively shared by the Hell Creek faunal-facies and Trochu, and eight species are shared exclusively by the Hell Creek faunal-facies and Lance.

It is intuitively obvious from the data in Table 12–1 that the Lance, Trochu, and Flat Creek local faunas are similar taxonomically. This degree of resemblance can be quantified through use of Simpson's (for example, 1947, 1965) index 100 $C/N_1$, (see Table 12–2 for definition of terms). Table 12–2 shows the results of these comparisons at the species level. Data in Tables 12–1 and 12–3 were used for the computations. Each taxonomic listing in these tables was assumed to represent a separate species, even if unnamed. Several of the taxa qualified by the term

| | Lance local fauna | Flat Creek local fauna | Trochu local fauna | Bug Creek Anthills local fauna | Paleocene "relatives" |
|---|---|---|---|---|---|
| Mesodma hensleighi | x | x | x | | † |
| Mesodma formosa (& cf. M. formosa) | x | x | x | | † |
| Mesodma thompsoni | x | x | x | | † |
| ?Neoplagiaulax n. sp. | | x | | | † |
| ?Neoplagiaulacidae gen. et sp. indet. | | x | | | ?† |
| Cimolodon nitidus | x | x | x | † | † |
| Kimbetohia campi | x | | | | † |
| Ptilodontidae gen. et sp. indet. | | | | x | ? |
| Cimolomys gracilis | x | x | x | † | |
| Cimolomys trochuus | | | | x | |
| Meniscoessus robustus | x | x | | † | |
| Essonodon browni | x | x | | † | |
| Cimexomys minor | x | 0 | | † | † |
| "Cimexomys" priscus (n. gen.) | | x | x | | |
| Alphadon marshi | x | x | x | † | † |
| Alphadon wilsoni | x | x· | x | † | |
| Alphadon lulli | x | x | | | |
| Alphadon rhaister | x | x | x | | |
| Alphadon cf. A. rhaister | ?x | | | | |
| Glasbius intricatus | x | | | | |
| Glasbius n. sp. | | x | | † | |
| Pediomys elegans | x | x | x | † | |
| Pediomys cooki (& cf. P. cooki) | x | x | | † | |
| Pediomys krejcii | x | x | x | † | |
| Pediomys hatcheri | x | x | x | † | |
| Pediomys florencae | x | x | | † | |
| ?Pediomys cf. P. florencae | | x | | | |
| Didelphodon vorax | x | x | x | † | |
| Didelphodon padanicus | ?x | | | | |
| Gypsonictops hypoconus | x | 0 | x | † | † |
| Gypsonictops illuminatus (& cf. G. illuminatus) | | x | x | ?† | † |
| Cimolestes incisus | x | x | | † | † |
| Cimolestes cerberoides (& cf. C. cerberoides) | | 0 | x | | † |
| Cimolestes propalaeoryctes | | x | x | | † |
| Cimolestes stirtoni | x | x | | | ?† † |
| Cimolestes magnus | x | 0 | x | † | † |
| Batodon tenuis | x | x | x | † | † |
| Telacodon laevis | x | | | | ? |

| | No. | (%) | No. | (%) "-0" | No. | (%) "+0" | No. | (%) |
|---|---|---|---|---|---|---|---|---|
| Multituberculata | 9 | (32) | 10 | (37) | 11 | (35) | 8 | (38) |
| Metatheria | 13 | (46) | 12 | (44) | 12 | (39) | 7 | (33) |
| Eutheria | 6 | (21) | 5 | (19) | 8 | (26) | 6 | (29) |
| Total | 28 | (100) | 27 | (100) | 31 | (100) | 21 | (100) |

"cf." are included with the respective species because they probably are conspecific. Others qualified by "cf." or "aff." that might be distinct species are listed separately.

The left column in Table 12–2 is a pair-wise comparison of the Lance, Trochu, and Flat Creek local faunas and the Flat Creek and Bug Creek Anthills local faunas. The central column is a similar comparison, but all species reported from the Hell Creek faunal-facies are used, that is, the Flat Creek local fauna plus five additional taxa. It is not certain whether the comparisons in the left or central column more accurately reflect faunal resemblances. We favor the results listed in the central column because the five species absent from the Flat Creek local fauna are everywhere rare members of the Hell Creek faunal-facies. Thus, their absence from the Flat Creek local fauna may be accidental. However, there are unresolved problems regarding two of these taxa, *Gypsonictops* (various species) and *Cimexomys minor*. First, it is still not clear how many and which species of *Gypsonictops* are present in the Hell Creek Formation. Second, recognition of *C. minor* at any given locality is dependent upon which teeth are found. Among isolated teeth, only P4/R's and M1/ of *Cimexomys minor* can be distinguished from those of *Mesodma*. Thus M/1's and M2/2's of *C. minor* may be present at some localities, but have not been correctly identified.

The significance of the high level of taxonomic resemblance between these local faunas is open to various interpretations. Similarly, the small differences could be a reflection of ecological diversity among contemporaneous local faunas, attributable to differences in age, or the result of the interaction of both factors.

Most geological evidence suggests the Lance, Flat Creek, and Trochu local faunas existed in low-lying coastal environments, although sedimentation types and rates, distances to the mid-continental seaway, floras, and such, varied to some degree. The high resemblance of these

---

Table 12–1. Hell Creek faunal-facies. The first three columns show species of "typical latest Cretaceous" mammals in the Lance (southernmost), Flat Creek (Intermediate), and Trochu (northernmost) local faunas. "0's" under the Flat Creek local fauna designate taxa known from other Hell Creek faunal-facies localities, but not the Flat Creek local fauna. The fourth column lists Hell Creek faunal-facies mammals known from the Bug Creek Anthills local fauna (see Table 12–3). The fifth column lists Hell Creek faunal-facies mammals that either persist into the Paleocene, are ancestral to Paleocene taxa, or share close common ancestry with a Paleocene species. The figures at the bottom are the number and relative frequences of the three major groups of "typical" latest Cretaceous mammal species in each local fauna. The two columns below the Flat Creek local fauna are with (+ "0") and without (− "0") the five additional Hell Creek faunal-facies mammals.

| $100C/N_1$ | | $100C/N_1$ | | Distance between the local faunas |
|---|---|---|---|---|

| Lance l.f. / Flat Creek l.f. | 77% | Lance l.f. / Hell Creek faunal-facies | 82% | 530 km. |
| Trochu l.f. / Flat Creek l.f. | 76% | Trochu l.f. / Hell Creek faunal-facies | 90% | 680 km. |
| Lance l.f. / Trochu l.f. | 71% | Lance l.f. / Trochu l.f. | 71% | 1190 km. |
| Flat Creek l.f. / Bug Creek Anthills l.f. | 63% | Hell Creek faunal-facies / Bug Creek Anthills l.f. | 79% | 25 km. |

Table 12–2. Faunal resemblance between various Lancian local faunas (l.f.) at the species level using Simpson's (for example, 1947, 1965) index $100C/N_1$. Data for the calculations are from tables 1 and 3. C is the number of faunal units (in this case species) in common between the two regions. $N_1$ is the number of faunal units (also species) in the smaller of the two samples. See text for further information.

mammalian local faunas at the species level appears to be a reflection of this environmental homogeneity.

A comparison of the faunal resemblance and distance between local faunas (Table 12–2) shows, not unexpectedly, that the local faunas that are most distant from one another tend to show least resemblance. The major exception is shown in the comparison between the Bug Creek Anthills local fauna (Bug Creek faunal-facies) and Flat Creek local fauna (Hell Creek faunal-facies). Although only 25 kilometers apart, these faunas do not show the greatest resemblance. Simpson (for example, 1965, pp. 123–124) presented an example from the Late Paleocene to show that ecologically similar faunas from distant areas may resemble one another more than ecologically different faunas that are in closer proximity. Other data suggest that the differences between the local faunas of the Bug Creek Anthills and the Flat Creek probably are in large part the result of ecological differences.

Previous authors understandably have been cautious in trying to assess the relative ages of the Lance, Trochu, and Bug Creek Anthills local faunas. Lillegraven (1969, p. 33 and p. 50), utilizing biostratigraphic correlations based on the multituberculates and metatherians, could find no evidence for significant difference in the ages of the Trochu and Lance local faunas. However, he (1969, p. 33) agreed with Van Valen and Sloan (1966, p. 265, figure 1), "that the Bug Creek Anthills locality and those stratigraphically above it in the Hell Creek Formation are slightly younger than the upper part of the Edmonton Formation" (= Scollard Formation). Although this might still be a possibility, stratigraphic correlations within the Hell Creek Formation demonstrate that many Hell Creek faunal-facies localities are coeval with, and possibly younger than, the majority of the Bug Creek faunal-facies localities. Thus, if it can be shown that the Flat Creek, Trochu, and Lance local faunas are of similar age, then there is no basis for considering the majority of the Bug Creek faunal-facies localities to be younger.

Biostratigraphic correlations based on organisms other than mammals have also been attempted. For example, L. S. Russell (1975, p. 155) has suggested that the apparent differences in eutherian abundance and diversity among local faunas in the Lance, Hell Creek, and Scollard formations were not due to age differences; rather, the similarities in the reptilian faunas "suggest that the eutherian distribution indicates a southward incursion of these mammals, not yet completed in Lancian time."

In summary, the degree of resolution apparently required for determination of the relative ages of the Lance, Trochu, and various Hell Creek local faunas is greater than that possible with current biostratigraphic data and techniques. Other techniques of correlation such as paleomagnetic sequencing are just beginning to be employed. Without a better basis for correlation a "simplistic dictum" phrased by Savage (1977, p. 431) is employed ". . . [L]acking evidence to the contrary organisms appearing most alike are most closely related, and organisms which are alike lived during the same geochronological interval, no matter how extensive their geographic ranges." Based on this "dictum" we consider that the high degree of faunal resemblance (minimum = 71%) at the *species* level among the Lance, Trochu, and Flat Creek local faunas reflects a close similarity in age. The ages of most of the Bug Creek faunal-facies local faunas also fall within the same interval of time, despite the differences between the two faunal-facies.

There are a number of interesting and consistent patterns in numerical abundance and diversity of the mammals in the Lance, Trochu, and Flat Creek faunas. Relative abundances of the species from these local faunas have not been determined, except for the species of *Alphadon* and *Pediomys* (Clemens, 1966, p. 85) and the genera (Estes and Berberian, 1970, table 1) in the Lance local fauna. However, it is possible to make qualitative comparisons based on the comments and data presented by Clemens (1964, 1966, 1973), Lillegraven (1969), and Archibald (1980).

As has been noted by several authors (Lillegraven, 1969; Clemens, 1973; and others), the Lance local fauna is dominated by marsupials and multituberculates in both numerical abundance and diversity. Clemens

| | Bug Creek Anthills local fauna | Bug Creek West local fauna | Harbicht Hill local fauna | Hell's Hollow local fauna | Billy Creek I (V72211) | McKeever Ranch I (V72210) | Hell Creek faunal-facies | Paleocene "relatives" | References |
|---|---|---|---|---|---|---|---|---|---|
| Cimolomys gracilis | x | | | | | | † | | 1 |
| Essonodon browni | x | | | | | | † | | 1 |
| Alphadon marshi | x | | | | | | † | † | 1 |
| Alphadon wilsoni | x | | | | | | † | | 2 |
| Didelphodon vorax | x | | | | | | † | | 1 |
| Gypsonictops hypoconus | x | | | | | | † | † | 1 |
| Cimolestes incisus | x | | | | | | † | † | 1 |
| Cimolestes magnus | x | | | | | | † | † | 2 |
| Batodon tenuis | x | | | | | | † | † | 3 |
| Mesodma sp. | x | cf. | | | | | ?† | | 4 |
| Cimolodon nitidus | x | x | | | | | † | † | 1 |
| Cimexomys minor | x | x | | | | | † | † | 1 |
| Stygimys kuszmauli | x | x | | | | | † | | 1 |
| Pediomys elegans | x | x | | | | | † | | 1 |
| Pediomys krejcii | x | ? | | | | | † | | 2 |
| Pediomys hatcheri | x | x | | | | | † | | 1 |
| Pediomys cooki | x | ? | | | | | † | | 2 |
| Pediomys florencae | x | ? | | | | | † | | 2 |
| Glasbius n. sp. | x | x | | | | | † | | 5 |
| Gypsonictops "petersoni" (=G. illuminatus ?) | x | x | | | | | † | † | 2 |
| Protungulatum donnae | x | x | | | | | | † | 1 |
| Meniscoessus robustus | x | x | x | | | | † | | 1 |
| Catopsalis joyneri | x | x | x | | | | † | | 1 |
| Procerberus formicarum | x | x | x | x | | | † | | 1 |
| Protungulatum cf. P. donnae | cf. | cf. | x | | | | † | | 5 |
| Protungulatum sp. M | | x | x | x | | | † | | 6 |
| Cimexomys sp. | | | | x | | | † | | 1 |
| Eucosmodontidae (undesc. gen. et sp.) | | | | x | | | ? | | 1 |
| Marsupialia (undescribed) | | | | x | | | ? | | 7 |
| Purgatorius ceratops | | | | x | | | † | | 8 |
| Mesodma aff. M. thompsoni | | | cf. | x | | | ?† | | 5 |
| Stygimys aff. S. kuszmauli | | | cf. | x | | | ?† | | 5 |
| Protungulatum sp. E. | | | x | x | | | † | | 6 |
| ? Oxyclaeninae gen. et sp. indet. | | | cf. | x | | | † | | 6 |
| Protungulatum sp. H. | | | x | ? | | cf. | † | | 6 |
| Cimexomys n. sp. | | | | x | | | † | | 5 |
| ?Eucosmodontidae gen. et sp. indet. | | | | x | | | † | | 5 |
| Catopsalis sp. | | | | x | | | † | | 5 |
| Peradectes cf. P. pusillius | | | | x | | | † | | 5 |
| Protungulatum cf. P. sp. M | | | | | x | | † | | 6 |
| Periptychidae n. gen. et sp. | | | | | | x | - | | 5 |
| ?Periptychidae gen. et sp. indet. | | | | | | x | - | | 5 |
| Number of species - level taxa at each locality | 24 | 14-17 | 14 | 11-12 | 1 | 3 | | | |

(1973, p. 94) has estimated that in numerical abundance the eutherians comprise only 10 to 15% of this local fauna, whereas in terms of diversity the marsupials comprise almost 50% (13 of 28 species-level taxa, Table 12–1) of the mammals.

In contrast, Lillegraven (1969, p. 50) has found that in the Trochu local fauna, eutherians rival the marsupials both in taxonomic diversity and numerical abundance (Table 12–1). Eutherians and marsupials each comprise about one-third of the taxa known from this local fauna, with multituberculates contributing slightly more than one third.

The Flat Creek local fauna appears to resemble most closely the Lance local fauna in the diversity of the three mammalian groups (Table 12–1). When the three species of eutherians known from other Hell Creek faunal-facies are added, eutherians comprise about one-quarter of the taxonomic diversity. However this figure is deceptive. The Hell Creek faunal-facies localities, taken in total, are essentially like a mixture of the Lance and Trochu local faunas, which is not surprising when we consider the geographically intermediate position of these localities. Thus, the Hell Creek faunal-facies localities share not only the increased relative diversity of the Trochu eutherians, but also the diversity of the marsupials found in the Lance Formation.

The Flat Creek local fauna is also like an admixture of the other two local faunas in terms of relative numerical abundances. Clemens (1966, Table 16) has reported that *Pediomys* is by far the most common genus of marsupial in the Lance Formation. In contrast, Lillegraven (1969) has found two species of *Alphadon, A. marshi* and *A. wilsoni,* to be numerically the most abundant marsupials in the Trochu local fauna. Unlike

---

Table 12–3. Species-level taxa in Lancian Bug Creek fauna-facies local faunas in McCone County, and in a pre-mantuan local fauna (Hell's Hollow) and at mantuan localities (McKeever Ranch) in Garfield County. The local faunas and localities are arranged in biostratigraphic (and stratigraphic) order with the oldest to the left. Question marks indicate questionable occurrences or uncertainties as to the correct species assignment. The sixth column lists Hell Creek faunal-facies mammals found in Lancian Bug Creek faunal-facies or younger local faunas. The seventh column lists Lancian mammals (some also shown in Table 12–1) that either persist into the Paleocene, are ancestral to Paleocene taxa, or share ancestry with a Paleocene species. Column eight gives the taxonomic references used in this table:

    1) Sloan and Van Valen, 1965, table 1;
    2) Estes and Berberian, 1970, faunal list, p. 7–8;
    3) Clemens, unpublished data;
    4) Novacek and Clemens, 1977;
    5) Archibald, 1980;
    6) Van Valen, 1978;
    7) UCMP collections;
    8) Van Valen and Sloan, 1965.

Trochu, the Flat Creek local fauna includes all the species *Pediomys* recognized in the Lance; yet, resembling Trochu, *A. marshi* and *A. wilsoni* are more abundant. The Flat Creek fauna is unique in the numerical abundance of a new species of *Glasbius*. In this local fauna it is as abundant as either of the species of *Alphadon*, whereas the Lance local fauna the closely related species, *G. intricatus*, is rare, and the genus is totally absent from Trochu. Finally, although the eutherians in the Hell Creek faunal-facies are as taxonomically diverse as those from Trochu, they are numerically as rare as those in the Lance local fauna.

The multituberculates are a numerically abundant and taxonomically diverse group throughout all three local faunas. In abundance and diversity the neoplagiaulacid *Mesodma* is the dominant genus. Most of the other multituberculate genera and species occur in all three local faunas; the few exceptions probably are due to slight differences in age and/or ecology.

The Lance local fauna includes a rare ptilodontid, *Kimbetohia campi*. *Cimexomys minor* is also present. As noted earlier, it is sometimes difficult to recognize the latter species. However, if the recorded occurrences are correct, this species has an odd distributional pattern. Besides being present in the Lance, it is known from both the Hell Creek and Bug Creek faunal-facies in McCone County, only the pre-mantuan "level" in Garfield County, and is absent from Trochu. If this genus is as cladistically primitive as suggested by Archibald (1980), this may be a relict distributional pattern.

The Flat Creek local fauna includes the earliest record of neoplagiaulacid of Paleocene aspect, a new species of *?Neoplagiaulax*. This multituberculate has a high, arcuate P/4-blade previously known only in Tertiary representatives of the family. The sample of this local fauna also contains a P/4 of *Mesodma thompsoni* that is almost indistinguishable in size from *Mesodma ambigua* of the earliest Paleocene Mantua lentil, Wyoming. However, two other equally large P/4's of *M. thompsoni* were recovered from localities that are thought to be more than 30.5 meters lower in the section. Also, a questionable neoplagiaulacid with a high, arcuate blade was recovered from one of these stratigraphically low localities.

The Trochu local fauna is odd in that it lacks the ubiquitous species *Meniscoessus robustus*. It also includes a rare ptilodontid, other than *Kimbetohia*, and the unique species *Cimolomys trochus*. If Lillegraven (1969) is correct in his identification and assignment of the latter species to the Cimolomyidae, it would be the first representative of this family known to lack the small, peg-like P/3.

Although there are differences between the Flat Creek, Trochu, and

Lance local faunas in taxonomic diversity and relative species abundance, the overall pattern is remarkably uniform over the large geographic area that these local faunas sample.

If these three local faunas are compared with successively earlier mammals known from North American there is also a suggestion of relative uniformity or stasis in a temporal sense during the latest Cretaceous. A moderately good sequence of localities have been found in North America beginning with terrestrial sediments that are approximately correlative with the lower Campanian. Approximately one-half of the mammalian genera known from the Flat Creek, Trochu, and Lance local faunas combined have been recognized by R. C. Fox at localities that are regarded as Aquilan in age (approximately Early Campanian correlative; see Clemens and others, 1979, for references). The sequence of faunas through the Lancian are dominated by ptilodontoid multituberculates and marsupials. There definitely are changes within this sequence, such as the increase in eutherians, but in contrast to the explosive radiation that begins in the Bug Creek faunal-facies, the former sequence is one of relative stasis.

BUG CREEK FAUNAL-FACIES, PRE-MANTUAN, AND MANTUAN

The three local faunas, Bug Creek Anthills, Bug Creek West, and Harbicht Hill, that Sloan and Van Valen described in 1965, show striking differences from other Lancian local faunas (Clemens, 1964, 1966, 1973; Lillegraven, 1969; Archibald, 1980). In addition to "typical latest Cretaceous" mammals, these local faunas include a number of mammals of "Paleocene aspect."

Sloan and Van Valen (1965; Van Valen and Sloan, 1977) have demonstrated that a turnover in species occurred in their stratigraphic sequence—Bug Creek Anthills, Bug Creek West, and Harbicht Hill. This was attributed to either immigration or in situ evolution of the "Paleocene aspect" mammals and progressive reduction in diversity and numbers of "typical latest Cretaceous" mammals. Dinosaurs show a similar decline. They are present at Harbicht Hill, but the relative abundance of *Triceratops* is reduced to 10% of what it was at Bug Creek Anthills (Van Valen and Sloan, 1977, p. 44). These authors have noted (1977, figure 4) that this decline of "typical latest Cretaceous" mammals may be in part an artifact of sample size. Bug Creek Anthills has yielded a tremendously larger number of specimens than any of the other localities. This sample includes some species (for example, *Essonodon browni*) that are usually rare even at rich Hell Creek faunal-facies localities.

Although there may be some biases due to differences in sample size, our research conducted to the west in Garfield County lends support to the trends recognized by Van Valen and Sloan. The Bug Creek faunal-facies is represented in Garfield County by only one or possibly two poorly known localities. However, two lowermost Paleocene mammal "levels" in the uppermost Hell Creek and lowermost Tullock formations of Garfield County are a faunal continuation of the Bug Creek faunal-facies. These are the pre-mantuan and mantuan "levels" noted earlier. Both "levels" are older than the Puercan land mammal age which was first characterized in the San Juan Basin, New Mexico.

As discussed below, the mantuan "level" is approximately contemporaneous with the Mantua lentil locality, Wyoming. Van Valen (1978) has proposed the Mantuan as a new land mammal age. Although we agree that Mantua lentil is older than known Puercan localities, we feel that the recognition of a new, admittedly short land mammal age would only add to the present confusion associated with the North American land mammal ages. Referral to a more concisely defined Puercan age is preferred, but until this is done the distinctiveness of these "levels" will be recognized informally. As implied by the name, we regard the pre-mantuan "level" as older than mantual localities from Montana and Wyoming. The pre-mantuan is best known by the Hell's Hollow local fauna as was discussed in the stratigraphy section.

Table 12–3 lists the mammals for the three Bug Creek faunal-facies of Sloan and Van Valen (1965, with modifications and additions) and the pre-mantuan (Hell's Hollow local fauna) and mantuan (McKeever Ranch localities) "levels" in Garfield County (Archibald, 1980). These are arranged in biostratigraphic order with the youngest to the right. This arrangement is also in accord with stratigraphic sequencing, with the possible exception of the McKeever Ranch localities. However, as we suggested previously, it seems likely that the channel deposit which includes these localities cut more deeply into the underlying sediments than the base of the probably older Hell's Hollow channel deposit.

As can be seen, the Harbicht Hill and Hell's Hollow local faunas may have as many as half of their species in common. A closer comparison may show that more species are common to both local faunas. For example, it is not known whether the undescribed Harbicht Hill eucosmodon-tid of Sloan and Van Valen (1965) and the new genus and species of microcosmodontine eucosmodontid of Archibald (1980) are the same. Similarly, Sloan and Van Valen (1965) have not noted any marsupials at Harbicht Hill, but a University of California crew recovered a marsupial dentary from this locality. However, the specimen is too worn to determine whether it belongs to a Lancian genus, such as *Alphadon,* or the earliest Paleocene *Peradectes.*

Most of the differences between the Harbicht Hill and the Hell's Hollow local faunas suggest the latter is younger. The few specimens of the taeniolabidid *Catopsalis* at Hell's Hollow are larger and more than *C. joyneri* at Harbicht Hill. They most resemble the Puercan *C. foliatus* but may belong to a new species known from a lower Paleocene locality in Colorado (M. Middleton, pers. comm.). The didelphid *Peradectes* is known only from the Tertiary and is a common taxon at Hell's Hollow. This taxon is absent or extremely rare at Harbicht Hill. A new species of the arctocyonid *Ragnarok* is present at Hell's Hollow (Archibald, 1980). It is derived in comparison with *R. harbichti* from Harbicht Hill. A single molar from Hell's Hollow that was questionably referred to the species of *Ragnarok,* may also belong to *R. harbichti* or possibly the mantuan species, *R. nordicum.*

Three Hell Creek faunal-facies genera, *Meniscoessus, Gypsonictops,* and *Cimolestes,* that have been recorded from Harbicht Hill (Sloan and Van Valen, 1965; Sloan, 1976); have not been recovered from Hell's Hollow. The latter two genera are not included in Table 12–3 because species designations have not been published (Sloan, 1976). Only two genera of Hell Creek faunal-facies, *Mesodma* and *Cimexomys,* and one species, *Cimexomys minor,* are present at Hell's Hollow. *Mesodma, Cimexomys,* and *Cimolestes* are known from Puercan species, so the occurrence of the first two genera at Hell's Hollow is not surprising. Finally, dinosaurs are present at Harbicht Hill, but not at Hell's Hollow (except for two probably reworked hadrosaur teeth).

Only one taxon, the primate *Purgatorius ceratops,* reported from Harbicht Hill but not from Hell's Hollow, suggests a younger age for Harbicht Hill. However, subsequent work at Harbicht Hill has not uncovered material beyond the original lower molar reported by Van Valen and Sloan (1965), and use of this species for biostratigraphic correlation at least for now remains suspect.

The mantuan "level" is represented by small samples from the McKeever Ranch localities (Table 12–3). Since only four taxa, all condylarths, are known from the McKeever Ranch localities, the assignment to mantuan rests entirely on joint occurrences between the McKeever Ranch localities and the richer Mantua lentil locality, Wyoming, and the relatively more derived evolutionary state of the condylarths at McKeever as compared with the Hell's Hollow local fauna. The two species in common at McKeever and Mantua lentil are the anisonchine *Mimatuta minuial* and the arctocyonid *Ragnarok nordicum.* The first species, and possibly the second, are more derived than representative of these genera at Hell's Hollow. A new species of *Protungulatum* (Archibald, 1980) recovered from one of the McKeever Ranch localities seems to be slightly more derived than *Protungulatum* cf. *P. donnae* known from Hell's Hollow. Finally, a dentary fragment preserving the third molar was found at one of the McKeever Ranch localities. Although it can be referred only questionably to the Periptychidae, it is distinctly more derived than any condylarth at Hell's Hollow.

## COMPARISON OF HELL CREEK AND BUG CREEK FAUNAL-FACIES

The evidence given by Sloan and Van Valen (1965 and later papers), Archibald (1980), and as summarized in this paper, clearly indicates mammals were undergoing rapid evolutionary change *within* the Bug Creek faunal-facies, and *within* this faunal-facies the relative abundance of dinosaurs and Hell Creek faunal-facies mammals was decreasing. A stratigraphic sequence of well-sampled Hell Creek faunal-facies localities are not known, but all indications point to a low mammalian speciation (or immigration) compared with that in the Bug Creek faunal-facies. There are a few exceptions, such as the appearance of *?Neoplagiaulax* in the Hell Creek Formation and *Kimbetohia campi* in the Lance Formation. Also, palaeoryctids may have radiated early in the Lancian and begun to spread southward during this time (Lillegraven, 1969; L. S. Russell, 1975). However, these are minor changes compared to geologically rapid events that were occurring within the Bug Creek faunal-facies.

Van Valen and Sloan (1977, pp. 43–44) noted that the diversity and relative abundance of dinosaurs and Hell Creek faunal-facies mammals decrease in the upper part of the Hell Creek Formation. However, they did not draw a distinction between Hell Creek and Bug Creek faunal-facies localities. As has been pointed out, the evidence from Garfield County supports their idea of decrease in diversity and abundance of dinosaurs and Hell Creek faunal-facies mammals *within* the Bug Creek faunal-facies. However, the data from Garfield County does not support the idea of a general decrease of diversity of mammals within the Hell Creek faunal-facies. The Flat Creek local fauna localities are only 5.2 meters below the base of the Tullock Formation (the base of the lower Z coal) and about 2.2 meters below the Cretaceous-Tertiary boundary, and yet the diversity of Hell Creek faunal-facies mammals is as great as in the Lance and Trochu local faunas. Dinosaurs, represented primarily by isolated teeth, are also common.

Other Hell Creek faunal-facies localities, although not so rich as those represented by the Flat Creek local faunas, have been discovered at levels up to at least 3 meters below the top of the formation. Dinosaurs are common at these sites, which are largely concentrations of bones in small, silty channel fillings. Evidence from surrounding floodplain deposits is somewhat contradictory. Van Valen and Sloan (1977, p. 44) have noted that dinosaur remains become increasingly scarce up-section, at least in the Bug Creek area. Dinosaur remains are also rare in certain parts of the upper Hell Creek Formation in Garfield County. However, in other places they are common. For example, parts of what appear to be seven individuals of *Triceratops* were found within approximately 0.4 square kilometers about 10.7 meters below the base of the Tullock. The specimen of *Tyrannosaurus rex* collected for the Natural History Museum, Los Angeles County, in the 1960s was within 18.3 meters of the base of the Tullock Formation. Teeth of *T. rex* have been found within 7.6 meters of the contact. Clearly a general survey is needed to establish whether dinosaur remains decrease in relative abundance in the floodplain deposits of the upper Hell Creek Formation.

The extinction of dinosaurs and the demise of the Hell Creek faunal-facies as a recognizable unit both occur at approximately 3 meters below the Hell Creek–Tullock formational contact. This level is used to mark the Cretaceous-Tertiary boundary. The only larger lower vertebrates that have been recovered from this 3-meter interval at the top of the Hell Creek Formation are turtles, eosuchians, and crocodilians. The only two mammal-producing localities thus far discovered in this interval are most similar in composition to the pre-mantuan Hell's Hollow local fauna which occurs near the base of the overlying Tullock Formation. Although the pre-mantuan localities on either side of the formational contact are considered to be Paleocene, they are clearly a continuation of the Bug Creek faunal-facies.

## MAMMALIAN EVOLUTION NEAR THE CRETACEOUS-TERTIARY BOUNDARY

The evidence of latest Cretaceous mammalian evolution presented here agrees in many aspects with models proposed by Sloan (1970, 1976) and Van Valen and Sloan (1977).

The latest Cretaceous local faunas of a generally similar composition, characterized by the presence of a variety of dinosaurs and "typical latest Cretaceous" mammals, appear to have inhabited low-lying, coastal environments from Alberta southwart to Wyoming. In eastern Montana, after deposition of the sandy, lowest part of the Hell Creek Formation, the sediments became generally finer grained. At that time animals of the Hell Creek faunal-facies probably lived on the broad flood plains and along the margins of small, sluggish, possibly ephemeral streams that crossed them.

In the upper part of the Hell Creek Formation larger, sandier channel-fill deposits record the development of larger streams in the area. First records of invading Bug Creek faunal-facies mammals are preserved in the stratigraphically lowest of these channel-fill deposits where they are mixed with the remains of dinosaurs and mammals of the Hell Creek faunal-facies. Members of the Hell Creek faunal-facies become progressively reduced in diversity and relative abundance in each stratigraphically successive large channel-fill deposit. In contrast, the taxonomic diversity of mammals of the Bug Creek faunal-facies increased within this sedimentary facies as a result of evolutionary radiations and immigration.

While these changes were occurring in the environments peripheral to the larger streams, along the margins of the smaller streams and on intervening flood plains, Hell Creek faunal-facies mammals and dinosaurs continued to exist. To date there is no evidence of significant evolutionary change in the lineages of Hell Creek faunal-facies mammals. However, a few new species of mammals dispersed into such areas. At the time of the beginning of deposition of the last approximately 3 meters of the Hell Creek Formation and after they probably disappeared from local faunas of the Bug Creek faunal-facies, dinosaurs became extinct. At the same time several lineages of Hell Creek faunal-facies mammals, particularly lineages of marsupials, also became extinct, and thus this faunal-facies ceased to exist as a unit.

Study of the faunas of the Hell Creek and Tullock formations of eastern Montana is being continued. The data presented in this interim report allow evaluation of some of the many hypotheses concerning the causes of extinction of dinosaurs and other organisms used to mark the end of the Cretaceous. Clearly what has been discovered in eastern Montana does not support hypotheses invoking a sudden, cataclysmic event as the causal factor. Although the possibility that such an event set the stage cannot be ruled out, the extinction of several lineages of terrestrial vertebrates involved a recognizable period of time.

Drawing on botanical data, Van Valen and Sloan (1977) have suggested that the climate of eastern Montana changed during the deposition of the Hell Creek Formation and was characterized by increasingly cooler winter temperature. Correlated with this change was a modification of the flora with replacement of subtropical by temperate forests. These authors have suggested that in these increasingly rigorous environments, extinction of dinosaurs and other vertebrates was the result of what they term "diffuse" competition with invading and evolving members of the Bug Creek faunal-facies. One test of the climatic portion of this hypothesis will be to determine whether the floral change occurred first along the banks of the major streams leaving refugia on the flood plains and along smaller streams where members of the Hell Creek faunal-facies survived.

The origin and evolution of Paleocene mammalian faunas currently is being studied by several paleontologists. This work is demonstrating that most Bug Creek and many Hell Creek faunal-facies mammals gave rise to members of Early Paleocene, North American faunas. Other lineages became extinct. It is not yet possible accurately to assess the magnitude of mammalian extinction at this time. However, comparison of the date in Tables 12–1 and 12–3 gives a crude estimate and suggests that the change in the mammalian fauna was similar to that usually seen between any given succession of North American land mammal ages.

## ACKNOWLEDGMENTS

This paper reports the results of a continuing research project in eastern Montana jointly sponsored by the Museum of Paleontology, University of California, Berkeley; and the Natural History Museum, Los Angeles County. The financial and other kinds of support given by these organizations, and by the Annie M. Alexander Endowment and the National Science Foundation (grants GB 39789 and BMS 75–21017) are gratefully acknowledged. We also greatly appreciate the help in field and laboratory work provided by many colleagues, too numerous to cite individually. Harley Garbani discovered the first Paleocene mammals in Garfield County and in following summers made great contributions to the collections of Cretaceous and Paleocene vertebrates on which this work is based. Both Leigh Van Valen and Michael Middleton provided us with data and stimulating discussion. David Schindel and John Ostrom read various drafts of the paper. Finally, we thank the Engdahl family and many other residents of Garfield County without whose help this work would not have been possible.

## LITERATURE CITED

Archibald, J. D., 1979. Mammalia. In: *McGraw-Hill 1979 Yearbook of Science and Technology* (New York: McGraw-Hill), pp. 278–280.
———, 1977. Fossil Mammalia and testudines of the Hell Creek Formation and the geology of the Tullock and Hell Creek Formations, Garfield County, Montana. Doctoral, Univ. of California, Berkeley, California, 705 pp.
Archibald, J. D., and W. A. Clemens, 1977. The beginnings of the age of mammals. *Jour. Pal.,* 51 (suppl. to no. 2):1 (abstract).
Brown, B., 1907. The Hell Creek beds of the Upper Cretaceous of Montana. *Amer. Mus. Nat. Hist. Bull.,* 23:823–845.
Brown, R., 1952. Tertiary strata in eastern Montana and western North and South Dakota. *Billings Geol. Soc. Guidebk.,* 3:89–92.

Butler, R. F., and E. H. Lindsay, 1980. Magnetostratigraphy, biostratigraphy, and geochronology of Cretaceous-Tertiary boundary sediments Red Deer Valley (a comment on Lerbekmo and others, 1979). *Nature*, 248:375.

Butler, R. F., E. H. Lindsay, and L. L. Jacobs, 1977. Magnetostratigraphy of the Cretaceous-Tertiary boundary in the San Juan Basin, New Mexico. *Nature*, 267:318–323.

Calvert, W. R., 1912. Geology of certain ignite fields in eastern Montana. Contributions to economic geology (short papers and preliminary reports). *U. S. Geol. Surv. Bull.*, 471:187–201.

Clemens, W. A., 1964. Fossil mammals of the type Lance Formation, Wyoming. Part I. Introduction and multituberculata. *Calif. Univ. Publ. Geol. Sci.*, 48:1–105.

———, 1966. Fossil mammals of the type Lance Formation, Wyoming. Part II. Marsupalia. *Calif. Univ. Publ. Geol. Sci.*, 62:1–222.

———, 1973. Fossil mammals of the type Lance Formation, Wyoming. Part III. Eutheria and summary. *Calif. Univ. Publ. Geol. Sci.*, 94:1–102.

———, 1974. *Purgatorius*, an early paromomyid primate (Mammalia). *Science*, 184:903–905.

Clemens, W. A., and J. D. Archibald, 1980. Evolution of terrestrial faunas during the Cretaceous-Tertiary transition. In: Écosystèmes continentaux de Mésozoïque. *Soc. Géol. France Mém.*, n. ser., t LIX:67–74.

Clemens, W. A., J. A. Lillegraven, E. H. Lindsay, and G. G. Simpson, 1979. Where, when, and what—a survey of known Mesozoic mammal distribution. In: J. A. Lillegraven, Z. Kielan-Jaworowska, and W. A. Clemens (eds.), *Mesozoic Mammals: the First Two-Thirds of Mammalian History* (Berkeley, Calif.: Univ. Calif. Press), pp. 7–58.

Collier, A. J., and M. Knechtel, 1939. The coal resources of McCone County, Montana. *U. S. Geol. Surv. Bull.*, 905:1–80.

Estes, R., and P. Barberian, 1970. Paleoecology of a Late Cretaceous vertebrate community from Montana. *Breviora*, 343:1–35.

Feldman, R. M., 1972. Stratigraphy and paleoecology of the Fox Hills Formation (Upper Cretaceous) of North Dakota. *North Dakota Geol. Surv. Bull.*, 61:1–65.

Folinsbee, R. E., H. Baadsgaard, and G. L. Cumming, 1970. Geochronology of the Cretaceous-Tertiary boundary of the western plains of North America. *Eclogae Geol. Helvetiae*, 63:91.

Folinsbee, R. E., H. Baadsgaard, G. L. Cumming, J. Mascimbene, and M. Shafigullah, 1965. Late Cretaceous radiometric dates from the Cypress Hills of western Canada. *Alberta Soc. Petrol. Geol. Ann. Field Conf. Guidebk.*, 15:161–174.

Fox, S. K., and R. K. Olsson, 1969. Danian planktonic foraminifera from the Cannonball Formation in North Dakota. *Jour. Pal.*, 43:1397–1404.

Gibson, D. W., 1977. Upper Cretaceous and Tertiary coal-bearing strata in the Drumheller-Ardley region, Red Deer River Valley, Alberta. *Geol. Surv. Canada Paper*, 76-35:1–41.

Gill, J. R., and W. A. Cobban, 1973. Stratigraphy and geologic history of the

Montana Group and equivalent rocks, Montana, Wyoming, and North and South Dakota. *U. S. Geol. Surv. Prof. Paper,* 776:1–37.

Gradziński, R., Z. Kielan-Jaworowska, and T. Maryańska, 1977. Upper Cretaceous Djadokhta, Barun Goyot and Nemegt formations of Mongolia, including remarks on previous subdivisions. *Acta Geol. Polonica,* 27:281–318.

Grambast, L., M. Martinez, M. Mattauer, and L. Thaler, 1967. *Perutherium altiplanense,* nov. gen., nov. sp., premier Mammifère Mésozoïque d'Amérique du Sud. *Acad. Sci. Paris C.R. Ser. D,* 264:707–710.

Hall, J. W., and N. J. Norton, 1967. Palynological evidence of floristic change across the Cretaceous-Tertiary boundary in eastern Montana. *Palaeogeogr. Palaeoclimatol. Palaeoecol.,* 3:121–131.

Jensen, F. S., and H. D. Varnes, 1964. Geology of the Fort Peck area, Garfield, McCone, and Valley counties, Montana. *U.S. Geol. Surv. Prof. Paper,* 414-F:1–49.

Kielan-Jaworowska, Z., 1974. Multituberculate succession in the Late Cretaceous of the Gobi Desert (Mongolia). In: Z. Kielan-Jaworowska (ed.), Results of the Polish-Mongolian Palaeontological Expedition, pt. 5. *Pal. Polonica,* 30:23–44.

Kielan-Jaworowska, Z., and A. V. Sochava, 1969. The first multituberculate from the uppermost Cretaceous of the Gobi Desert (Mongolia). *Acta Pal. Polonica,* 14:355–367.

Ledoux, J.-C., J.-L., Hartenberger, J. Michaux, J. Sudre, and L. Thaler, 1966. Découverte d'un Mammifère dans le Crétacé supérieur à Dinosaures de Champ-Garimond près de Fons (Gard). *Acad. Sci. Paris C.R., Ser. D,* 262:1925–1928.

Lerbekmo, J. F., M. E. Evans, and H. Baadsgaard, 1979. Magnetostratigraphy, biostratigraphy, and geochronology of Cretaceous-Tertiary boundary sediments, Red Deer Valley. *Nature,* 279:26–30.

———, 1980. Magnetostratigraphy, biostratigraphy, and geochronology of Cretaceous-Tertiary boundary sediments, Red Deer Valley (a reply to Butler and Lindsay, 1980). *Nature,* 284:376.

Lillegraven, J. A., 1969. Latest Cretaceous mammals of upper part of Edmonton Formation of Alberta, Canada, and a review of marsupial-placental dichotomy of mammalian evolution. *Kansas Univ. Contrib. (vert)* art. 50 (12):1–122.

Lindsay, E. H., L. L. Jacobs, and R. F. Butler, 1978. Biostratigraphy and magnetostratigraphy of Paleocene terrestrial deposits, San Juan Basin, New Mexico. *Geology,* 6:425–429.

Norton, N. J., and J. W. Hall, 1969. Palynology of the Upper Cretaceous and Lower Tertiary in the type locality of the Hell Creek Formation, Montana. *Palaeontographica, Abt. B,* 125:1–64.

Novacek, M. J., 1977. A review of Paleocene and Eocene Ieptictidae (Eutheria:Mammalia) from North America. *PaleoBios.,* 24:1–42.

Novacek, M. J., and W. A. Clemens, 1977. Aspects of intrageneric variation and evolution of *Mesodma* (Multituberculata, Mammalia). *Jour. Pal.,* 51:701–717.

Obradovich, J. D., and W. A. Cobban, 1975. A time-scale for the Cretaceous of the western interior of North America. *Geol. Assoc. Canada Spec. Paper*, 13:119–136.

Oltz, D. F., 1969. Numerical analyses of palynological data from Cretaceous and early Tertiary sediments in east central Montana. *Palaeontographica, Abt. B.*, 128:39–42.

Rogers, G. S., and W. Lee, 1923. Geology of the Tullock Creek coal field, Rosebud and Big Horn counties, Montana. *U.S. Geol. Surv. Bull.*, 749:1–181.

Russell, D. A., and C. Singh, 1978. The Cretaceous-Tertiary boundary in southcentral Alberta—a reappraisal based on dinosaurian and microfloral extinctions. *Canadian Jour. Earth Sci.*, 15:284–292.

Russell, L. S., 1975. Mammalian faunal succession in the Cretaceous System of western North America. *Geol. Assoc. Canada Spec. Paper*, 13:137–161.

Savage, D. E., 1977. Aspects of vertebrate paleontological stratigraphy and geochronology. In: E. G. Kauffman and J. E. Hazel (eds.), *Concepts and Methods of Biostratigraphy* (Stroudsburg, Penn.: Dowden, Hutchinson, and Ross, Inc.), pp. 427–442.

Shafiqullah, M., R. E. Folinsbee, H. Baadsgaard, G. L. Cumming, and J. F. Lerbekmo, 1968. Geochronology of Cretaceous-Tertiary boundary, Alberta, Canada. *Int. Geol. Congr., 22nd Session, India, 1964, Rep.*, 3:1–20.

Sigé, B., 1968. Dents de Micromammifères et fragments de coquilles d'oeufs de Dinosauriens dans la fauna de Vertébratés de Crétacé supérieur de Laguna Umayo (Andes péruviennes). *Acad. Sci. Paris C. R., Ser. D*, 267:1495–1498.

———, 1971. Les Didelphoidea de Laguna Umayo (formation Vilquichico, Crétacé supérieur, Pérou), et le peuplement marsupial d'Amerique du Sud. *Acad. Sci. Paris C. R., Ser. D*, 273:2479–2481.

———, 1972. La faunule de mammifères du Crétacé supérieur de Laguna Umayo (Andes péruviennes). *Mus. Natl. Hist. Nat. Bull.*, ser. 3, no. 99, Sciences de la Terre 19:375–409.

Simpson, G. G., 1927. Mammalian fauna of the Hell Creek Formation of Montana. *Amer. Mus. Novitates*, 267:1–7.

———, 1947. Holarctic mammalian faunas and continental relationships during the Cenozoic. *Geol. Soc. Amer. Bull.*, 58:613–688.

———, 1965. *The Geography of Evolution* (New York: Capricorn Books), 249 pp.

Sloan, R. E., 1970. Cretaceous and Paleocene terrestrial communities of western North America. In: E. L. Yochelson (ed.), *Proceedings of the North American Paleontology Convention (1969)* (Lawrence, Kansas: Allen Press, Inc.), pp. 427–453.

———, 1976. The ecology of dinosaur extinction. In: C. S. Churcher (ed.), Athlon. Essays on palaeontology in honor of Loris Shano Russell. *Roy. Ontario Mus. Life Sci. Misc. Publ.*, pp. 134–154.

Sloan, R. E., and L. S. Russell, 1974. Mammals from the St. Mary River Formation (Cretaceous) of southwestern Alberta. *Roy. Ontario Mus. Life Sci. Contrib.*, 95:1–20.

Sloan, R. E., and L. Van Valen, 1965. Cretaceous mammals from Montana. *Science*, 148:220–227.

Tedford, R. H., 1970. Principles and practices of mammalian geochronology in North America. In: E. L. Yochelson (ed.), *Proceedings of the North American Paleontology Convention (1969)* (Lawrence, Kansas: Allen Press, Inc.), pp. 666–703.

Thom, W. T., and C. E. Dobbin, 1924. Stratigraphy of Cretaceous-Eocene transition beds in eastern Montana and the Dakotas. *Geol. Soc. Amer. Bull.,* 35:481–505.

Tschudy, P. H. 1970. Palynology of the Cretaceous-Tertiary boundary in the northern Rocky Mountain and Mississippi Embankment regions. *Geol. Soc. Amer. Spec. Paper,* 127:65–111.

Trofimov, B. A., 1975. New data on *Buginbaatar* Kielan-Jaworowska et Sochava, 1969 (Mammalia, Multituberculata) from Mongolia. In: N. N. Kramarenko, et al. (eds.), Fossil fauna and flora of Mongolia. *Moscow, Acad. Nauk, Joint Soviet-Mongolian Pal. Exped. Trans.,* 2:7–13 (Russian with English summary).

Van Valen, L., 1978. The beginning of the age of mammals. *Evol. Theory,* 4:45–80.

Van Valen, L.,and R. E. Sloan, 1965. The earliest primates. *Science,* 150:743–745.

———, 1966. The extinction of the multituberculates. *Syst. Zool.,* 15:261–278.

———, 1977. Ecology and the extinction of the dinosaurs. *Evol. Theory,* 2:37–64.

Waage, K. M., 1968. The type Fox Hills Formation, Cretaceous (Maastrichtian), South Dakota. part 1. Stratigraphy and paleoenvironments. *Yale Univ. Peabody Mus. Nat Hist. Bull.,* 27:1–175.

Chapter 13

# TERMINAL CRETACEOUS EXTINCTIONS
# OF LARGE REPTILES

DALE A. RUSSELL

Paleobiology Division,
National Museum of Natural Sciences, Ottawa

## INTRODUCTION

Reptiles of unusual size, by modern standards, inhabited the seas, lands, and skies of our planet during the Mesozoic Era. As we have grown more familiar with the development of the great reptiles through Cretaceous time, so has our evolutionary sophistication and our deepened interest in the enigma of their sudden disappearance. It seems that nothing in their history prior to the end of the Cretaceous presaged their imminent extinction. The fossil record of the great reptiles as it pertains to the problem of their disappearance, is briefly reviewed in the first part of this presentation. Few would deny that these creatures vanished during an interval characterized by the extinction of many other species of organisms in a general biotic crisis. Some apparently paradoxical attributes of this crisis are therefore noted. Finally, I wish to present an apologia for my own inclination to regard the extinctions as a result of a brief, catastrophic environmental stress, and for my reluctance, in the present state of our knowledge, to reject a nearby supernova as a possible cause of the catastrophe.

## THE FOSSIL RECORD

Remains of large marine reptiles, including those of ichthyosaurs (McGowan, 1978), mosasaurs, and plesiosaurs have been collected in association with those of other marine organisms in strata of Maastrichtian age at localities as widely separated as Arctic Canada, Patagonia, Bulgaria and New Zealand (Welles, 1962; Russell, 1967, 1975b; Mabesoone

373

et al., 1968, Welles and Gregg, 1971; Azzaroli et al., 1975; Chong Dias, 1976; and references cited therein). It is not possible to define boreal and austral limits to the distribution of mosasaurs and plesiosaurs during Maastrichtian time, although plesiosaurs appear to be relatively more abundant in higher latitudes (see Valle and others, 1977 for a Campanian plesiosaur occurrence in Antarctica). In Belgium (studies in progress by the author) and the United States (Russell, 1975a), where sequences of relatively large assemblages of mosasaurs are known of Maastrichtian and Campanian-Maastrichtian age, respectively, there is no evidence of a decline in generic diversity as the Cretaceous-Tertiary transition is approached.

Isolated remains of the two great orders of large terrestrial reptiles popularly known as dinosaurs have been recovered from Upper Cretaceous localities scattered around the globe. These relatively minor occurrences contain information pertaining to the geographic distribution of dinosaurs and the relative frequency of localities that have produced dinosaur bone high in the Cretaceous. Among them may be cited a pedal phalanx of a Maastrichtian bird or theropod and the caudal vertebra of a peculiar dinosaur of uncertain affinities (Scarlett, pers. comm., 1976; Mölnar, pers. comm., 1980), from New Zealand, a Cenomanian sauropod from Queensland, Australia (Mölnar, pers. comm., 1977), and Senonian sauropods and duckbilled dinosaurs from Laos (Hoffet, 1943a, 1943b). Poorly preserved bones of theropod, sauropod, and ornithischian dinosaurs are known from several localities in the central provinces of India (Lapparent, 1957; Prasad, 1968; Coombs, 1971), and are presumed to be of late Upper Cretaceous age (see Jeletzky, 1962). Theropod and sauropod bones have been recovered in association with Maastrichtian guide fossils in southern India (Prasad, 1968: Govindan, 1972). Campanian theropods, sauropods, and orinthischians occur in Madagascar (Russell et al., 1976) and at several localities in northwestern Africa (Lapparent, 1957; Taquet, 1976). Two minor occurrences of duckbilled dinosaurs are known from Maastrichtian marine sediments in Belgium and the Crimea (Paris and Taquet, 1973).

Skeletal remains of duckbilled dinosaurs, some of which were found articulated, typically dominate sites of Cenomanian through Santonian age in Kazakhstan and Soviet Central Asia, where more than two dozen localities have been discovered over a vast region (Rozhdestvensky, 1967, 1973). Elements of other ornithischians and theropods of different sizes occur in subordinate amounts, and sauropod bones and dinosaur egg fragments have also been recovered. Hydraulic accumulations dominated by duckbilled dinosaur bones have been sampled along the Amur River in eastern Siberia, and one articulated skeleton was collected on the island of Sakhalin. These localities have been considered by Rozhdestvensky (1967, 1973) to be of Senonian age. The Wangshih Beds in the province of Shantung, China, probably of Campanian-Maastrichtian age, have produced generally disarticulated bones of several varieties of duckbilled dinosaurs, as well as those of theropods, armored dinosaurs, sauropods and dinosaur eggs (Young, 1958). Finally, skeletal material of dinosaurs is known from shallow marine or nearshore terrestrial sediments, of Santonian through Maastrichtian age, along the Atlantic and Pacific coasts of North America (Langston, 1960; Miller, 1967; Morris, 1967, 1973: Kaye and Russell, 1973; Mölnar, 1974; Gaffney, 1975: Baird and Horner, 1977).

None of the foregoing localities have yielded skeletal fragments in sufficient abundance to be considered adequate samples of the dinosaurs which were living in the vicinity of the site of deposition. These small samples do demonstrate the world-wide distribution of dinosaurs late in Cretaceous time, and the continued presence of sauropods as an important element in terrestrial faunas.

## DIVERSITY TRENDS

There are at least five regions on earth where two or more Upper Cretaceous dinosaur assemblages of different ages occur in general proximity to each other, and which have produced, or have the potential to produce, dinosaurian fossils in sufficient quantity to document diversity trends during this period.

A remarkable sequence of horizons has yielded dinosaur bones in Niger, on the southwestern edge of the Sahara (Taquet, 1976). The bones were evidently not articulated, and have not yet been identified below a subordinal level. It is hoped that future studies will generate information on dinosaurian diversity in then-equatorial Africa through Cenomanian to Maastrichtian time.

In Europe, two dinosaur assemblages from different horizons in Upper Cretaceous strata are sufficiently complete to be usefully compared. The early Campanian Gosau and Ajka Beds (Jeletzky, 1962) in Austria and Hungary, respectively, contain a peculiar fauna of small ornithischians and sauropods (Nopcsa, 1929; Huene, 1932; Coombs, 1971). Their relatively small size has been attributed to their occurrence on a small emergent area in central Europe. At least six genera have been recognized and distributed among five families, and more may be represented in collected material. The fauna could profitably be restudied. Late Maastrichtian (Rognacian) strata in southern France had produced a similar assemblage of relatively small dinosaurs in addition to large quantities of dinosaur eggs (Lapparent, 1947, 1967; Dughi and Sirugue, 1968; Paris and Taquet, 1973). Six genera belonging to five families have also been identified here. The two regions were evidently prospected with equal diligence and it can be concluded that no apparent decline in diversity exists.

In Argentina, dinosaurian bones and skeletal fragments (Huene, 1929) from the Neuquén and Chubut Groups are dominated by sauropods to an extent unequalled in other Cretaceous assemblages and comparable to conditions in some quarries of Late Jurassic age in Tanzania and the United States. Material of theropods, horned dinosaurs, and duckbilled dinosaurs has also been identified. The Neuquén Group is considered to lie within Coniacian to Maastrichtian limits (Leanza, 1972) and is overlain in northern Patagonia by Maastrichtian sediments of terrestrial and coastal facies containing associated material of duckbilled dinosaurs and eroded skeletal parts of other large reptiles (Casamiquela 1964, pers. comm., 1977; Volkheimer, 1973). The Chubut Group in southern Patagonia has produced dinosaurian remains at several levels. Higher strata in the Group, which also contain the youngest remains of dinosaurs known in the region, interdigitate with marine strata of Maastrichtian age (Volkheimer, 1969). The Upper Cretaceous of Argentina has already yielded a biogeographically and ecologically interesting sequence of dinosaurian occurrences. Future collecting and stratigraphic studies will probably produce as much information as is contained in comparable North American successions (see Bonaparte et al., 1977). Existing evidence appears to show no indication of a decline in dinosaurian diversity toward the end of Cretaceous time (Casamiquela, pers. comm., 1977).

In Mongolia, well-sampled and excellently preserved dinosaur assemblages have been collected from three successive horizons of Upper Cretaceous age (Kielan-Jaworowska, 1968-1975). The position of the assemblages within the global sequence of Upper Cretaceous marine stages is uncertain (Fox, 1978), although they are generally considered to fall within the latter part of this period. Collections in the Soviet, Polish, and Mongolian academies of science, and in the American Museum of Natural History, suggest that at least eight genera of dinosaurs and seven families are represented from the older Djadochta Formation, seven genera and seven families from the intermediate Barun Goyot Formation, and twelve genera and nine families from the later Nemegt Formation. Diversity differences between the Djadochta and Nemegt Formations are probably real, as more dinosaur specimens have been collected from the older unit.

Western Canada has produced Upper Cretaceous dinosaurs in greater number and variety than any other region on the globe. Twenty-seven valid species of dinosaurs have been recognized among the 320 specimens collected from the Oldman Formation in Dinosaur Provincial Park, Alberta, and they represent the largest number of dinosaurian species known from any single area of outcrop of terrestrial Mesozoic strata. Curiously, however, sauropod dinosaurs have not been identified in Canada, and Sloan (1970) has suggested that the animals preferred relatively warmer environments to the south. Four Campanian-Maastrichtian faunas from the Canadian prairies, as well as important assemblages from equivalent strata on the eastern flanks of the Rocky Mountains in the United States, were examined for diversity trends by me in a previous paper (Russell, 1975a). No significant temporal trend was found, although an obvious correlation was noted between diversity and sample size.

Van Valen and Sloan (1977) have proposed a model linking the extinctions to a global temperature decline wherein meridional terrestrial communities ("Cretaceous" facies) are gradually replaced by communities typical of higher latitudes ("Tertiary" facies). Evidence of dinosaurian decline is based on screenings from three microvertebrate localities in faunally and sedimentologically heterogeneous sediments separated by as much as 22 kilometers in central Montana. Existing collections of articulated dinosaur skeletons contain few well-preserved specimens from this region (only one skeleton of *Tyrannosaurus,* more than one-third complete, has been described). The relatively large number of incomplete and poorly preserved skeletal fragments suggests that, even though large areas of badlands have exposed the strata, sedimentological environments may not have been uniformly favorable for bone preservation. As has been noted by Brown (1907), dinosaurian macrofossils tend to be better preserved and therefore more often collected from sandstone units that lie near the base and center of the terrestrial terminal Cretaceous sequence. Field records of the American Museum of Natural History show no unusual decline in the frequency of dinosaurian skeletal fragments collected from the highest levels in the unit. It should also be noted that there is no indication of "Tertiary" or temperate forest communities lacking dinosaurs in high latitudes which are demonstrably of Cretaceous age (see Krassilov, 1978), and that the "Danian" duckbilled dinosaur from the Southern Hemisphere actually occurs in Maastrichtian strata (Casamiquela, pers. comm., 1977).

## THE MAASTRICHTIAN-DANIAN TRANSITION

It may be concluded that, whatever the course of other groups of organisms, large marine reptiles and dinosaurs show no evidence of a general decline prior to their extinction. Although diversity contained in collections from most sites is low, this is at least in part due to the great surface area of exposures needed to locate specimens and the great cost of excavating and cleaning skeletal fragments. Large dinosaurs were clearly more diversified in the Cretaceous of Alberta than are large mammals in Africa today (Béland and Russell, 1978). Remains of the great reptiles continue to occur in normal abundance and diversity in the highest biostratigraphic divisions of the Cretaceous, and they abruptly vanish at the level of the general biotic crisis which corresponds biostratigraphically to the Maastrichtian-Danian boundary (Jeletzsky, 1962: Béland et al., 1977; and many other authors). The nature of the crisis can be more economically and accurately evaluated through the study of fossils of less than dinosaurian dimensions. What, then, are some of the biologic incongruities which appear to have occurred at the Maastrichtian-Danian boundary?

1) On a global basis, about half as many genera of organisms have been described from strata above the boundary of early Tertiary age as from strata below the boundary of late or terminal Cretaceous age (Russell in Béland et al., 1977).

2) Freshwater organisms are more exposed to temperature changes, but extinctions of mollusks, fishes, and reptiles were much more severe in marine environments (Russell in Béland et al., 1977).

3) Dinosaurs and coral reefs existed in tropical environments during Upper Cretaceous time. Paleobotanists have never postulated a climatic deterioration to less than temperate conditions at the Cretaceous-Tertiary boundary in Cretaceous mid-latitudes (Jarzen in Béland et al., 1977) or even in Cretaceous polar regions (Rahmani and Hopkins, 1977). Coral reefs persisted in tropical refugia through Quaternary glacial maxima when many of these palynofloral sites were covered perennially by ice. Would a temperature decline that permitted bald cypress to survive in polar regions during basal Paleocene time have eliminated tropical refugia in equatorial regions?

4) Several different, severe, and geologically rapid sources of environmental stress (including alternating glacial and interglacial epochs, eustatic changes in sea level, and spread of human cultures) combined to produce extinctions during the Pleistocene. These stresses appear to have been on a scale at least as great as observed temperature and strandline fluctuations at the end of the Cretaceous, and have been operative during a one-million-year span. The resulting extinctions, both direct and as a result of trophic ramifications, have been much less severe (Russell, 1976) and more staggered in time (Webb, 1969) than those at the Cretaceous-Tertiary boundary.

A few of the many hypotheses which have been advanced to account for the terminal Cretaceous extinctions have been reviewed by me elsewhere (Russell, 1979). A brief case is here presented for not *a priori* rejecting a hypothesis involving an extraterrestrial source of biotic stress.

Life flourishes on earth within narrow physical limits. Electromagnetic energy from the sun drives weather systems and ocean currents and feeds the biosphere through photosynthesis. The solar wind sweeps low-energy cosmic radiation from our region of space. Colossal stellar explosions called supernovae are known to emit enormous quantities of electromagnetic and particulate ("cosmic") radiation (Tucker in Béland et al., 1977). Their frequency in our region of the galaxy can be measured through several lines of observational evidence and it is estimated that several supernovae should have been close enough to significantly affect terrestrial environments during the past half-billion years.

Tucker (in Béland et al., 1977) has shown that two intervals of biotic stress may result as a consequence of a nearby supernova, one within 30 years, from the arrival of a relativistic shock wave, and another between 3,000 and 30,000 years later, from the expanding shell of the supernova remnant. Reid (in Béland et al., 1977) and Reid and others (1978) have shown that the thermal properties and transparency of the atmosphere to solar radiation could be altered by high energy radiation from a supernova. Of direct relevance to the biosphere would be a cosmic-ray induced rise in background radiation at the base of the atmosphere to about 300 roentgens per year, and exposure of marine and terrestrial plants to relatively unattenuated ultraviolet radiation from the sun as a result of the removal of the ozone layer (Russell, 1979). Several other interesting extraterrestrial models are currently under consideration (see Alvarez et al., 1980). The sedimentary processes of our planet are strongly influenced by the biosphere. Tappan's thought-provoking synthesis (1968) suggests that if the biosphere were thrown into a state of acute disequilibrium by extinctions of organisms in essential, low trophic levels, it is conceivable that sedimentary and indeed planetary environments in general would be altered for thousands of years following an extinction event.

These and other sedimentological phenomena associated with the Cretaceous-Tertiary transition (Tappan, 1968; Percival and Fischer, 1977; Béland et al., 1977; and references cited therein) suggest to me that this relatively brief interval of geologic time presents problems which cannot be understood in terms of geological processes normally operative on the surface of our planet. Paleomagnetic studies will be of great assistance in locating the transition more precisely in time than by biostratigraphic means alone. Observations by Lowrie and Alvarez (1977) and cited in Percival and Fischer (1977) imply that the boundary occurs in an interval of reversed polarity in calcareous strata in northern Italy and Spain. Butler and others (1977) have confined the terminal Cretaceous extinction event to a part of a polarity column, derived from terrestrial sediments in

New Mexico, which they interpret to have been deposited 0.5 to 1.5 million years later than the extinction datum in marine facies in western Europe. Evidence of this diachroneity should appear as a substantial vertical offset of extinction levels where marine and continental strata interdigitate. This clearly does not seem to be the case in North Dakota (Fox and Olsson, 1969; Frye, 1969). The sedimentological context of the boundary at different localities will have to be carefully examined for possible short-term hiatuses (Foster in Béland et al., 1977).

It does not seem possible to me at present to eliminate extraterrestrial models from the complex of hypotheses under current consideration, nor can I recommend that others eliminate them from their thinking. The Cretaceous-Tertiary transition will remain a challenging and exciting area of research for many years to come. I look forward to joining my colleagues in different disciplines in seeking to better understand these mysterious extinctions and their place in the history of the earth.

## ACKNOWLEDGMENTS

I am grateful to Dr. Norman Newell, of the American Museum of Natural History, and to Drs. Denise Sigogneau-Russell and Donald E. Russell, of the Muséum National d'Histoire Naturelle, Paris, for their constructive comments which have substantially improved the content of this paper. The brisk exchange of ideas before, during, and after the North American Paleontological Convention with other members of the panel on the Cretaceous-Tertiary boundary problem has been most helpful in defining the scope and orientation of the approach adopted here. Drs. Barry Cox, of King's College, London, Eugene S. Gaffney, of the American Museum of Natural History, and John H. Ostrom, of Yale University, generously reviewed the manuscript and also made recommendations which have enhanced the clarity and content of the text. Finally, I wish to express my deep and sincere thanks to my wife, Dr. Janice Alberti Russell, for her patient and helpful counsel on a problem that has greatly vexed her husband. No one is more aware than she is that the last word has not yet been written on this fascinating subject.

## LITERATURE CITED

Alvarez, L. W., W. Alvarez, F. Asaro, and H. V. Michel, 1980. Extraterrestrial cause for the Cretaceous-Tertiary extinction. *Science,* 208:1095–1108.
Azzaroli, A., C. De Giuli, G. Ficcarelli, and D. Torre, 1975. Late Cretaceous mo-

sasaurs from the Sokoto district, Nigeria. *Firenze, Accad. Lincei, Atti, Cl. Sci. Fisiche, Ser. 8,* 13 (2):21–34.

Baird, D., and J. R. Horner, 1977. A fresh look at the dinosaurs of New Jersey and Delaware. *New Jersey Acad. Sci. Bull.,* 22:50.

Béland, P., P. Feldman, J. Foster, D. Jarzen, G. Norris, K. Pirozynski, G. Reid, J.-R. Roy, D. Russell, and W. Tucker, 1977. Cretaceous-Tertiary extinctions and possible terrestrial and extra-terrestrial causes. *Canada Nat. Mus. Syllogeus,* 12, 162 pp.

Béland, P., and D. A. Russell, 1978. Paleoecology of Dinosaur Provincial Park (Cretaceous), Alberta, interpreted from the distribution of articulated vertebrate remains. *Canadian Jour. Earth Sci.,* 15:1012–1024.

Bonaparte, J. F., J. A. Salfity, G. Bossi, and J. E. Powell, 1977. Hallazgo de dinosaurios y aves cretacicas en la Formacion Lecho de El Brete (Salta), proxima a limite de Tucuman. *Acta Geol. Lilloana,* 14:5–17.

Brown, B., 1970. The Hell Creek Beds of the Upper Cretaceous of Montana. *Amer. Mus. Nat. Hist. Bull.,* 23:823–845.

Butler, R. F., E. H. Lindsay, L. L. Jacobs, and N. M. Johnson, 1977. Magnetostratigraphy of the Cretaceous-Tertiary boundary in the San Juan Basin, New Mexico. *Nature,* 267:318–323.

Casamiquela, R. M., 1964. Sobre un dinosaurio hadrosaurido de la Argentina. *Ameghiniana,* 3:285–308.

Chong Diaz, G., and Z. Bradoni de Gasparini, 1976. Los vertebrados Mesozoicos de Chile su aporte geopaleontologico. *Argentina, Acta VI Congr. Geol.,* 26 pp.

Coombs, W. P., Jr., 1971. *The Ankylosauria.* Columbia Univ., Ph.D dissertation, 487 pp.

Dughi, R., and F. Sirugue, 1968. Marnes à oeufs d'Oiseaux du Paléocène de Basse-Provence. *Soc. Géol. France Bull.,* ser. 7, 10:542–548.

Fox, R. C., 1978. Upper Cretaceous terrestrial vertebrate stratigraphy of the Gobi Desert (Mongolian People's Republic) and western North America. *Geol. Assoc. Canada Spec. Paper,* 18:577–594.

Fox, S. K., Jr., and R. K. Olsson, 1969. Danian planktonic foraminifera from the Cannonball Formation in North Dakota. *Jour. Pal.,* 43:1397–1404.

Frye, C. I., 1969. Stratigraphy of the Hell Creek Formation in North Dakota. *North Dakota Geol. Surv. Bull.,* 54:1–65.

Gaffney, E. S., 1975. A revision of the sidenecked turtle *Taphrosphys sulcatus* from the Cretaceous of New Jersey. *Amer. Mus. Nat. Hist. Novitates,* 2571:1–24.

Govindan, A., 1972. Upper Cretaceous planktonic foraminifera from the Pondicherry area, south India. *Micropaleontology,* 18 (2):160–193.

Hoffet, J. H., 1943a. Description de quelques ossements de titanosauriens du Sénonien du Bas-Laos. *Con. Rech. Sci. Indochina, Compt. Rendu Seances Année 1942,* pp. 49–57.

⸻, 1943b. Description des ossements les plus caractéristiques appartenant à avipelviens du Sénonien du Bas-Laos. *Con. Rech. Sci. Indochina, Compt. Rendu Seances Année 1943,* pp. 1–8.

Huene, F., 1929. Los saurisquios y ornitisquios del Cretaceo argentino. *Asunción, Mus. La Plata, Ann.*, 3 (2):1–196.

⸻, 1932. Die fossile Reptil-Ordnung Saurischia, ihre Entwicklung und Geschichte. *Geol. Pal. Monats.*, 4:1–361.

Jeletzky, J. A., 1962. The allegedly Danian dinosaur-bearing rocks of the globe and the problem of the Mesozoic-Cenozoic boundary. *Jour. Pal.*, 36 (5):1005–1018.

Kaye, J. M., and D. A. Russell, 1973. The oldest record of hadrosaurian dinosaurs in North America. *Jour. Pal.*, 47:91–93.

Kielan-Jaworowska, Z. (ed.), 1968–1975. Results of the Polish-Mongolian Palaeontological Expeditions, parts 1–4. *Pal. Polonica*, nos. 19, 21, 25, 27, 30, 33.

Krassilov, V. A., 1978. Late Cretaceous gymnosperms from Sakhalin and the terminal Cretaceous event. *Palaeontology*, 21:893–905.

Langston, W., 1960. The vertebrate fauna of the Selma Formation of Alabama, part 6. the dinosaurs. *Chicago Nat. Hist. Mus., Fieldiana Geol. Mem.*, 6:313–363.

Lapparent, A. F., 1947. Les dinosauriens du Crétacé supérieur du Midi de la France. *Soc. Géol. France Mém.*, 56:1–51.

⸻, 1957. The Cretaceous dinosaurs of Africa and India. *Pal. Soc. India Jour.*, 2:109–112.

⸻, 1967. Les dinosaures de France. *Sciences*, 51:4–19.

Leanza, A. F. (ed.), 1972. Geologia regional de Argentina. *Acad. Nac. Cienc. Cordoba*, 869 pp.

Lowrie, W., and W. Alvarez, 1977. Upper Cretaceous magnetic stratigraphy. *Geol. Soc. Amer. Bull.*, 88:74–377.

Mabesoone, J. M., I. M. Tinoco, and P. N. Coutinho, 1968. The Mesozoic-Tertiary boundary in northeastern Brazil. *Palaeogeogr. Palaeoclimatol. Palaeoecol.*, 4:161–185.

McGowan, C., 1978. An isolated ichthyosaur coracoid from the Maastrichtian of New Jersey. *Canadian Jour. Earth Sci.*, 15:169–171.

Miller, H. W., 1967. Cretaceous vertebrates from Phoebus Landing, North Carolina. *Philadelphia Acad. Nat. Sci. Proc.*, 119:219–235.

Molnar, R. E., 1974. A distinctive theropod dinosaur from the Upper Cretaceous of Baja California (Mexico). *Jour. Pal.*, 48:1009–1017.

Morris, W. J., 1967. Baja California: late Cretaceous dinosaurs. *Science*, 155:1539–1541.

⸻, 1973. A review of Pacific coast hadrosaurs. *Jour. Pal.*, 47:551–561.

Nopcsa, F., 1929. Dinosaurierreste aus Siebenburgen, V. *Geol. Hungarica, Pal. ser.*, 4 (1):1–76.

Paris, J. P., and P. Taquet, 1973. Découverte d'un fragment de dentaire d'Hadrosaurien dans le Crétacé supérieur des Petites Pyrennes. *Mus. Natl. Hist. Nat., Bull., ser. 3* (130):18–27.

Percival, S. F., and A. G. Fischer, 1977. Changes in calcareous nannoplankton in the Cretaceous-Tertiary biotic crisis at Zumaya, Spain. *Evol. Theory*, 2:1–35.

Prasad, K. N., 1968. Some observations on the Cretaceous dinosaurs of India. *Geol. Soc. India Mem.*, 2:248–255.

Rahmani, R. A., and W. S. Hopkins, Jr., 1977. Geological and palynological interpretation of Eureka Sound Formation on Sabine Peninsula, northern Melville Island, District of Franklin. *Geol. Surv. Canada Paper*, 77-1B:185–189.

Reid, G. C., J. R., McAfee, and P. J. Crutzen, 1978. Effects of intense stratospheric ionization events. *Nature*, 275:489–492.

Rozhdestvensky, A. K., 1967. Late Mesozoic land vertebrates of the Asian part of the U. S. S. R. (in Russian). *Moscow, Acad. Nauk, U. S. S. R., Lab Kont. Obraz.*, pp. 82–92.

———, 1973. The study of Cretaceous reptiles in Russia (in Russian). *Pal. Zhur.*, 2:90–99.

Russell, D. A., 1967. Systematics and morphology of American mosasaurs. *Yale Univ. Peabody Mus. Nat. Hist. Bull.*, 23:1–51.

———, 1975a. Reptilian diversity and the Cretaceous-Tertiary transition in North America. *Geol. Assoc. Canada Spec. Paper*, 13:119–136.

———, 1975b. A new species of *Globidens* from South Dakota and a review of globidentine mosasaurs. *Chicago Nat. Hist. Mus., Fieldiana Geol*, 33 (13):235–256.

———, 1976. Mass extinctions of dinosaurs and mammals. *Nat. Canada*, 5 (2):18–24.

———, 1979. The enigma of the extinction of the dinosaurs. *Ann. Rev. Earth Planet. Sci.*, 7:163–182.

Russell, D. S., D. E. Russell, P. Taquet, and H. Thomas, 1976. Nouvelles récoltes de vertébrés dans les terrains continentaux du Crétacé supérieur de la région de Majunga (Madagascar). *Soc. Geol. Fr., C. R. somm.*, 5:205–208.

Sloan, R. E., 1970. Cretaceous and Paleocene terrestrial communities of western North America. *Proceedings of the North American Paleontological Convention, Section E*, pp. 427–453.

Tappan, H., 1968. Primary production, isotopes, extinctions and the atmosphere. *Palaeogeogr. Palaeoclimatol. Palaeoecol.*, 4:187–210.

Taquet, P., 1976. Géologie et Paléontologie du Gisement de Gadoufaoua. *Cahiers Pal.*, 191 pp.

Valle, R. A., F. Medina, and Z. Gasparini de Brandoni, 1977. Nota preliminar sobre el hallazgo de reptiles fósiles marions del suborden *Plesiosauria* en las islas James Ross y Vega, Antártida. *Inst. Antarct. Argentino Contrib.*, 212:1–13.

Van Valen, L., and R. E. Sloan, 1977. Ecology and the extinction of the dinosaurs. *Evol. Theory*, 2:37–64.

Volkheimer, W., 1969. Problemas del Grupo Chubut. *Ameghiniana*, 6 (2):173–180.

———, 1973. Observaciones geologicas en el area del Ingeniero Jacobacci y adyacencias. *Assoc. Geol. Argentina Rev.*, 28 (1):13–36.

Webb, S. D., 1969. Extinction-origination equilibria in Late Cenozoic mammals of North America. *Evolution*, 23 (4):688–702.

Welles, S. P., 1962. A new species of elasmosaur from the Aptian of Columbia and a review of the Cretaceous plesiosaurs. *Calif. Univ. Publ. Geol. Sci.,* 44 (1):1–96.

Welles, S. P., and D. R. Gregg, 1971. Late Cretaceous marine reptiles of New Zealand. *Canterbury Mus. Rec.,* 9 (1):1–111.

Young, C. C., 1958. The dinosaurian remains of Laiyang, Shantung. *Pal. Sinica,* (whole no. 142) n. ser. C, 16:49–138.

# Part III

# CATASTROPHIC PROCESSES IN THE
# GEOLOGICAL RECORD

Chapter 14

# LOW SEA LEVELS, DROUGHTS, AND MAMMALIAN EXTINCTIONS

NILS-AXEL MÖRNER

Geological Institute, Stockholm University

## INTRODUCTION

Cenozoic periods of drastic environmental changes and mammalian extinctions have been documented. Two major periods of changes are recognized: one beginning with the Eocene-Oligocene boundary at about 37 myr and one in the Late Miocene (Messinian) at about 5 to 6 myr. Both these periods were found to have corresponded with major sea level regressions.

These changes, however, must have been caused by mechanisms that were quite different from those of the Late Cenozoic that were caused by major global climatic changes of a fairly short duration. One of the best examples may be the 10,000-BP change in global climate (The Pleistocene-Holocene boundary) by Behrensmayer (1977), now also shown to correspond to a faunal change and a sudden appearance of a microlithic culture in East Africa.

The eustatic question has been thoroughly discussed in a separate paper in this volume (Mörner, Chapter 15). In this respect, reference is specifically made to Figure 15-3 in that paper. A eustatic change in the oceans will also change the geoid under the continents and hence affect the ground water table and the crustal movements (Mörner, Chapter 15 and 1977b). This obvious effect of a eustatic regression on the ground water table has not been discussed previously, although it seems to provide a solution to major Cenozoic changes in local environment and fauna without having to infer drastic global climatic changes of Pleistocene type, as well as for the dryness in equatorial and low-latitude regions during Ice Ages (Science, 1976).

## THE EOCENE-OLIGOCENE EVENT

Ingle (1977) has noted that the Oligocene began with a major cooling, sea level fall, and decrease in diversity in several faunal elements. Van Couvering (1977) has shown that the Eocene-Oligocene transition corresponds to changes from general forest to woodlands both in North America and in Central Asia, Bakker (1977) has demonstrated that the Eocene-Oligocene boundary corresponds to a major evolutionary discontinuity during which the top predator subguilds were emptied by extinction due to some environmental change initiated by a world-wide regression.

Well, what could have caused the sea level drop and the corresponding environmental and faunal changes? There are no records of Antarctic glaciation at this time. And even if there were, an Antarctic icecap increase is not likely to be linked to global climatic changes of the type we know from the Pleistocene.

The regression recorded must have been caused by orogeny and/or plate tectonics decreasing the ocean basin volume which led to a tectono-eustatic drop in sea level with a corresponding lowering of the geoid under the continents (Mörner, 1977b). A lowering of the geoid under a continent means lowering the ground water table and hence producing drought and other local environmental changes. This is exactly what Van Couvering (1977) has reported from the Eocene-Oligocene boundary. These environmental changes, in turn, explain the drastic faunal changes reported by Bakker (1977) and Ingle (1977). The change in vegetation affects the albedo and may therefore even lead to some general change in climate. Such an effect, together with minor glaciers in Antarctica, may explain the temperature drop recorded in some $\delta$ $O^{18}$ curves from the South Pacific (Shackleton and Kennett, 1975a).

Figure 14-1 illustrates the suggested mechanism and causal relationship for the Eocene-Oligocene event (see Mörner, 1977b).

## THE LATE MIOCENE (MESSINIAN) EVENT

The expansion of the Antarctic Ice Cap to the Queen Maud Glaciation (Mayewski and Goldthwait, 1977), the cooling in the South Pacific (Kennett and Vella, 1975; Shackleton and Kennet, 1975b) and the regression in New Zealand (Kennett, 1967) have been correlated with the regression in the Mediterranean during the Messinian (Berggren and Haq, 1976). Although this correlation may be questioned (Mörner, Chapter 15) and may need further confirmation, there can be little doubt that the Miocene

# Eocene/Oligocene Boundary Model

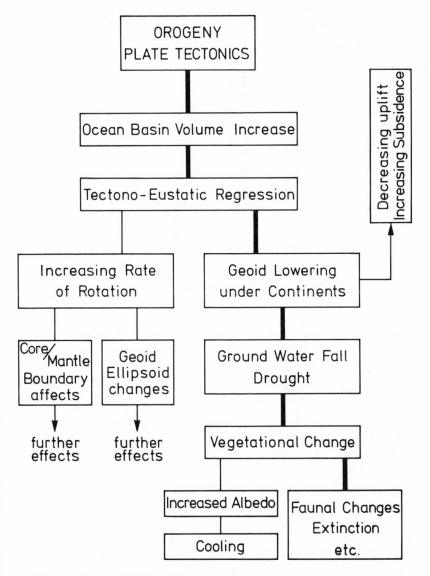

Figure 14-1. Multiple effect of a tectono-eustatic regression.
Left side is not complete. It may be noted, however, that "further effects" (like those in Figure 14–2) include possible geomagnetic field and geoidal-eustatic changes (Mörner, 1977). Thick line gives "straight" line from earth movements to extinction.

# Late Miocene (Messinian) Model

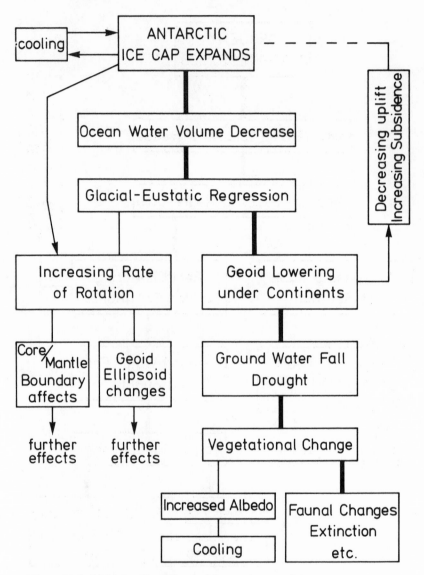

Figure 14–2. Multiple effects of a glacial-eustatic regression.
Left side is not complete. Thick line gives "straight" line from icecap formation to extinction.

ended with a world-wide glacial-eustatic regression. Ryan (1977) has presented solid geological-geophysical data on the distinct regression and erosion in the Mediterranean in relation to the formation of the huge salt layers of the Messinian. Van Couvering (1977) has shown that the land vegetation underwent a major change at about 5 myr (for example, during the Messinian); in both North America and Central Asia this marks the sudden beginning of prairie formation. Webb (1977) has pointed out that the North American vertebrate fauna records a mass extinction somewhere between 7 and 5 myr without any corresponding increase in the number of immigrants, which indicates an environmental origin of this extinction. From Florida, he has reported a simultaneous regression of at least 40 to 50 meters (followed by a transgression of at least 60 meters) Adams and others (1977) have been given a synthesis of the available data.

The suggested model for this Late Miocene (Messinian) regression-drought-extinction is given in Figure 14-2. The pattern is similar to that in Figure 14-1. However, in the latter the regression is caused by glacial-eustasy. No world-wide cooling, such as in the Pleistocene is suggested,

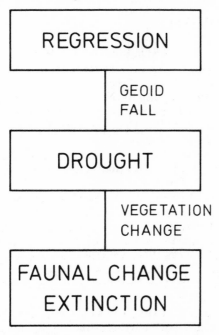

Figure 14-3. Main causal chain behind some of the major non-competitive extinction levels during the Cenozoic (for example, those recorded at the Eocene-Oligocene boundary and in the Late Miocene).

only a chain reaction on the regression, namely, the lowering of the geoid causing a general ground water fall giving rise to drought and vegetational changes that affected the fauna to such an extent that mass extinction is recorded (see Mörner, 1977b).

## SUMMARY

The present models offer a simple and logical explanation of the established correlation between regressions, vegetational changes, and extinctions during the Cenozoic without having to imply drastic and worldwide changes in climate like those characterizing the later part of the Pleistocene (the "Glacial Pleistocene"). The new factor that makes the chain logical is simply the lowering of the geoid under the continents and its effect on the ground water and the drought. Figure 14-3 summarizes the causal relationship and the main chain reaction: Regression→ Drought→Extinction→.

## LITERATURE CITED

Adams, C. G., R. H. Benson, R. B. Kidd, W. B. F. Ryan, and R. C. Wright, 1977. The Messinian salinity crisis and evidence of Late Miocene eustatic changes in the World Ocean. *Nature,* 269:383–386.

Bakker, R., 1977. "Directionalism vs. Non-competitive Catastrophe in Cenozoic Mammals." Woods Hole Symposium, oral presentation.

Behrensmeyer, A. K., 1977. "The Human Effect in the Pleistocene." Woods Hole Symposium, oral presentation.

Berggren, W. A., and B. ul Haq, 1976. The Andalusian State (Late Miocene): biostratigraphy, biochronology and paleoecology. *Palaeogeogr. Palaeoclimatol. Palaeoecol.,* 20:67–129.

Ingle, J., 1977. "The Oligocene Hiatus." Woods Hole Symposium, oral presentation.

Kennett, J. P., 1967. Recognition and correlation of the Kapitean State (Upper Miocene, New Zealand). *New Zealand Jour. Geol. Geophys.,* 10:1051–1063.

Kennett, J. P., and P. Vella, 1975. Late Cenozoic planktonic foraminifera and paleooceanography at DSDP Site 284 in the cool subtropical South Pacific. *Initial Reports of the Deep Sea Drilling Project* (Washington, D.C.: U.S. Govt. Printing Office), vol. 21, pp. 769–799.

Mayewski, P. A., and R. P. Goldthwait (in press). The glacial history of the Transantarctic Mountains: a record of the East Antarctic ice cap. *Jour. Glaciol.,* vol. 18.

Mörner, N.-A., 1977a. "Eustasy, Geoid Changes and Multiple Geophysical Interaction." Woods Hole Symposium, oral presentation.

————, 1977b. Palaeogeoid changes and palaeoecological changes in Africa with respect to real and apparent palaeoclimatic changes. *Palaeoecol. Africa,* 10–11:1–12.

Ryan, W. B. F., 1977. *The Messinian Event.* Woods Hole Symposium, oral presentation.

Science, 1976. Paleoclimate: Ice Age earth was cool and dry. *Science,* 191:455.

Shackleton, N. J. and J. P. Kennett, 1975a. Paleotemperature history of the Cenozoic and the initiation of Antarctic glaciation: oxygen and carbon isotope analyses in DSDP Sites 277, 279, and 281. *Initial Reports of the Deep Sea Drilling Project* (Washington, D. C.: U. S. Gov. Printing Office), vol. 24, pp. 743–755.

————, 1975b. Late Cenozoic oxygen and carbon isotopic changes in DSDP Site 284: implication for glacial history of the northern hemisphere and Antarctica. *Initial Reports of the Deep Sea Drilling Project* (Washington, D. C.: U. S. Govt. Printing Office), vol. 24, pp. 801–807.

Van Couvering, J., 1977. "The Grassland Biome." Woods Hole Symposium, oral presentation.

Webb, S. D., 1977. "Faunal Equilibria and Extinctions." Woods Hole Symposium, oral presentation.

# Chapter 15

# EUSTASY, GEOID CHANGES,
# AND MULTIPLE GEOPHYSICAL INTERACTION

NILS-AXEL MÖRNER

Geological Institute, Stockholm University

## INTRODUCTION AND METAPHORS

The effects on eustasy of geoidal changes have been discussed by Mörner (1975, 1976, 1977a, 1977b, 1979a), Reyment and Mörner (1977) and Newman and others (1977, 1979). The effects on eustasy of glacial mass attraction of water and of the visco-elastic response of the earth to deglaciation have been discussed by Clark (1976), Farrell and Clark (1976) and Clark and others (1978). This has made the cause and nature of ocean level changes complicated, and has called for a redefinition of the old concept of "eustasy" (Mörner, 1976).

First, two metaphors will explain the new situation:

1) The oceans can be compared with a rubber cup of water (Figure 15-1). The size of the cup (= ocean basin volume) can be changed by temper (= earth movements) compressing and expanding the cup and causing the water level to rise and fall correspondingly (= tectono-eustasy). The water volume in the cup can be changed by drinking and refilling (= climate), giving rise to corresponding rises (refilling = glacial melting) and falls (drinking = glacial accumulation) in the water level (= glacial-eustasy). Consequently, temper and thirst (= earth movements and climate) determine the level of the water in the cup (= the eustatic ocean level). However, in this strange cup, the water table is not a flat surface but a rough and uneven one (= the geoid or geodetic sea level): nothing but witchcraft (= gravity). Any change in the witchcraft gives rise to redistribution of the irregularities in the water surface (= geoidal-eustasy).

2) The earth as a whole can be compared with an amoeba that changes

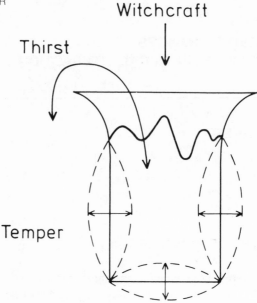

Witchcraft

Thirst

Temper

Figure 15-1. Metaphor 1: the ocean as a rubber cup of water.

its body shape with respect to active forces. If it is pressed on one side, the other side projects correspondingly. The visco-elastic earth exhibits a similar amoeboid sensitivity to changes in load, mass redistribution, and rotation.

## EUSTASY AND ITS NEW DEFINITION

Eustasy was first defined as the oceans' own changes in opposition to crustal (tectonic and isostatic) movements. Because the eustatic ocean level changes were always thought to affect the oceans globally, eustasy was also defined as "world-wide simultaneous changes in sea level" as distinguished from local (relative) sea level changes.

With the introduction of geoid changes (Mörner, 1976), the old concept of eustasy and its global validity must be accordingly changed. The best definition of eustasy is today simply "ocean level changes" (Figure 15-2) regardless of causation and implying vertical—global or local—movements of the ocean surface at a particular point (excluding local meteorological, hydrological, and oceanographic changes). Three factors determine the eustatic ocean level changes:

1) climate changing the ocean water volume and hence giving rise to glacial-eustatic changes that would have been of equal size all over the globe had the attraction and rotation potentials been affected and made the ocean surface changes irregular over the globe;

2) earth movements changing the ocean basin volume, hence giving rise to tectono-eustatic changes of equal size all over the globe had the attraction and rotation potentials not been affected and made the ocean surface changes irregular over the globe;

3) gravitational (and gravitational-rotational) changes of the ocean level distribution giving rise to geoidal-eustatic changes that are of totally different size, and even sign, over the globe (though they may occur simultaneously).

The relative sea level changes (that is, the field data we investigate) are the function of crustal and eustatic changes, as shown in Figure 15-2.

Figure 15-3 gives an integrated scheme of the "processing" of input changes in:

1) external gravity (of the universe, the galaxies or the solar system),
2) rotation (the rate or the tilt), and
3) mass redistribution, and the output changes in "eustasy," "hydrology," and "isostasy." The scheme illustrates and summarizes the compli-

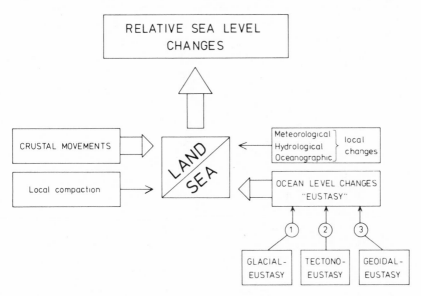

Figure 15–2. Metaphor 2: the earth as an amoeba.

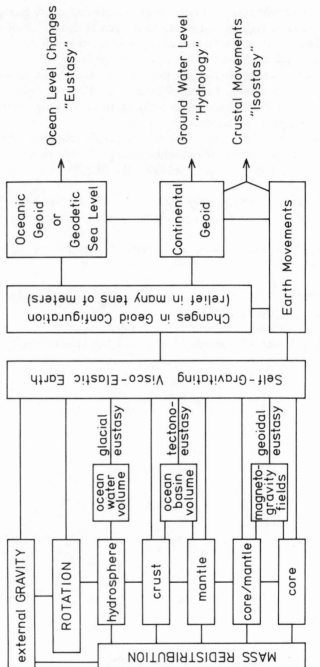

Figure 15-3. The influence of eustatic sea level change on ground water table.

cated multiple geophysical interaction behind ocean level changes (and its relation to ground water movements and crustal movements).

## THE GEOID AND ITS INSTABILITY

The geoid is an equipotential surface determined by the attraction and rotation potentials. The ocean geoid is often termed the geodetic sea level. The geoid relief amounts to several tens of meters with a 180-meter sea level difference between the geoid hump at New Guinea and the geoid depression of the Maldive Islands (a distance of only 50 to 60° longitude). Figure 15–4 gives geoid profiles for every 10° latitude of the globe (as calculated from the SSE III geoid map of Gaposchkin, 1973).

The present geoid configuration cannot, of course, have remained stationary back in time. It must have changed with changes in the attraction and rotation potentials (Figure 15–5). The main question is not whether the geoid has changed, but over what time units geoid changes have played a significant role (Mörner, 1976).

While the earth's ellipsoid is described by the second harmonic, the geoid configuration (relief) is described by the higher harmonics. A power spectrum of the gravity field models possesses the highest degree of variance for the lower seven harmonics (Marsh and Marsh, 1976). The geoid pattern of the lower harmonics must mainly derive from other, deeper-seated, sources than compositional and thermal differences in the crust and upper mantle. According to Hide and Malin (1970), there is a correlation between the earth's gravity field and its non-dipole magnetic field, suggesting the presence of humps-depressions and eddies at the core-mantle interface (see Jacobs, 1975, p. 170). Mörner (1976), therefore, has suggested that the present geoid configuration (lower harmonics) to a great extent originates from the core-mantle interface, and that any change in the core-mantle coupling and interface would also affect the configuration of the geoid (giving rise to geoidal-eustatic changes) at the same time as these changes would directly affect the geomagnetic field. Figure 15–6 illustrates this double (geoidal and magnetic) effect of changes in the core-mantle coupling and interface. It implies that one would expect to find a correlation between geomagnetic and geoidal changes, which is exactly what the field data seem to indicate (see below).

The rate of changes offers a problem, however, as it is generally assumed that core-mantle motion is a slow process. Recent recording of rapid geomagnetic reversals, events, exclusions, and distinct short-period secular magnetic variations indicate that rapid motions, in fact, do occur, just as general long-term changes do (for example, Mörner, 1978, 1979b).

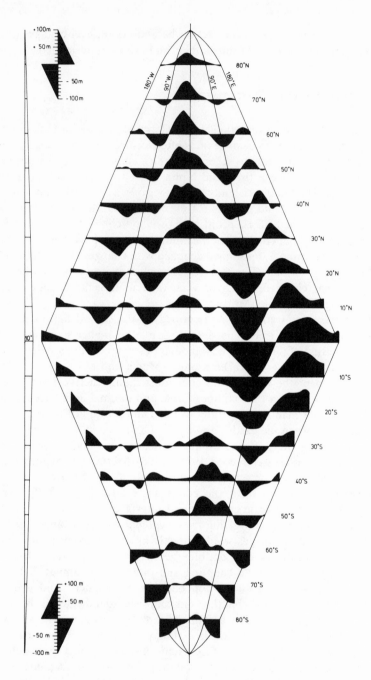

Figure 15–4. Geoid relief profiles after Gaposchkin (1973).

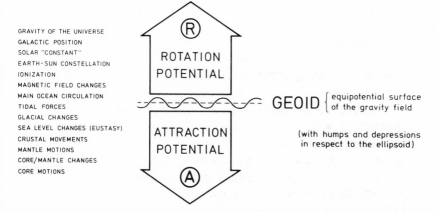

GRAVITY OF THE UNIVERSE
GALACTIC POSITION
SOLAR "CONSTANT"
EARTH-SUN CONSTELLATION
IONIZATION
MAGNETIC FIELD CHANGES
MAIN OCEAN CIRCULATION
TIDAL FORCES
GLACIAL CHANGES
SEA LEVEL CHANGES (EUSTASY)
CRUSTAL MOVEMENTS
MANTLE MOTIONS
CORE/MANTLE CHANGES
CORE MOTIONS

® ROTATION POTENTIAL

GEOID { equipotential surface of the gravity field

ATTRACTION POTENTIAL Ⓐ

(with humps and depressions in respect to the ellipsoid)

Figure 15-5. Influences on geoid configuration.

When I analyzed known variables that may have affected the geoid configuration directly or indirectly and geological records that may be interpreted as registering geoid changes (1976), I arrived at the conclusion that geoid changes (geoidal-eustasy) must have occurred not only during the Phanerozoic and Late Quaternary time units, but also during the Holocene and short-period time units. There are several known variables (earth movements, rotational changes, magnetic changes, Milanko-

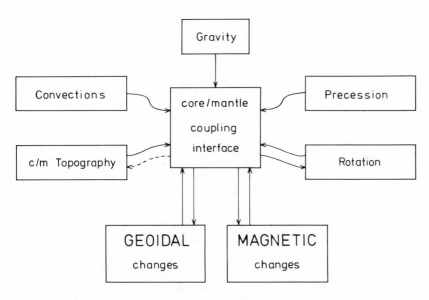

Gravity

Convections

core/mantle coupling interface

Precession

c/m Topography

Rotation

GEOIDAL changes

MAGNETIC changes

Figure 15-6. Relationships of geoidal and magnetic change.

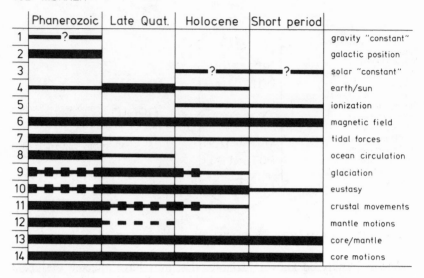

| | Phanerozoic | Late Quat. | Holocene | Short period | |
|---|---|---|---|---|---|
| 1 | ?  | | | | gravity "constant" |
| 2 | | | | | galactic position |
| 3 | | | ? ? | | solar "constant" |
| 4 | | | | | earth/sun |
| 5 | | | | | ionization |
| 6 | | | | | magnetic field |
| 7 | | | | | tidal forces |
| 8 | | | | | ocean circulation |
| 9 | | | | | glaciation |
| 10 | | | | | eustasy |
| 11 | | | | | crustal movements |
| 12 | | | | | mantle motions |
| 13 | | | | | core/mantle |
| 14 | | | | | core motions |

Figure 15–7. Variable terms in geoid distortion.

vitch parameters, glacial changes, eustatic changes, and such) that must have affected the geoid relief or must be linked to changes that also affected the geoid. Figure 15–7 lists some potential variables that may affect the rotation or attraction potentials (and hence the geoid), and their probable significance per time units on the changes in geoid configuration.

Mass redistribution leading to geoidal-eustatic changes in the order of $10^3$ years (or even $10^2$) can probably be generated by changes in the hydrosphere, crust, asthenosphere, core-mantle boundary, and outer core.

## SOME EXAMPLES OF GEOIDAL-EUSTASY

1) Reyment and Mörner (1977) have analyzed the Cretaceous transgressions and regressions in the South Atlantic region with respect to their global significance and possible influence of geoidal-eustatic changes. The data are synthesized in Figure 15–8. Tectono-eustasy (with global significance) was found to dominate during the Mercanton Normal Geomagnetic Epoch, while geoidal-eustasy (unequal over the globe) was found to dominate during the Beringov Mixed Geomagnetic Epoch (Figure 15–8, column E). The correlation between magnetism ( Figure 15–8, column B) and geoidal-eustatic activity (Figure 15–8, column E) agrees well with the theory of changes in the core-mantle interface. Furthermore, there is a correlation between tectono-eustasy (Figure 15–8C) and

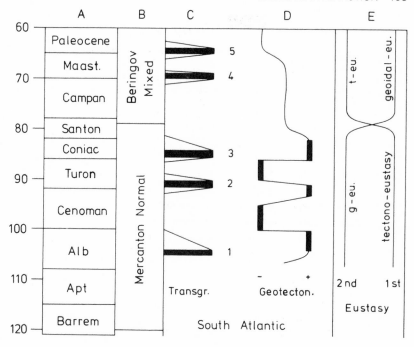

Figure 15-8. Cretaceous transgressions and regressions in the South Atlantic.

the activity of ridge growth and sea floor spreading (Figure 15-8D). The analysis has been considerably improved in a recent paper (Mörner, 1980).

2) The increased Antarctic glaciation (the Queen Maude Glaciation) at about 4.7 to 4.3 myr ago (Shackleton and Kennett, 1975; Mayewski and Goldthwait, 1977) must—besides a general glacial-eustatic lowering (estimated at about 37 meters by Mayewski and Goldthwait, 1977)—have led to geoidal-eustatic changes via mass redistribution and rotation changes (see Mörner, Chapter 14). Regressions and transgressions that are out of phase with each other are therefore to be expected during this time interval. Sea level changes must not, therefore, be used as a base of global correlations. Berggren and Haq (1976) have interpreted the regression initiating the Messinian Event as glacial-eustatic and therefore correlated it with the above-mentioned Antarctic glaciation (and corresponding glacial-eustatic regression), although available radiometric and paleomagnetic data, in fact, have suggested the opposite. The recorded changes at around the Miocene-Pliocene boundary, when strictly plotted against their suggested radiometric and paleomagnetic age, are given in Figure 15-9. The black area at the right margin indicates a period during which the following main geophysical changes took place: a) a

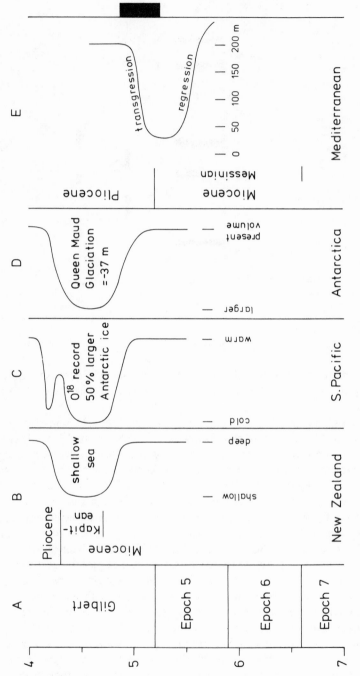

Figure 15–9. The Miocene-Pliocene boundary event.

major geomagnetic reversal (A), b) the onset of a glacial increase in Antarctica (D) with corresponding cooling (C) and glacial-eustatic lowering (B), and c) a major, obviously geoidal-eustatic, transgression in the Mediterranean region (E). If the available radiometric and paleomagnetic data thus are correct, which is not at all certain at this stage, it is not the regression but the subsequent transgression in the Mediterranean region that corresponds to the glacial increase in Antarctica (and the corresponding glacial-eustatic regression), and, if so, would provide evidence of a high-amplitude geoidal-eustatic transgression in the Mediterranean region. This would not be surprising, because an Antarctic glaciation must have affected the mass distribution over the globe and the rate of rotation which acts differently on the core and the mantle-crust and hence also affects the core-mantle coupling and interface (Figure 15–6).

3) The Blake Geomagnetic Event is reported from deep-sea sediment (in the middle of the X zone) estimated at an age of about 108,000 to 114,000 BP (Smith and Foster, 1969). The sea level records for the same period are controversial (Mörner, 1976, 1977c), viz.: a 70 meter regression in New Guinea (Chappell, 1974), a 50 meter transgression in Hawaii (Stearns, 1974), a very slight regression in the Mediterranean (Butzer and Cuerda, 1962), and no significant change at Barbados (Matthews, 1976) after age-correction of an earlier reported regression of 70 meters (Steinen and others, 1973). These inconsistent sea level changes really seem to record geoidal-eustatic changes of large amplitude. A geoidal-eustatic change at the same time as a geomagnetic field change is to be expected, as both are the effects of changes in the core-mantle coupling and interface (Figure 15–6). During the same period, the Milankowitch parameters show major changes in all three variables. As discussed by Mörner (1976), major changes in the earth-sun constellation must have affected the core-mantle coupling (see Figures 15–3, 15–5, 15–6). Consequently, there are geophysical reasons to expect major core-mantle changes at around 110,000 to 115,000 BP. Paleomagnetic studies (Mörner, 1977c, 1979b) of the palynological "standard section" at Grande Pile in France (Woillard, 1975) indicate that the Blake Event occurred right in the middle of the Eemian pollen zone (with an expanded non-normal magnetism ranging for about 128,000 to 75,000 BP, and hence agreeing well with Denham's giant core records of 1976). Consequently, the Grande Pile data exhibit no continental glaciation at the time of the Blake Event to support the theory that the regression in New Guinea was glacial-eustatically caused (Chappell, 1974); instead, it supports the author's geoidal-eustatic explanation. Figure 15–10 summarizes the variations during the period 70,000 to 130,000 BP. Although the available data seem to indicate that rotational changes, brought about by

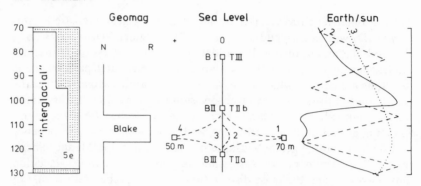

Figure 15-10. Core metal changes associated with the Blake event.

astronomical variables, gave rise to major geomagnetic (the Blake Event) and gravitational (geoidal-eustasy) changes, we need further field data of high dating precision in order to be able really to solve this fundamental question.

4) Mörner (1976, figure 14) analyzed major points of agreement and disagreement between various "eustatic" curves for the last 150,000 years and arrived at the conclusion that they seemed to record "geoid changes in a cyclic manner very much resembling the precession cycle." The 30,000 BP record is a major point of disagreement (Mörner, 1971c; Thom, 1974). It is also the period of the Lake Mungo Excursion (Barbetti and McElhinny, 1972) a magnetic excursion which is also very well established in the Grande Pile record (Mörner, 1978, 1979b). Consequently, it seems probable that the precession cycle affects the core-mantle boundary in such a way that magnetic and geoidal changes are generated (see Figure 15-6).

5) The anomalously low sea levels at −200 to −240 meters discussed by Schwartz (1972) and Pratt and Dill (1974) are supposed to be about 13,000 years old. If these shorelines are correctly interpreted as to genesis and age, they give evidence of high-amplitude geoidal-eustatic changes. It should be noted that the suggested age falls within the Gothenburg Magnetic Excusion (Mörner, 1977d, 1978). As magnetic and geoidal-eustatic changes are both caused by core-mantle changes (Figure 15-6), it is to be expected that both occur together. Mass redistribution and glacial attraction of water in connection with major glacial readvances in North America (for example, the Port Huron Readvance) and in Northern Europe (the Low Baltic Readvance) at around 13,000 BP may also have caused some changes in the geoid configuration.

6) A quite different mechanism of local "eustatic" changes (that is, changes in the geoid configuration) is represented by the local sea level

fluctuations caused by mass redistribution and glacial attraction of water in relation to Late Glacial changes of the icecaps (Clark, 1976; Farrell and Clark, 1976). This mechanism may explain some of the rapid local sea level records in Washington State and the Baltic (Mörner, 1977a, 1977b).

7) Mörner (1971a, 1971b, 1979a) noted that all supposedly eustatic curves converge at about 7,000 to 7,500 BP, although they differ very much for the interjacent period. He called this "the real Holocene eustatic problem" and suggested that it must be due to some unknown factor affecting the ocean level distribution differently over the globe between the present and about 7,000 radiocarbon years BP. The recorded differences in the "eustatic" curves must correspond to a cyclic geoidal-eustatic change. The length of this cycle is about 5,250 years (or ¼ of the precession cycle). It corresponds to the archaeomagnetic cycle of Bucha (1970) and the atmospheric $C^{14}$ production cycle of Suess (1970), which indicates a mutual origin in core-mantle changes according to Figure 15–11.

8) Although the rate problem makes it doubtful, there are many indications that even short period changes (at least some of them) in magnetic polarity, magnetic intensity, atmospheric $C^{14}$ production (ionization), climate, "eustasy," and volcanism may have an origin in core-mantle changes (Mörner, 1977e, 1977f, 1978).

9) After they had analyzed the frequency and time domains in the paleoclimatic record of some deep-sea cores, Hays and others (1976) arrived at the conclusion that "changes in the earth's orbital geometry are the fundamental cause of the succession of Quaternary ice ages." Besides a direct effect on the insolation (especially the annual insolation distribution) of the earth's orbital elements, they do also affect the core-mantle coupling and interface hence generating magnetic and geoidal-eustatic changes (Figure 15–15). It may even be that their effect on climate is not only a direct insolation effect, but also (and maybe even predominantly) the secondary effect of magnetic and geoidal-eustatic changes (Mörner, 1978). Long-term paleoclimatic records (for example, Imbrie, 1977) indicate that the amplitude-decreases drastically below the Brunhes-Matuyama boundary and becomes very low below the Gauss-Gilbert boundary. This paleoclimatic amplitude dependence of the geomagnetic field indicates that even the climatic effect of the orbital cycles to a great extent (in this case the amplitude) is controlled by core-mantle changes (Mörner, 1978). This is consistent with the suggested correlations between orbital variables and paleomagnetic and paleoclimatic changes (for example, Wollin et al., 1971; Kawai, 1976; Kent and Opdyke, 1977; Wollin et al., 1977).

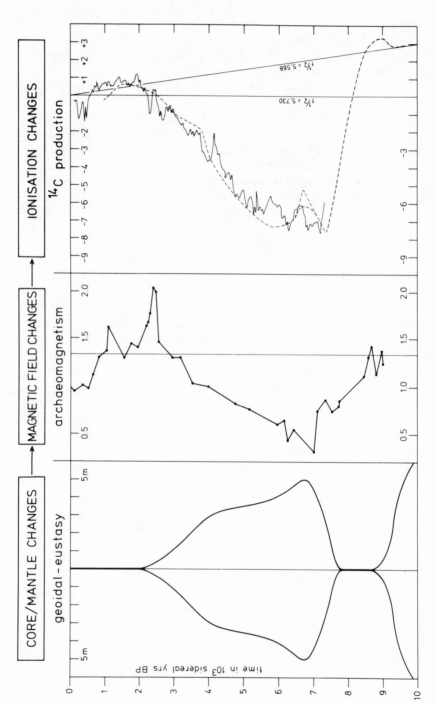

Figure 15–11. Holocene eustatic event.

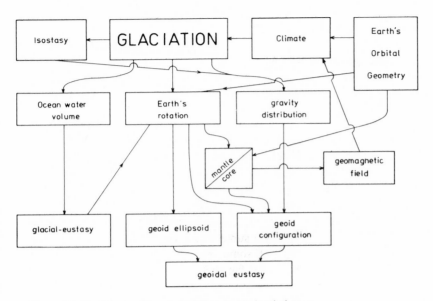

Figure 15-12. Relationship of glaciation to sea level change.

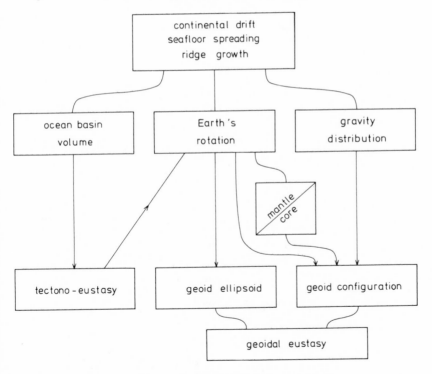

Figure 15-13. Relationship of continental drift to sea level change.

## MULTIPLE EUSTATIC EFFECTS

Major geophysical changes such as motions within the earth, sea floor spreading and ridge growth, and continental glaciations lead to multiple eustatic effects according to different schemes of causation and inter-action (Mörner, 1978), where only some lines can, as yet, be quantified (Dicke, 1966; Olausson and Svenonius, 1975; Farrell and Clark, 1976). Figure 15–12 shows the multiple effects of a glaciation on the "eustatic" ocean level changes. Figure 15–13 gives a corresponding scheme of the effects of continental drift, sea floor spreading, and ridge growth. Even the closing of the circum-equatorial circulation during the late Cenozoic may lead to similar effects (Figure 15–14).

An analysis of the sea level changes throughout the Cretaceous has revealed that the earth is also subject to large-scale latitudinal geoidal-eustatic wavelike changes or some sort of gravitational drop motions (Mörner, 1980). This is a novel factor in sea level analyses.

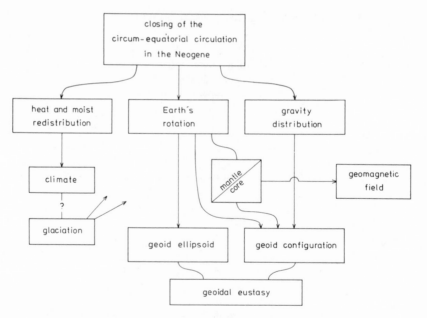

Figure 15–14. Relationship of ocean circulation to sea level change.

## PALEOCLIMATIC IMPLICATIONS

1) With the introduction of geoidal-eustasy (Mörner, 1976), Late Cenozoic transgressions and regressions must no longer be taken as a priori evidence of corresponding glacial volume changes.

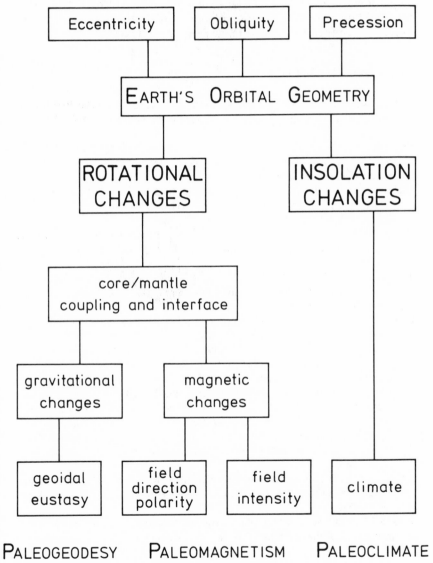

Figure 15-15. Interaction of orbital changes and earth history.

2) The earth's orbital variables affect both the insolation and the core-mantle coupling and interface. Paleoclimatic correlations with the orbital cycles may therefore be due not only to direct insolation changes but also to secondary effects of the changes in the geomagnetic and gravitational fields brought about by core-mantle changes (Figure 15-15).

3) Geoid lowering under the continents leads to a corresponding ground water lowering (Figure 15-3) that may have drastic effects on the flora and fauna and the extension of deserts. Direct paleoclimatic reconstructions can therefore not be made from these data.

## LITERATURE CITED

Barbetti, M. F., and M. W. McElhinny, 1972. Geomagnetic reversal 30,000 years ago from aboriginal fireplaces in Australia. *Nature (Phys. Sci.)*, 239:327–330.

Berggren, W. A., and B. ul Haq, 1976. The Andalusian Stage (Late Miocene): biostratigraphy, biochronology and paleoecology. *Palaeogeogr. Palaeoclimatol. Palaeoecol.*, 20:67–129.

Bucha, V., 1970. Influence of Earth's magnetic field on radiocarbon dating. In: I. Olsson (ed.), *Radiocarbon Variations and Absolute Chronology* (New York: Wiley & Sons), pp. 501–511.

Butzer, K. W., and J. Cuerda, 1962. Coastal stratigraphy of southern Mallorca and its implication for Pleistocene chronology of the Mediterranean Sea. *Jour. Geol.*, 70:398–416.

Chappell, J., 1974. Geology of coral terraces, Huon Peninsula, New Guinea: a study of Quaternary tectonic movements and sea level changes. *Geol. Soc. Amer. Bull.*, 85:553–570.

Clark, J. A., 1976. Greenland's rapid postglacial emergence: a result of ice-water gravitational attraction. *Geology*, 4:310–312.

Clark, J. A., W. E. Farrell, and W. R. Peltier, 1978. Global changes in postglacial sea level: a numerical calculation. *Quat. Res.*, 9:265–287.

Denham, C. R., 1976. Blake polarity episode in two cores from the Greater Antilles Outer Ridge. *Earth Planet Sci. Lett.*, 29:422–434.

Dicke, R. H., 1966. The secular acceleration of the Earth's rotation and cosmology. In: B. G. Marsden and A. G. W. Cameron (eds.), *The Earth-Moon System* (New York: Plenum Press), pp. 98–164.

Farrell, W. E., and J. A. Clark, 1976. On postglacial sea level. *Geophys. Jour.*, 46:647–668.

Gaposchkin, E. M., 1973. Satellite dynamics. In: E. M. Gaposchkin (ed.), Smithsonian standard earth (III). *Smithsonian Astronom. Obs., Spec. Rept.*, 353:85–192.

Hays, J. D., J. Imbrie, and N. J. Shackleton, 1976. Variations in the earth's orbit: pacemaker of the Ice Ages. *Science*, 194:1121–1132.

Hide, R., and S. R. C. Malin, 1970. Novel correlations between global features of the earth's gravitational and magnetic fields. *Nature,* 225:605–609.

Hinte, J. E. van, 1976. A cretaceous time scale. *Amer. Assoc. Petrol. Geol. Bull.,* 60:498–516.

Imbrie, J., 1977. On the underlying causes of Quaternary climatic changes. *INQUA 10th Congr. Birmingham 1977, Abstr.,* p. 221.

Jacobs, J. A., 1975. *The Earth's Core.* (New York: Academic Press, International Geophysical Series no. 20), 254 pp.

Kawai, N., 1975. Geomagnetic events and climate in the Quaternary. *IAGA Bull.,* 36:174.

Kennett, J. P., 1967. Recognition and correlation of the Kapitean Stage (Upper Miocene, New Zealand). *New Zealand Jour. Geol. Geophys.,* 10:1051–1063.

Kennett, J. P., and P. Vella, 1975. Late Cenozoic planctonic foraminifera and paleooceanography at DSDP Site 284 in the cool subtropical South Pacific. In: R. E. Burns, J. E. Andrews, et al. (eds.), *Initial Reports of the Deep Sea Drilling Project* (Washington, D. C.: U. S. Govt. Printing Office), vol. 21, pp. 769–799.

Kent, D. V., and N. D., Opdyke, 1977. Paleomagnetic field intensity variation recorded in a Brunhes Epoch deep-sea sediment core. *Nature,* 266:156–159.

Marsh, B. D., and J. G. Marsh, 1976. On global gravity anomalies and two-scale mantle convection. *Jour. Geophys. Res.,* 81:5267–5280.

Matthews, R. K., 1976. Sea level records from coral reefs in tectonically active areas. *Geol. Soc. Amer. Abstr. Progr.,* 8 (6):1000.

Mayewski, P. A., and R. P. Goldthwait, 1977. The glacial history of the Transantarctic Mountains: a record of the East Antarctic ice cap. *Jour. Glaciol.,* 18.

Mörner, N.-A., 1971a. The Holocene eustatic sea level problems. *Geol. Mijnbouw,* 50:699–702.

———, 1971b. Relations between ocean, glacial and crustal changes. *Geol. Soc. Amer. Bull.,* 82:787–788.

———, 1975. Eustasy and geoid changes. *Int. Union Geol. Geophys. Sci., 16th General Assembly, Grenoble 1975, Abstr.,* p. 245.

———, 1976. Eustasy and geoid changes. *Jour. Geol.,* 84:123–152.

———, 1977a. Eustasy and instability of the geoid configuration. *Geol. Fören, Stockholm, Förh.,* pp. 369–376.

———, 1977b. Eustatic changes and geoidal-eustasy. *Earth Rheology Late Cenozoic Isostatic Movements Sympos., Stockholm, 1977 Abstr.,* pp. 92–101.

———, 1977c. The Grande Pile records and the 115,000 BP events. *Int. Geol. Correl. Project 24, Quat. Glac. North. Hemisph. Rept.,* 4:47–52.

———, 1977d. The Gothenburg Magnetic Excursion. *Quat. Res.,* 7:413–427.

———, 1977e. Paleoclimatic records from South Scandinavia, global correlations, origin and cyclicity. *Paleolimnol. Lake Biwa Japan,* 4:499–528.

———, 1977f. Paleoclimate and short period changes of the core/mantle coupling and interface. *Jour. Interdiscipl. Cycle Res.,* 8:207–210.

———, 1978. Paleoclimatic, paleomagnetic and paleogeoidal changes: interaction and complexity. *CNES Colloq. Int. "Evolution of Planetary Atmospheres and Climatology of the Earth," Nice, 1978*, pp. 221–232.

———, 1979a. Eustasy and geoid changes as a function of core/mantle changes. In: N.-A., Mörner (ed.), *Earth Rheology, Isostasy and Eustasy* (New York: Wiley & Sons), pp. 535–553.

———, 1979b. The Grande Pile paleomagnetic/paleoclimatic record and the European glacial history of the last 130,000 years. *Int. Proj. Paleolimnol-Late Cenozoic Climate*, 2:19–24.

———, 1980. Relative sea level changes, tectono-eustasy, geoidal eustasy and geodynamics during the Cretaceous. *Cretaceous Res.*, 1 (2).

Newman, W. S., L. F. Marcus, R. R. Pardi, J. A. Paccione, and S. M. Tomecek, 1977. Eustasy and deformation of the geoid: 1000–6000 radiocarbon years B.P. *Earth Rheology Late Cenozoic Isostatic Movements Sympos., Stockholm, 1977 Abstr.*, pp. 103–106.

———, 1979. Eustasy and deformation of the geoid: 1000–6000 radiocarbon years B.P. In: N.-A. Mörner (ed.), *Earth Rheology, Isostasy and Eustasy* (New York: Wiley & Sons), pp. 555–568.

Olausson, E., and B. Svenonius, 1975. Past changes in the geomagnetic field caused by glaciations and deglaciations. *Boreas*, 4:55–62.

Pratt, R. M., and R. F. Dill, 1974. Deep eustatic terrace levels: further speculations. *Geology*, 2:155–159.

Reyment, R. A., and N.-A., Mörner, 1977. Cretaceous transgressions and regressions exemplified by the South Atlantic. *Pal. Soc. Japan Spec. Papers*, 21:247–261.

Schwartz, M. L., 1972. Seamounts as sea-level indicators. *Geol. Soc. Amer. Bull.*, 83:2975–2980.

Shackleton, N. J., and J. P. Kennett, 1975. Late Cenozoic oxygen and carbon isotopic changes of DSDP Site 284: implications for glacial history of the northern hemisphere and Antarctica. *Initial Reports of the Deep Sea Drilling Project* (Washington, D.C.: U.S. Govt. Printing Office), vol. 21, pp. 801–807.

Smith, J. D., and J. H. Foster, 1969. Geomagnetic reversal in Brunhes Normal Polarity Epoch. *Science*, 163:565–567.

Stearns, H. T., 1974. Submerged shorelines and shelves in the Hawaiian Islands and a revision of some of the eustatic emerged shorelines. *Geol. Soc. Amer. Bull.*, 85:795–804.

Steinen, R. P., R. S. Harrison, and R. K. Matthews, 1973. Eustatic low stand of sea level between 125,000 and 105,000 B.P.: evidence from the subsurface of Barbados, West Indies. *Geol. Soc. Amer. Bull.*, 84:63–70.

Suess, H. E., 1970. The three causes of the secular C14 fluctuations, their amplitude and time constants. In: I. Olsson (ed.), *Radiocarbon Variations and Absolute Chronology* (New York: Wiley & Sons), pp. 595–604.

Thom, B. G., 1974. The dilemma of high interstadial sea levels during the last glaciation. *Progr. Geogr.*, 5:167–246.

Woillard, G., 1975. Recherches palynologiques sur le Peistocene dans l'Est de la Belgique et dans les Vosges Lorraines. *Acta Geogr. Lovaniensia*, 14:1–118.

Wollin, G., D. B., Ericson, W. B. F. Ryan, and J. H. Foster, 1971. Magnetism of the earth and climatic changes. *Earth Planet Sci. Lett.*, 12:175–183.

Wollin, G., W. B. F. Ryan, D. B. Ericson, and J. H. Foster, 1977. Paleoclimate, paleomagnetism and the eccentricity of the earth's orbit. *Geophys. Res. Lett.*, 4:267–270.

Chapter 16

## ON TWO KINDS OF RAPID FAUNAL TURNOVER

S. DAVID WEBB

Florida State Museum

INTRODUCTION

From a paleontological vantage point, faunal change appears to proceed in two modes: one gradual, the other one rapid or even cataclysmic. In the gradual mode, biotic diversity remains essentially constant, and the evolutionary rates of included species are mainly horotelic or bradytelic. At times, however, this stately mode is disturbed. Then, over relatively short periods of time, major restructuring of whole biotas and tachytelic evolution of diverse taxa occur. These are the two modes to which Derek Ager's (1973) stratigraphic simile aptly refers: "The history of any one part of the earth, like the life of a soldier, consists of long periods of boredom and short periods of terror."

Paleontologists are confronted by history's irregularities at all levels; "herky-jerky" patterns are evident both in the history of particular taxa and in the chronicles of whole biotas. The current debate over "punctuated equilibria" (Eldredge and Gould, 1972; Gould and Eldredge, 1977) concerns the irregular rates of single species evolution; yet, curiously, the same term would be altogether appropriate (were it not preoccupied) in the present context wherein we consider evolutionary changes in whole faunas. One suspects that these are different masks of a single nature, but for the moment we shall continue to separate them.

The observation that whole communities change and that they change at irregular rates is not new. Particularly in the context of terrestrial vertebrates, E. C. Olson (1977) has perceptively considered the modes of community evolution. He restricted the meaning of "Community Evolution" to "changes that occur within integrated complexes of organisms that maintain direct continuity through time by persistence of their basic ecological structure." Such persistent integrated complexes are "chronofaunas" as he defined them earlier (Olson, 1952). On a longer time scale

417

he recognized "community succession" as a series of discretely separated chronofaunas, and he supposed that "most of the major events of evolutionary history took place as such new communities formed during rapid modifications of the environments." The present essay begins with this assumption that episodes of rapid community reorganization constitute a major aspect of evolutionary history.

What follows is an attempt to analyze in some detail two case histories of major community reorganizations. The two cases are selected from the relatively complete record of Late Cenozoic land mammals. They fall within a relatively well-documented stratigraphic framework, with radiometric control to about the nearest half million years. The analysis is restricted to the larger mammals, above about 5 kilograms,[1] because their record is essentially complete. The small mammals, while potentially more numerous, are inadequately known due to sampling difficulties and greater regional differentiation (Behrensmeyer, 1975). Thus a favorable part of the fossil record is chosen.

The two case histories to which I refer exemplify two very different kinds of rapid faunal turnover. The first concerns the demise of the richest large mammal fauna in North America. Known as the Clarendonian Chronofauna, its diverse ungulate fauna compared closely with that of the great African savannas in their heyday. The breakdown of this great fauna seems to have resulted from a major environmental crunch. The second case history concerns the dramatic invasion of South America by North American land mammals and the consequent major reorganization of that continent's fauna. The breakdown appears to be primarily the consequence of a horde of new immigrants. These two cases represent the two fundamental mechanisms that can generate rapid faunal turnover.

## THE MACARTHUR-WILSON EQUILIBRIUM HYPOTHESIS

If the punctuational view of earth history is correct, then during any short period of time one may expect to find stable communities over most of the earth. Study of the Recent biota has produced just such a view, well expressed by the MacArthur-Wilson Equilibrium Hypothesis (Figure 16-1). In this view, species number in a given area (usually an island) is a dynamic equilibrium between the species that immigrate and the species that become extinct.[2] Over decades or longer the equilibrium number remains stable. Even if all the taxa are extirpated (as in Krakatoa in 1883),

[1] As Bourliere (1975) has shown, a fundamental gap involving adaptive strategies, separates mammals below 3 and above 5 kilograms.
[2] An important conceptual result of this hypothesis is that it brings extinction out of the realm of mystery and assigns it a routine role among ecological processes.

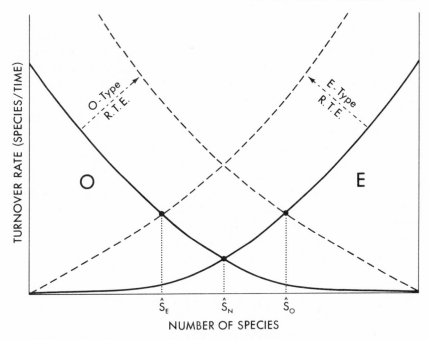

Figure 16–1. Faunal turnover rates.

the biota soon (within about three decades on Krakatoa) returns to its equilibrium number (MacArthur and Wilson, 1963, 1967; Simberloff and Wilson, 1969, 1970).

Paleontologists have been quick to note that, if valid, the equilibrium hypothesis would be applicable to greater than island areas, including continents, ocean basins, or even the whole earth's biota (Webb, 1969; Simberloff, 1972, 1974; Schopf, 1974). The real advantage of thus extending the species equilibrium concept was that it permitted the unification of neontological and paleontological data or at least an attempt at such unification. Early in the discussion, Wilson (1969) suggested that in evolutionary time (in contrast to ecological time) equilibrium numbers might be expected to drift upwards, as adaptive strategies and coadaptations are refined in an evolving community. Both cases cited below provide some evidence of such a maturation effect. Much the same consideration has led Webb (1969) and Simberloff (1972) to rename the "immigrations curve" as the "originations curve" so that in paleontological studies the equilibrium could incorporate new taxa that evolved within the community. Thus the equilibrium hypothesis was extended into the fossil record.

In Figure 16–1 the two curves governing the equilibrium are assumed to be determined by independent sets of factors. The originations (or immigration) curve is thought to be controlled by the distance from the barriers to immigrants, the diversity of potential species pools, and in general the availability of the environment to new taxa. The extinctions curve, on the other hand, generally represents the hostility of nature toward the species living in the area studied; it is inversely related to such factors as habitat diversity, climatic equability, and habitable area. The extinctions and originations curves are expected to be concave upward (MacArthur and Wilson). It can be grasped intuitively that in an unpopulated island, most of the first wave of immigrant taxa might find favorable situations in which to live, but as the island becomes more fully occupied and as pioneer populations spread and interact more, fewer subsequent immigrant taxa are likely to find favorable circumstances. By similar reasoning, the extinction curve is expected to rise geometrically, as more diverse taxa are piled ever more densely into finite areas. The interactions of taxa in such an ecosystem produce a dynamic equilibrium.

## O-TYPE AND E-TYPE RAPID TURNOVER EPISODES

The equilibrium model predicates control of species number either by shifts of the *originations curve* or by shifts of the *extinctions curve*. Raising either curve increases the turnover rate, that is, the equilibrated rates of both originations and extinctions. The principal difference is that a rapid turnover episode controlled by a steepening *originations curve* will increase the net species number, whereas an episode produced by a steepening *extinctions curve* will decrease the net species number. Thus the model predicts two kinds of rapid turnover episodes in the history of life.

In the following case histories I hope to demonstrate that the equilibrium model enlightens real paleontological problems and that both classes of rapid turnover episodes do occur. Hereafter I refer to the two kinds of change as *O-type* and *E-type* rapid turnover episodes.

## THE FALL OF THE CLARENDONIAN CHRONOFAUNA

An extraordinarily rich fauna of large mammals evolved in North America during the Late Miocene (Figure 16–2). At its peak some 10 myr ago it included about 50 genera of ungulates and large carnivores. It was closely comparable to the present savanna fauna of Africa both in faunal richness and in the array of adaptive types, yet phylogenetic ties between any of the North American and African large mammal taxa were remote.

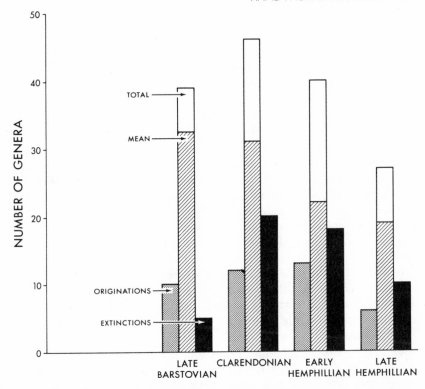

Figure 16–2. Faunal turnover during middle and late Miocene in North America.

The principal groups, each contributing numerous taxa to this North American savanna fauna, were hyaenoid dogs (Borophaginae), hypsodont horses, the Camelidae, and the Antilocapridae (native pronghorns of diverse sorts). This rich savanna fauna has been termed the Clarendonian Chronofauna (Webb, 1969b; Tedford, et al., in press), named for the mammal age during which it reached its acme.

During an interval of more than 10 myr (beginning before the Barstovian) the same principal groups of North American savanna mammals held sway. From *Parahippus* (of Hemingfordian age) stemmed diverse hypsodont horses, and the Protolabinae sired an array of progressive camels. The integrity of the evolving chronofauna is evident from the general continuity of these and other lineages and from the absence of major waves of new immigrants. The most notable exception were the two allochthonous genera of Proboscidea, *Gomphotherium* and *Miomastodon,* which were early integrated into the chronofauna, and toward the end, a string of large carnivores, none of which persisted long. While the net diversity of this chronofauna slowly swelled, the generic turnover rates were remarkably low.[3] In short, the gradual mode of faunal evolution prevailed.

Then at the peak of its diversity in the Late Clarendonian (9 myr ago) the chronofauna was shaken (Figure 16-3). Some 20 large mammal genera became extinct. The turnover rate accelerated by a factor of three. New autochthonous lineages appeared, but not so rapidly as the old groups disappeared. Only one allochthonous genus (the proboscidean *Platybelodon*) immigrated; thus, the accelerated turnover rate represents essentially internal faunal turmoil. And then about 3 myr later, during the mid-Hemphillian, the greatest extinction event in the history of North American land mammals took place. Some two dozen large mammal genera disappeared (Figure 16-3). This is nearly twice the number of large North American mammal genera than that which disappeared during the widely discussed Late Pleistocene episode (Martin and Wright, 1965). A detailed review of Hemphillian chronology by Richard H. Tedford shows that most of the Hemphillian extinctions clustered near the middle of that stage, although several others appear to have occurred at about the end of that stage. Thus, as the chronology becomes more precise, the extinctions are seen to be synchronized rather than scattered through time.

During the mid-Hemphillian extinctions the turnover rate continued to accelerate: it averaged nearly twice that of the Clarendonian, and five to seven times that of the Barstovian. Extinctions continued to mount and autochthonous new originations could not keep pace. A slightly larger contingent of immigrants helped bolster receding diversity, but a majority (three genera) were short-lived carnivoran taxa. Two ground sloth genera (*Pliometanastes* and *Thinobadistes*) from South America were more enduring additions. Nevertheless the net diversity of large mammals fell from about 40 genera in the Early Hemphillian to less than 30 in the Late Hemphillian.

The Late Hemphillian was the third and final phase in the fall of the Clarendonian Chronofauna. Although it was the least devastating phase, it is important because it demonstrates the persistence or recurrence of the same processes of faunal deterioration for still another 2 myr. Turnover rates slowed to only two or three times their Barstovian level, while the extinction rate continued at about the same level as in the preceding two phases of the fall. Two more short-lived carnivoran taxa appeared as

[3] It should be noted that I have not counted successive members of one evolving lineage with different names (for example, Pseudohipparion and its apparent descendant *Griphippus*), thus eliminating purely taxonomic turnover.

immigrants. By the end of the Hemphillian, large mammal diversity was half of what it had been 5 myr earlier at the peak.

ENVIRONMENTAL CHANGES

The demise of the Clarendonian Chronofauna was clearly an E-type rapid turnover episode (Figure 16–3). The Barstovian revealed a rich fauna with remarkably low turnover rates. During the Clarendonian, the fauna became still richer, but soon displayed markedly higher extinction rates. Thereafter, through Hemphillian time, the persistent pressure of high extinction rates is evident. In each successive phase extinctions drove diversity downward, while the few immigrant groups could not redress the balance. Independent evidence strongly suggests that these patterns resulted from an environmental crunch.

The large mammals that had produced this rich North American sa-

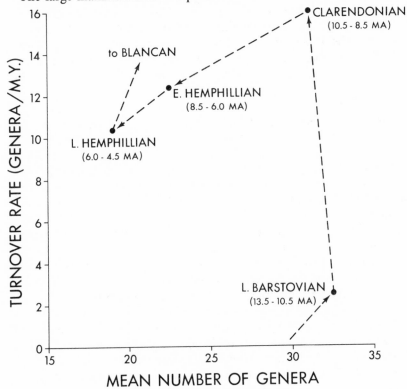

Figure 16-3. North American faunal turnovers.

vanna fauna did not all become extinct when it collapsed, and the survivors have produced an instructive pattern. Among the large carnivores, *Osteoborus* and then *Borophagus* persisted, and in many faunal samples were represented by two different-sized species. Most of the arctoid carnivores were lost, and then a series of broadly similar bearlike immigrants each met the same fate.

Among the Equidae the first genera to drop out were the browsers. Only grazers survived the Clarendonian, and three groups of them were lost then. Groups related to modern asses and zebras survived and later spread abroad, as did the basic stock of the peculiar South American groups *Onohippidion* (MacFadden). The only other persistent equid was the gazelle-like *Nannippus*, the most hypsodont and one of the smallest of the Late Miocene grazers.

Among nonruminant artiodactyls the pattern of reductions is revealing. Of the several kinds of oreodonts and peccaries in the Barstovian, only two survived the Clarendonian, and only one peccary came through the Hemphillian. This surviving genus, *Platygonus*, developed lophodont (shearing) cheek teeth and thus differed markedly from the earlier forms, such as *Prosthennops*, with bunodont (crushing) dentitions. The apparent shift is toward a coarser, more fibrous diet and more open country living (Lundelius).

The pattern of survivors among the cameloids resembles that among the equids. The first major losses were among brachydont, presumably browsing groups with unfused metapodials, such as the miolobine camels and the horned protoceratids. The most dramatic loss is that of *Aepycamelus*, the 12-foot giraffe-camel, rare among Late Clarendonian samples and gone after the Early Hemphillian. As in the horses, three or four progressively hyposodont lineages survived and some of them later dispersed to the Old World and South America.

Among ruminants the pattern of extinctions is similar. Small brachydont groups, such as leptomerycids and blastomerycines, dropped out early. Only *Pseudoceras* lingered on as a rare animal in the Early Hemphillian. The larger more deerlike dromomerycids, also brachydont, survived until the Late Hemphillian, but were rare after the Clarendonian, despite a marked hypsodont increase in *Pediomeryx*. The most successful native ruminants were the hypsodont merycodontines, from which were derived various antilocaprines (pronghorns). Many local faunas of Late Hemphillian and Blancan age include a large and a small genus of surviving antilocaprines. Thus the only group of native ruminants that survived was the most hypsodont one.

Clearly the common feature of the surviving native ungulates was adaptation to coarser fodder and more open country. Obvious browsers in each group had dropped out by the end of the Clarendonian; a few, such as *Aepycamelus* and *Pseudoceras,* lingered into the Early Hemphillian. The few persistent browsers or mixed feeders, notably the peccary *Platygonus* and the dromomerycid *Pediomeryx,* exhibited notably increased lophodonty and hypsodonty. Even the hypsodont ungulates experienced repeated decimations, but the greatest number of surviving lineages were among the horses, camels, and pronghorns, the premier hypsodont grazers.

This pattern of decreased diversity among grazing ungulates and wholesale loss of browsers and mixed feeders presumably represents a shift from park savanna habitats to pure grassland (steppe) conditions (Gregory, 1971; Webb, 1977). Most of the juicy ecotonal areas were lost or severely restricted. As a result of such a shift, the rich array of browsers and mixed feeding ungulates was devastated. Only the most progressive grazers, and often the smallest of them, persisted in the more restricted environments. This view requires partial revision of the classic story of how the great herds of the Clarendonian Chronofauna evolved in the midcontinental grasslands. Instead those herds are now seen to have received succor in park savannas (more like their analogues in Africa), as befits the many browsers and the giraffe-camel. Botanical evidence produced by MacGinitie (1962) from the Barstovian Kilgore Flora in Nebraska in fact led to this revised scenario, for his work directly documents park savanna on the interfluves and subtropical forest along the streams. Thus the scene before the great demise is well-documented. However, throughout the High Plains there are extensive caliche horizons, usually forming a resistant "cap rock" in the Hemphillian parts of the Ogallala Group. Evidently these caliche zones represent evaporative concentration of lime salts in former soil horizons. If so, no more extensive evidence of recurrent phases of persistent aridity can be sought.

Two complementary environmental hypotheses may explain this presumed vast shift from park savanna to steppe. One is that regional uplift in the midcontinent produced a less equable climate. Axelrod and Bailey (1976) have clearly documented such an effect in the Late Miocene (among other times) on the basis of floral evidence from the Rio Grande Rift, and major uplift during the same period is indicated more generally in the southern Rocky Mountains by Scott (1975) and Taylor (1975).

The second hypothesis that may explain this drastic environmental shift, is the onset of a Late Cenozoic glacial climate. Such a change might be expected to produce the first and therefore most devastating episodes of cooling and drying in the midcontinent, resembling episodes that continually recurred during the Pleistocene. This hypothesis gains support from the close correlation between the largest extinction event in the mid-Hemphillian and the eustatic events associated with the end of the Messinian in the Mediterranean Basin (Ryan, et al.). The great mid-Hemphillian extinctions occurred between 5 and 6 myr ago. Coffee Ranch Quarry in the type section of the Hemphillian Stage produces a Late Hemphillian fauna which is depauperate not because of poor sampling nor because of stenotypic ecology (Shotwell, 1958), but because the major extinctions had already occurred. The age based on fission tracks in zircon crystals taken just above the quarry is 5.4 Ma (Boellstorff, 1976). Older pre-extinction sites such as FT-40 occur in the "Hemphill Beds" with basal scoria derived from the Raton Basalt dated at 6.0 Ma (Tedford et al., 1977). Thus the Hemphillian extinctions seem to cluster in the interval between 5.4 and 6.0 Ma. This is also the interval of the Messinian "salinity crisis" in the Mediterranean Basin and elsewhere (Ryan, et al.).

Direct evidence of the correlation between the Hemphillian land mammal succession and the Messinian eustatic cycle can be found in Florida (Figure 16-4). There, a rich set of fluvio-estuarine sites occupies the western flank of the tectonically stable peninsula. Late Clarendonian through Early Hemphillian sites with at least some marine influence occur between 38 and 50 meters above markedly lower sea levels: Withlacoochee 4A produces nonmarine deposits below 10 meters and the Manatee County Dam Site produces estuarine material at present sea level (Webb and Tessman, 1969a, 1969b). By the late Hemphillian, estuarine sites once more occupy elevations of 40 to 50 meters. The Bone Valley Formation regularly consists of two members, separated by an unconformity that locally attains considerable relief, and this unconformity is sandwiched between the Early and Late Hemphillian. Thus the Florida evidence indicates that a eustatic regression and transgression of at least 40 meters occurred between medial and Late Hemphillian time, that is between 5 and 6 myr ago. It seems probable that this represents the same world-wide eustatic event that occurs during the same time interval (at the end of the Messinian) in the Mediterranean region.

In summary, the combined effects of a major world-wide sea level regression and regional uplift may have produced the cumulatively increasing E-type rapid turnover episode that marked the demise of the Clarendonian Chronofauna. The more direct cause of the major vertebrate extinctions was probably a broad shift in the midcontinent from a park-savanna biome to a steppe or grassland biome.

## THE GREAT AMERICAN INTERCHANGE: AN O-TYPE EPISODE

Quite different from an E-type episode would be a set of events arising from an Originations Curve. The South American fauna during the Great American Interchange provides an example of such a pattern. The

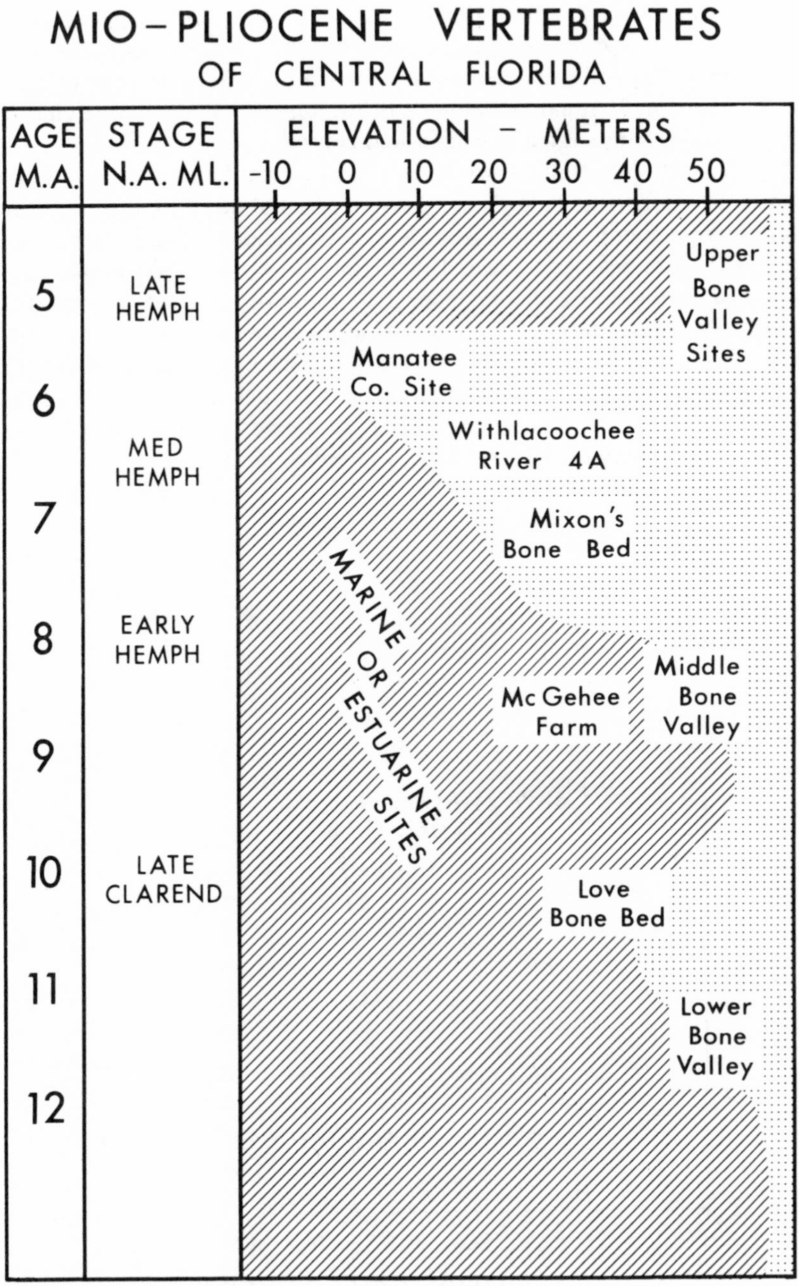

# MIO-PLIOCENE VERTEBRATES
## OF CENTRAL FLORIDA

| AGE M.A. | STAGE N.A. ML. | ELEVATION – METERS |
|---|---|---|
| | | -10  0  10  20  30  40  50 |
| 5 | LATE HEMPH | Upper Bone Valley Sites |
| 6 | MED HEMPH | Manatee Co. Site / Withlacoochee River 4A |
| 7 | | Mixon's Bone Bed |
| 8 | EARLY HEMPH | McGehee Farm / Middle Bone Valley |
| 9 | | MARINE OR ESTUARINE SITES |
| 10 | LATE CLAREND | Love Bone Bed |
| 11 | | Lower Bone Valley |
| 12 | | |

Figure 16–4. Mio-pliocene vertebrates of central Florida.

history of this fauna conforms to the definition of an O-type punctuation in showing a marked increase in new lineages, a rising but relatively lower extinction rate, and consequently a net increase in number of taxa.

The principal faunal movements of the Great American Interchange began about 3 myr ago and had slowed considerably by about 7 myr ago (Webb, 1976; Marshall, in press). It is appropriate here to focus only on the South American side of the story. During about 2 myr, nearly two dozen large mammal genera from North America established themselves in South America. As indicated in Figure 16–5, the turnover rate quadrupled during the Chapadmalalan (Late Pliocene) when the first considerable group of northern immigrants appeared and continued at about the same rate during the Uquian (Early Pleistocene) at the acme of the interchange. Although the extinction rate also increased, it did not keep pace, and brought the number of genera in the post-interchange fauna to about 150 percent of those in the pre-interchange fauna.

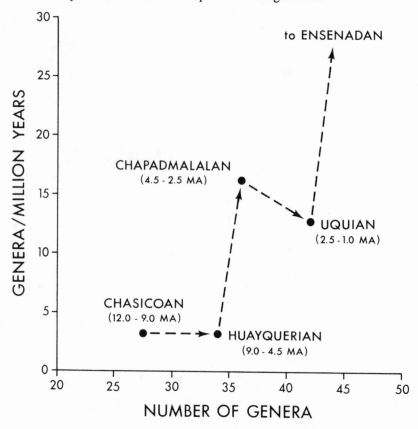

Figure 16–5. South American faunal turnovers.

The cause of this rapid turnover episode can be largely attributed to the influx of northern immigrants. Suddenly, introducing a score of new genera into a fauna that had remained wholly isolated for at least 30 myr surely had a major impact. This wholly biological effect had its geological cause in the formation of the isthmian land connection as part of the Cordilleran Orogeny. It is possible however, that these same geological events had important direct effects on the South American biota, for example, by enlarging the area of semiarid habitats. And to the extent that such environmental changes influenced faunal changes, this case is not purely an O-type episode, but includes some E-type effects.

At least two factors suggest that the South American faunal turnover is largely an O-type episode. The considerable net increase in number of taxa suggests that positive (origination) effects far outweigh negative extinction effects. This view is strengthened by the observation that a majority of originations during the Uquian are northern immigrants, and thereafter (in Ensenadan and Lujanian) are their derivatives (Webb, 1975). Most of the major environmental factors such as Andean uplift had begun to act during the Miocene and were well under way before the faunal interchange of the Late Pliocene (Marshall, in press; Webb, 1975). Thus, the E-type effects on this episode appear to be far less important than the O-type effects.

## RECOGNITION OF E-TYPE AND O-TYPE EPISODES

E-type and O-type episodes produce remarkably similar patterns: each type yields accelerated extinction rates and accelerated turnover rates.[4] It would not be surprising if in the fossil record such patterns were brigaded together as "mass extinctions," "turnover episodes," and "artifacts due to stratigraphic gaps." Any kind of rapid turnover episode punctuates the predominating gradual mode of chronofaunal evolution.

In view of these resemblances, it is essential to provide practical criteria that will permit distinction between O and E Punctuations. The most powerful criterion is $\Delta$ S, the change in number of taxa. An episode producing a net increase in taxa is an O-type episode; an episode that produced a net decrease is an E-type episode. The relationships can be visualized in Figure 16-1.

Some difficulty with the criterion of $\Delta$ S may arise in poorly sampled parts of the fossil record. It then becomes difficult to specify the boundary

---

[4] These resemblances are to be expected of a dynamic equilibrium which translates one perturbation to another by faunal feedback. Such feedback represents the summed evolutionary responses of diverse species to major changes either in resources or in competitors. One of the possible responses of each species is extinction.

conditions beyond which this criterion may not be applied. Nevertheless, if the fossil sample appears to be randomly drawn from a complete sample, then the proportions of first appearing, continuing, and last appearing taxa should remain the same and the criterion can still be applied. One may expect a systematic sampling bias against populations on the verge of extinction because they may be small and/or localized. On the other hand, a similar bias may be expected to affect the sample of newly originating taxa as well. Therefore, unless the suspected biases unequally affect first and last appearances, $\Delta$ S may be retained as the primary criterion.

Rate and mode of recovery may be other useful distinctions between E-type and O-type episodes. After a major environmental crunch one expects reconstruction of a new community to be an arduous evolutionary task requiring fundamental new faunal interactions. Opportunities should be great both for serendipitous immigrants and for enterprising natives. This rate of rebuilding would be dramatic at the beginning, but even so would span a considerable period of time. Thus, recovery from an E-type episode would involve fundamental rebuilding and long recovery time.

These criteria can be recognized in the E-type case history here considered. Recovery time from the Clarendonian Chronofaunal crash was slow. After the Late Hemphillian (Early Pliocene), low, land mammal diversity rose during the Blancan (Late Pliocene) about halfway to its former peak, and then during the Irvingtonian (Early Pleistocene) surpassed its Clarendonian apogee. Thus the recovery time was just over three myr (about 4.5 to 1.5 Ma).

This Late Cenozoic recovery time can be compared with recovery times from Recent island studies. MacArthur and Wilson (1967) and Simberloff (1972) have developed the following equation for time to restore 90 percent of the equilibrium number following defaunation:

$$t_{0.90} = 1.15 \left( \frac{\hat{S}}{\hat{X}} \right)$$

where $\hat{S}$ is the equilibrium number of taxa and $\hat{X}$ the equilibrium turnover rate. Employing the Clarendonian values for large land mammals from Figure 16–3:

$$\hat{S} = 31 \text{ genera}$$
$$\hat{X} = 16 \text{ genera/myr}$$

gives the expected recovery time for 90 percent of the equilibrium as $t_{0.90} = 2.23$ myr. The turnover rates for land mammal genera on whole continents are several orders of magnitude slower than those for avian and insect species on small islands, and therefore the recovery times are likewise several orders of magnitude longer. For our purposes the important consequence of these calculations is that recovery times for large biotas are measurable in the fossil record.

A second predicted feature of recovery from an E-type episode is a fundamental reorganization of community structure. In the present example the most obvious feature is a wholesale importation of Old World taxa across the Bering Bridge (Repenning, 1965). During the Late Pliocene and into the Pleistocene, major immigrant contingents were eurytopic carnivores (such as bears and hyenas) and r-selected microtine rodents. In the course of the Pleistocene these immigrants became more and more rigorously adapted to steppe and then cold-steppe and tundra conditions. By the Late Pleistocene a new holarctic chronofauna had arisen Phoenix-like from the ashes of the Clarendonian Chronofauna, "as great in admiration as herself."

The aftermath of an O-type episode may be expected to be less dramatic. Communities enriched by insinuation and accommodation of immigrant taxa might readily relax to lower equilibrium levels once the source of new immigrants became occluded or wholly redundant. On the other hand, some new opportunities might lead to further origination pressure over evolutionary time on the part of successful immigrant groups. Thus the relaxation of equilibrium number and turnover rate predicted in ecological time might be superseded in evolutionary time by integrative enrichment. In either event, the results would be less remarkable than those following an E-type episode.

There is no evidence in the South American O-type episode of a relaxation time lag. Simpson (1975) studying Andean Paramo "island" floras and Diamond (1977) working on New Guinean avifaunas have described some supersaturated biotas which have not relaxed to lower equilibrium levels for more than 10,000 years. Nonetheless, these differences endure only in faunas of large islands that have been considerably reduced from a previous very wide extent.

Quite the opposite in this particular case, with change in faunal area there is a continued diversity increase. Once integrated, the newly mingled fauna ascended to even higher diversity levels than before the Great American Interchange. The descendants of northern immigrants produced most subsequent origination during the Ensenadan (medial Pleistocene) and Lujanian (Late Pleistocene). Thus, much of the impetus for a continued high diversity level seems to have flowed from the O-type episode (Webb, 1975).

## The Gamma-Diversity Paradox

As defined above, an O-type rapid turnover episode produces an increased number of taxa in a given area, and is presumably triggered by an increased pulse of immigrant taxa. On a world-wide scale, however, rather than "in a given area," this relationship may change. An O-type episode could conceivably reduce world heterogeneity, and thereby produce an interesting paradox. Cody's (1976) distinction between *Beta* (regional) diversity and *Gamma* (world-wide) diversity is essential to resolving this paradox which may be called the *Gamma-Diversity* Paradox.

Imagine that the world's seven continents all become closely connected and pass major sets of immigrants to one another. As a result the terrestrial biota of each continent (Beta diversity) is enriched by an O-type episode. On the other hand, Gamma diversity drops to almost the same level as that of any one continent. Similarly, Valentine (1973) has shown that provinciality has a first-order control of world diversity.

A clear example of such a pattern can be observed in steppe mammals of the Late Cenozoic. As climatic deterioration vastly expanded the extent of steppe and semi-desert, concomitant eustatic and tectonic events multiplied the intercontinental connections between such habitats, thus combining formerly separate steppe faunas. Camels and horses from North America reached South America, Eurasia, and Africa; bison and other bovids, elephants, and cheetahs from the Old World reached North America. Pioneer or tramp taxa, notably cricetine and microtine rodents, spread ubiquitously. While each continent is the richer for it, Gamma diversity has dropped. The loss has been particularly severe on large k-selected vertebrates, whereas smaller r-selected vertebrates such as microtine rodents have gained in numbers.

## E-O COMPOUND EPISODES: THE ULTIMATE CATACLYSMS

In this essay, I have tried to distinguish two possible kinds of rapid turnover episodes. For any meaningful analysis of paleoequilibria, this distinction ought to be made. As Rosenzweig (1977) has stated, "environmental changes must be affecting only the extinction or only the origination rates, and we must know which." Each type of episode can be produced quite independently by distinct classes of environmental effects, as discussed above. By the same token, however, both kinds of events could be produced concurrently by separate sets of causes. It is of some interest to speculate on the nature of such compound episodes.

The results of such compound episodes would be easy to predict if the two curves rose symmetrically and simultaneously. The net diversity at any given time would not change, but the turnover rate would increase nearly twice as much as it would for either effect alone. And because of the high turnover rate, diversity accumulated over any long period of time (say a million years), would appear vastly inflated. Extinctions and originations, though balanced, would be unusually high.

What if the two kinds of punctuation events came at approximately but not exactly the same time? Such dischroneity might be somewhat more probable than perfect synchroneity. The biotic effects in such cases would be far more interesting. In an O-E compound punctuation a diverse biota with rapidly differentiating taxa and intense competition would then be dramatically wiped out. The cataclysm probably would be followed by a long recovery period.

Even more interesting would be an E-O compound episode. Following an extinction event, with diversity at a very low ebb, a major new set of allochthonous taxa, predominantly opportunists, would enter an under-saturated environment. The taxa building toward a new equilibrium would be largely new to the region and would evolve rapidly from r-strat-egists to seize the diverse opportunities available to k-strategists. The early equilibrium phase would be superseded by a mature or assortative equilibrium of perhaps twice the diversity and a wholly new community would rapidly evolve. To a large degree the full history of Late Cenozoic mammals in North America follow an E-O scenario. The Late Creta-ceous-Early Cenozoic history of land vertebrates is an even greater ex-ample (Sloan, 1974). It is not difficult to imagine even more complex pat-terns. For example, an oscillating environmental pattern, such as the glacial ages, might produce numerous E-O pairs cyclically on many con-tinents. Such compound E-O episodes are probably associated with the great cataclysms in the history of life.

## SUMMARY

The MacArthur-Wilson Equilibrium Hypothesis predicts two distinct kinds of rapid turnover episodes, an *E-type* triggered by some environ-mental crunch and resulting in a decreased number of taxa, and an *O-type* triggered by more immigrants and resulting in an increased num-ber of taxa. An example of each is given.

In North America the Clarendonian Chronofauna, a richly diversified savanna fauna that flourished from about 15 to about 8 Ma, was disman-tled during the next few million years by severe climatic deteriorations that produced a depauperate steppe fauna. Probable causes included regional uplift and the Messinian eustatic-climatic events about 5 Ma. Recovery to about the former equilibrium level took some three myr (through Blancan into Irvingtonian).

The South American mammal fauna was markedly enriched by northern immigrants during the Great American Interchange. The mean number of genera increased by 30 to 50% between pre-interchange (Chasicoan and Huayquerian) and interchange stages (Chapadmalalan and Uquian). The equilibrium number never again dropped to its former low level.

Since these two modes of rapid turnover are attributable to different sets of causes, they can by chance coincide or variously interact. The most powerful interaction is probably an E-type followed by an O-type episode, the compound E-O episode thereby producing first a large opportunity which was then intensively filled by many newly interacting taxa. Presumably the great revolutions in the history of life represent such compound episodes.

## ACKNOWLEDGMENTS

I wish to thank Richard H. Tedford, Daniel S. Simberloff and Bruce J. MacFadden for their helpful comments on these and related subjects. Work on Florida vertebrate sites was partially supported by NSF grants GB 3862 and 33500.

## LITERATURE CITED

Ager, D. V., 1973. *The Nature of the Stratigraphical Record* (New York: John Wiley), 114 pp.

Axelrod, D. I., and H. P. Bailey, 1976. Tertiary vegetation, climate, and altitude of the Rio Grande depression, New Mexico–Colorado. *Paleobiology,* 2:235–255.

Behrensmeyer, A. K., 1975. The taphonomy and paleoecology of Plio-Pleistocene vertebrate assemblages east of Lake Rudolf, Kenya. *Harvard Univ. Mus. Comp. Zool. Bull.,* 146:477–587.

Boellstorff, J., 1976. The succession of Late Cenozoic ashes in the Great Plains: a progress report. *Kansas Geol. Surv. Guidebk.,* 1:37–71.

Bourliere, F., 1963. Observations on the ecology of some large African mammals. In: R. C. Howell and F. Bourliere (eds.), *African Ecology and Human Evolution* (Chicago, Ill.: Aldine Publishing Co.), pp. 43–54.

Breyer, J., 1975. The classification of Ogallala sediments in western Nebraska. *Michigan Univ. Mus. Pal. Paper,* 12:1–8.

Dalquest, W. W., 1969. Pliocene carnivores of the Coffee Ranch local fauna. *Texas Mem. Mus. Bull.,* 15:1–44.

Diamond, J., 1975. Assembly of species communities. In: M. L. Cody and J. M. Diamond (eds.), *Ecology and Evolution of Communities* (Cambridge, Mass.: Harvard Univ. Press), pp. 342–444.

Eldredge, N., and S. J. Gould, 1972. Punctuated equilibria: an alternative to phyletic gradualism. In: T. J. M. Schopf (ed.), *Models in Paleobiology* (San Francisco, Calif.: Freeman, Cooper & Co.), pp. 82–115.

Gregory, J. T., 1971. Speculations on the significance of fossil vertebrates for the antiquity of the Great Plains of North America. *Hess. Landesamtes Bodenforsch. Abh. (Tobien Fest.),* 60:64–72.

Gould, S. J., and N. Eldredge, 1977. Punctuated equilibria: the tempo and mode of evolution reconsidered. *Paleobiology,* 3:115–151.

Lundelius, E. L., 1960. *Mylohyus nasutus,* long-nosed peccary of the Texas Pleistocene. *Tex. Mem. Mus. Bull.,* 1:1–41.

MacArthur, R. H., and E. O. Wilson, 1963. An equilibrium theory of insular zoogeography. *Evolution,* 17:373–387.

————, 1967. The theory of island biogeography. *Princeton Univ. Popul. Biol. Int. Monogr.,* pp. 1–203.

Martin, P. S., and H. E. Wright, Jr., 1967. *Pleistocene Extinctions: The Search for a Cause* (New Haven, Conn.: Yale Univ. Press), 453 pp.

MacGinitie, H. D., 1962. The Kilgore flora. A Late Miocene flora from northern Nebraska. *Calif. Univ. Publ. Geol. Sci.,* 35 (2):67–158.

MacFadden, B. J., and M. F. Skinner (in press). Diversification of the one-toed horses *Onohippidium* and *Hippidion. Postilla, Peabody Mus.*

Marshall, L., 1980. In-M. Nitecki (ed.), *Biotic Crises in Ecological and Evolutionary Time* (New York: Academic Press).

Olson, E. C., 1978. Taphonomy and vertebrate paleoecology: with special reference to the Late Cenozoic of Sub-Saharan Africa. *Burg-Wartenstein Sympos.,* 69.

Raup, D. M., 1977. Probabilistic models in evolutionary paleobiology. *Amer. Sci.,* 65:50–57.

Repenning, C. A., 1967. Palearctic-Nearctic mammalian dispersal in the Late Cenozoic. In: D. M. Hopkins (ed.), *The Bering Land Bridge* (Palo Alto, Calif.: Stanford Univ. Press), pp. 288–311.

Rosenzweig, M. L., 1975. On continental steady states of species diversity. In: M. L. Cody and J. M. Diamond (eds.), *Ecology and Evolution of Communities* (Cambridge, Mass.: Belknap Press), pp. 121–140.

Ryan, W. B. F., et al., 1973. *Initial Reports of the Deep Sea Drilling Project* (Washington, D.C.: U. S. Govt. Printing Office), vol. 13, 144 pp.

————, 1977. On interpreting the results of perturbation experiments performed by nature. *Paleobiology,* 3:322–324.

Schopf, T. J. M., 1974. Permo-Triassic extinctions relation to sea floor spreading. *Jour. Geol.,* 82:129–143.

Scott, G. R., 1975. Cenozoic surfaces and deposits in the southern Rocky Mountains. *Geol. Soc. Amer. Mem.,* 144:227–248.

Sepkoshi, J. J., Jr., 1976. Species diversity in the Phanerozoic: species-area effects. *Paleobiology,* 2:298–303.

Simberloff, D. S., 1972. Models in biogeography. In: T. J. M. Schopf (ed.), *Models in Paleobiology* (San Francisco, Calif.: Freeman-Cooper), pp. 161–191.

――――, 1974. Permo-Triassic extinctions: effects of area on biotic equilibrium. *Jour. Geol.,* 82:267–274.

Shotwell, J. A., 1958. Intercommunity relationships in Hemphillian Mid-Pliocene mammals. *Ecology,* 39:271–282.

Simpson, G. G., 1953. *The Major Features of Evolution* (New York: Columbia Univ. Press), pp. 1–434.

Sloan, R. E., 1976. The ecology of dinosaur extinction. In: C. S. Churcher (ed.), Athlon: Essays in paleontology in honour of Loris Shano Russell. *Roy. Ontario Mus. Life Sci. Misc. Publ.,* pp. 134–154.

Taylor, R. B., 1975. Neogene tectonism in southcentral Colorado. *Geol. Soc. Amer. Mem.,* 144:211–226.

Tedford, R. H., et al. (in press). Faunal succession and biochronology of the Arikareean through Hemphillian Interval of North America. Calif. Univ. Publ. Geol. Sci.

Valentine, J. W., 1973. Phanerozoic taxonomic diversity: a test of alternate models. *Science,* 180:1078–1079.

Van Valen, L., 1969. Evolution of communities and Late Pleistocene extinctions. *Proceedings of the North American Paleontological Convention, Section E,* pp. 469–485.

Webb, S. D., 1969a. Extinction-origination equilibria in Late Cenozoic land mammals of North America. *Evolution,* 23:688–702.

――――, 1969b. The Burge and Minnechaduza Clarendonian mammalian faunas of north-central Nebraska. *Calif. Univ. Publ. Geol. Sci.,* 78, pp. 191.

Webb, S. D., and N. Tessman, 1968. A Pliocene vertebrate fauna from low elevation in Manatee County, Florida. *Amer. Jour. Sci.,* 266:777–811.

――――, 1977. A history of savanna vertebrates in the new world. Part 1: North America. *Rev. Ecol. Syst. Ann.,* 8:355–380.

Wilson, E. O., 1969. The species equilibrium. *Brookhaven Symposia in Biology,* 22:38–47.

# Chapter 17

# THE PHANEROZOIC "CRISIS" AS VIEWED FROM THE MIOCENE

RICHARD H. BENSON

Smithsonian Institution

## INTRODUCTION

A crisis is an event in the history of a system when stress, usually originating externally, causes the alteration of its principle structures to be imminent; and through the absorption of this stress into the subsystems, the system survives (see Benson, Chapter 2, this volume). It becomes catastrophic when deformation of the reaction pathways of the subsystem forces new arrangements to be formed suddenly. The old system collapses and is replaced by a new one composed of this new arrangement. A cataclysm is the total destruction of the system and the subsystems. Replacement, if it takes place, is from somewhere outside.

There have been two almost cataclysmic events in geological history, one at the end of the Permian called the "Phanerozoic Crisis" (McAlester, 1973); and the other, much more geographically restricted, called the "Messinian Salinity Crisis." I have worked on the latter Late Miocene "crisis" (Benson and Sylvester-Bradley, 1971; Benson, 1972–78) for several years. How would this younger, smaller one compare with the one that ended the Paleozoic? Does the assumption that "the present is the key to the past" include the possibility that the present has not yet recovered from a "crisis"? It is worth taking a look backward from the Neogene, however imperfect that view may be.

The authors of the names of these Permian and Miocene events (Ruggieri, 1967; McAlester, 1973) were notably, if not understandably timid in their choice of terms to describe the conditions of survival of these times as "crisis." Any time that all or almost all kinds of inhabitants perish and the oceanographic and geologic systems are completely restructured, there has been more than just a crisis. Children pass through "crises" when their fever gets high from the chicken pox. If one of them dies and a

437

new child is adopted to take its place, you wouldn't say your family had gone through a "crisis." It would have suffered a "catastrophe." However, this criticism is not likely to change the names.

Concerning an issue of more substance, I do not agree with the work of Lantzy and others (1977) that concluded that the extinctions at the end of the Permian were largely due to a "catastrophic" reduction of salinity. I believe that they misused the catastrophe-theory model through assuming that salinity changes and changes in shelf habitat areas are in any way counteracting forces in the general marine environment. The combined stress must yield a result that can find redundant reaction pathways in the subsystems. Their model had one set of stresses acting through competition for space within populations against a general physiological effect. Either, conceivably, could have separately resulted in extinction, but any special acceleration effect by their combination has yet to be demonstrated.

I also question that a world-wide salinity reduction of the magnitude required can happen fast enough to cause mass extinction throughout a whole class or order of marine animals. This would require the isolation of evaporite sinks on a scale that is unrealistic in comparison to the volume of water in the World Ocean. The results of the present comparison of the Miocene crisis with that at the end of the Paleozoic makes an effective reduction even more unlikely.

## THE PERMIAN AND SALINITY

The end of Permian time is marked by a profound change in the history of life. The reduction in the diversity of marine and land animals and land plants near the Permian-Triassic boundary is one of the most severe in the fossil record (called the "Phanerozoic Crisis" by McAlester, 1973; see also Rhodes, 1967; Valentine, 1969). The standing diversity of Permian families dropped from 228 in the Leonardian to 161 in the Djulfian (Schopf, 1974).

Explanations for this relatively sudden deterioration have included trace element poisoning (Cloud, 1959); reduction of habitat space (Newell, 1963; Schopf, 1974); radiation (Hatfield and Camp, 1970); trophic resource fluctuations (Valentine, 1973); high temperatures (Waterhouse, 1973); changes in salinity (Fischer, 1963; Bowen, 1968; Lantzy et al., 1977; Stevens, 1977) and other causes. The most interesting explanation from the point of view of one working on the Messinian "salinity crisis" (Benson, 1973a, 1976b, 1976c) is the suggestion that a sudden reduction in World Ocean salinity, combined with a lowering of sea level, was in some way responsible for the extinction of Permian faunas. There can be no doubt that for some groups the effect was catastrophic.

In 1963, Fischer proposed a brine reflux model and a deep-sea brine sink, in part similar to the present Mediterranean, in order to account for a lowering of salinity of the upper levels of Permian oceans from a normal level of 35 to 30°/00 through the withdrawal of about $3\times10^6$km$^3$ of dissolved salts. The actual amount of halite that could have been withdrawn can never be known. However, Stevens (1977) has estimated that the total amount now entrapped in Permian deposits is $1.59\times10^6$km$^3$. This is about half of that required to lower the salinity to 30°/00, one-tenth of the volume of salts now in solution in the World Ocean.

## THE MESSINIAN SALINITY CRISIS

The final event in the history of Tethys came in the Late Miocene when its last connection with the World Ocean was severed. The severance took place when the Betic and Riff orogenic complexes were formed closing the Iberian Portal (possibly a Moroccan one also) by transcurrent and convergent plate movement along a line from the Atlantic into the Mediterranean (Benson, 1976d). During the Burdigalian and Serravallian (18 to 12 Ma), the Atlantic had nourished a psychrospheric deep-sea fauna in the western basins of the shrinking Tethys Ocean complex (Benson, 1978).

To the east, the formerly wide Tethys had been gradually sealed off from the Indian Ocean by progressive plate movement northward, forming the great Paratethyan lakes of Sarmatian-Pontian age from the Caspian (which is a Paratethyan brackish relic) to the Vienna Basin. By about 6 Ma, Tethys had become a warm deep sea occupying an isolated complex of deep basins. This restriction was accompanied by the lowering of its sea level and the concentration of more than $1.2\times10^6$km$^3$ of evaporitic salts in a series of lakes (Hsü, Cita and Ryan, 1973: Ryan, 1973; and others) causing the total extinction of the stenohaline marine faunas (Ruggieri, 1967; Benson, 1976b).

Only those Tethyan species that lived outside on the narrowing shelves of the Atlantic survived as the World Ocean sea level was also lowered as much as 70 meters (Adams, et al., 1977). The basins of the Persian Gulf and the Red Sea also became isolated and at least an additional $1.2\times10^6$km$^3$ of evaporites were formed in these areas (estimate by W.B.F. Ryan, pers. comm.).

The crisis, which had catastrophic effects in Tethys and extended the Paratethyan fauna from the Caspian region to Gibraltar, ended with a sudden breaking of the dam that separated the vast shallow and multiple level lake system (known as "lago mare") from the Atlantic. The sudden invasion of oceanic waters was strong enough to carry in elements of the North Atlantic deep-sea fauna and to sustain them in the newly formed Mediterranean throughout the Pliocene.

The deeper species of this fauna became extinct sometime during the Pleistocene, and only the upper bathyal fauna survives on the floor of the Mediterranean today. However, with the sudden invasion, whose history and paradox in faunas can be seen within only a few centimeters of marl (Zanclean) along the southern shores of Sicily, the widespread "lago mare" fauna (mostly Caspian-like, brackish-water ostracodes) was extinguished. It survives only in isolated marginal lagoons and on the fringes of the Black Sea (especially in the Sea of Azov) today.

## PERMOTRIASSIC FAUNAS AND THE OSTRACODES

The critical faunas, those most affected by the "Phanerozoic Crisis" that survived, were, for the most part, the stenohaline groups including the brachiopods, echinoids, ostracodes, and bryozoans. Some continue, in modified form of course, until today. It is possible to suggest, therefore, a uniformitarian extrapolation from the present limits of tolerance of these forms, to retrodict a reduced salinity to the end of the Permian sufficient to have caused large-scale extinction. That is, it is possible, if the tolerances of the survivors have remained the same; also, if today's World Ocean salinity is a reliable datum for normalcy. I will examine a few of the special characteristics of the ostracodes that survived and question the datum in terms of questioning today's normalcy.

Triassic ostracodes are anything but common. In fact they were, for the most part, unknown until the 1960s (Kollmann, 1963; Sohn, 1968; Kozur, 1969, 1971, 1972: Kristan-Tollman, 1969; Bolz, 1971; Ulrichs, 1971). Most of the faunas occur in the Upper Triassic (Carnian and Rhaetic). I estimate that from among 2000 described ostracode genera, less than 1 percent are known from the Triassic, and many of these are fresh-water forms.

Indeed, the event that ended the Paleozoic was more than critical for the 24 families of ostracodes known from the Permian. Only six of these survived, and of these six, four families survive today; the bythocytherids, the bairdiids, the darwinulids, and the cytherideids. To these, however, might be added the platycopids, the cyprids, and the macrocyprid-para-cyprid group, which have either closely related or analogous forms coming from somewhere in the Paleozoic, but whose exact status near the

Paleozoic-Mesozoic boundary is in doubt. Of course, there are some mysteries like the origin of the cytherids, the most populous of today's ornate ostracode groups, which may have originated in the Paleozoic but whose record is incomplete. In short, it is reasonable to state that 75 percent of the known Permian ostracode fauna became extinct (some data taken from Bate and others, 1967).

The crisis survivors can be divided into two groups, those whose predominant membership today lives in fresh-water or marginal environments (darwinulids, cyprids, cytherideids) and those whose marine distribution is either very broad or typical of the deep sea. Kozur (1972) has in fact postulated that the Triassic of Hungary was psychrospheric (cold, deep oceanic) on the basis of the similarity of the bythocytherids to living psychrospheric forms. This is especially significant, as there are few oceanic faunas known from this time interval and even fewer phyletic links to Permo-Triassic oceans. However, it should be noted that *Cytherella, Bairdia, Macrocypris,* and *Paracypris* are commonly found in today's oceans at depths exceeding 2000 meters and these genera are among the first to be found in the newly evolving South Atlantic during the Late Cretaceous (Peypouquet and Benson, in preparation). Was the Triassic Ocean a refuge? We just do not know. But, it seems unlikely that this fauna would have survived had a salt brine existed, as Fischer (1963) has suggested. It does not fare well even today in the warm, slightly saltier depths of the Mediterranean.

One of the possible causes of the "Phanerozoic Crisis" event was the formation of the Laurasia-Gondwana complex by the continental "suturing" of Schopf (1974) and the beginning of the Tethys Ocean-Sea. Most ocean systems begin and end as seas and with salinity crises. The Tethys began in the Permian and ended in the Miocene (5 to 6 Ma), or perhaps the tectonic relic is still ending; but the Tethyan fauna was destroyed, except for bits and pieces now along the equatorial African coast.

## MIOCENE EXTINCTIONS

Besides the catastrophic aspect of the Messinian event, in which there was more than a 40 percent changeover in marine species from the Miocene to the Pliocene within a million years (Lyell noted a change of about 25 percent between the middle of the Miocene and the middle of the Pliocene), there is the fact that the evaporite deposits, mostly gypsum with some halite, still remain entombed in the bottoms of these Tethyan gulf regions. How much of this was removed from the oceans? How much did it lower the geologically normal salinity level—the osmotic threshold, which is used as the baseline for estimating the Permian "salinity draw-down" and Phanerozoic Crisis extinctions?

Information concerning the brachipods and echinoids involved in the Late Miocene "salinity crisis" is scant. Data concerning the foraminifera is not relevant, as the ranges of the larger forms and planktonic species are too short for comparison with the Permian-Triassic. Corals are rare or absent. The Bryozoa are generally too rare in the deeper facies and are therefore not so useful as the remaining major group, the ostracodes.

It is a curious fact that among the first invaders of the Mediterranean, after the Late Miocene destruction of the Tethyan marine and the Para-tethyan brackish or hypersaline fauna, were the ostracode genera (*Cytherella* (*C. vulgata*, a platycopid) and *Bythoceratina* (*B. scaberrima*, a bythocytherid). The bairdiids are not abundant, but they are found higher in the section. The remaining surviving ostracodes, the cytheri-deids, were the dominant faunal element of the Paratethyan "lago mare." These are many of the same genera that appear first after the Phaneozoic Crisis.

## COMPARING THE OSTRACODES

Making the perhaps unwarranted assumption that the processes that pro-duced the fossil ostracode evidence in both the Neogene and the Triassic remained uniform through time; we might make the following prelimi-nary observations:

1) Of the invertebrate faunas that left fossil remains from the Phanero-zoic and the Miocene Crises, the ostracodes have left as abundant a record as any single group common to both events.

2) The ostracode survivors of the "Phanerozoic Crisis" and the first in-vaders after the Miocene "salinity crisis" are many of the same ostracodes or closely related.

3) These forms are stenohaline, commonly found in stressed condi-tions, often deep (bathyal or abyssal), with few other species present.

4) The range in geologic age of all these families is among the longest in the fossil record (in the case of the bairdiids, 500 myr; the bythocyth-erids, 400 myr).

5) By comparison, *Cyprideis*, the dominant cyprideid of the Late Mio-cene hypersaline "lago mare," is a very euryhaline ostracode whose sa-linity tolerance allows it to persist in waters with nearly double the "nor-mal" marine saline or alkaline concentrations. The cytherideids begin in the Permian.

## CONCLUSIONS

Therefore, considering the model that the terminal phases of the Permian and those of the Miocene were catastrophic for the faunas involved (one world-wide; the other "ocean"-wide), with a similarity between the groups of survivors that appeared afterward, and further considering that some "salinity draw-down" may have affected both groups, I believe that we must tentatively conclude:

1) If today's marine salinity is a baseline for a measure of stenohaline restriction, the postulated Permian salinity reduction was not enough to cause a general destruction of the normal marine faunas.

2) If the post-Miocene salinity ranges have been depressed by 3 to 4 °/00, "normal" as seen through the whole of post-Cambrian time, the postulate of "death by dilution" of the Permian seas becomes even less credible.

The minimum threshold of tolerance of the most isohaline marine taxa to lowering of "normal" marine salinity is about 25 to 28 °/00. Obviously, if the salinity of 35 percent is presently depressed as a consequence of the Miocene crisis, then more than four times the evaporite deposits now known from the Permian would be required.

Thus viewed from the Miocene, or even the Recent, with knowledge of the Miocene "salinity crisis" and using the analogy of possible extinction by salinity reduction, we must conclude that although the mechanisms of crisis may be similar, the historical effects are not. The catastrophe of the end of the Tethys caused local extinction only, with a world-wide reduction in sea level and possible changes in ocean salinity. There was no world-wide faunal crisis, or at least I have not seen it. Mörner (1978) has suggested that the continental ground-water table was also affected causing widespread vegetation changes and mammalian extinctions. I would opt for another explanation of the extinctions of the Permian, perhaps a restriction of habitat space on the shelves as was described by Schopf (1974), because the ostracode survivors are either continental or deep-sea forms.

## ACKNOWLEDGMENTS

Thanks are due to Richard E. Grant, with whom I had many discussions, William Oliver, Martin Buzas, and W. A. Berggren who reviewed the manuscript, and to Laurie Brennan and Marita Penny who helped to prepare it.

## LITERATURE CITED

Adams, C. G., et al., 1977. The Messinian salinity crisis and evidence of Late Miocene eustatic changes in the World Oean. *Nature*, 269:383-386.

Bate, R. H., et al., 1967. Arthropoda: Crustacea. In: W. B. Harland, et al. (eds.), *The Fossil Record* (London: Geol. Soc. London), 535 pp.

Benson, R. H., 1972. Ostracodes as indicators of threshold depths in the Mediterranean during the Pliocene. In: D. J. Stanley (ed.), *The Mediterranean Sea* (Stroudsberg, Penn.: Dowden, Hutchingson and Ross), pp. 63-70.

———, 1973a. An ostracodal view of the Messinian salinity crisis. In: D. W. Drooger (ed.), *Messinian Events in the Mediterranean* (Amsterdam: North-Holland Publ. Co.), pp. 235-242.

———, 1973b. Psychrospheric and continental Ostracoda from ancient sediments in the floor of the Mediterranean. In: W. B. F. Ryan, et al. (eds.), *Initial Reports of the Deep Sea Drilling Project* (Washington, D. C.: U. S. Govt. Printing Office), vol. 13, pp. 1002-1008.

———, 1975. Morphologic stability in Ostracoda. *Bulls. Amer. Pal.*, 65 (282):13-45.

———, 1976a. The evolution of the ostracode costa analysed by "Theta-Rho Differences." In: G. Hartmann (ed.), Evolution of post-Paleozoic Ostracoda. *Naturwiss. Ver. Hamburg Abh. Verh.*, 18/19:127-139.

———, 1976b. Biodynamics of the Messinian salinity crisis. *Palaeogeogr. Palaeoclimatol. Palaeoecol.*, 20:1-170.

———, 1976c. Changes in the ostracodes of the Mediterranean with the Messinian salinity crisis. *Palaeogeogr. Palaeoclimatol. Palaeoecol.*, 20:147-170.

———, 1978. The paleoecology of the ostracodes in DSDP Leg 42A. In: K. J. Hsü, et al. (eds.), *Initial Reports of the Deep Sea Drilling Project* (Washington, D.C.: U. S. Govt. Printing Office), vol. 42, pp. 777-787.

Benson, R. H., and P. C. Sylvester-Bradley, 1971. Deep-sea ostracodes and the transformation of ocean to sea in the Tethys. In: H. J. Oertli (ed.), Paléoécologie d'Ostracodes, Pau, 1970. *Soc. Natl. Petrol. Aquitaine, Cent. Rech. Pau, Bull.*, 5:63-92.

Bolz, H., 1971. Late Triassic Bairdiidae and Healdiidae. In: H. J. Oertli (ed.), Paléoécologie d'Ostracodes, Pau, 1970. *Soc. Natl. Petrol. Aquitaine, Cent. Rech. Pau, Bull.*, 5:717-745.

Bowen, R. L., 1968. Paleoclimatic and paleobiologic implications of Louaan salt deposition. *Amer. Assoc. Petrol. Geol. Bull.*, 52:1833 (abstract).

Cloud, P. E., Jr., 1959. Paleoecology—retrospect and prospect. *Jour. Pal.*, 33:926-962.

Fischer, A. G., 1963. Brackish oceans as the cause of the Permo-Triassic marine faunal crisis. In: A. E. M. Nairn (ed.), *Problems in Palaeoclimatology* (New York: John Wiley and Sons Ltd.), pp. 566-574.

Hatfield, C. B., and M. J. Camp, 1970. Mass extinctions correlated with periodic galactic events. *Geol. Soc. Amer. Bull.*, 81:911-914.

Hsü, J. J., M. B. Cita, and W. B. F. Ryan, 1973. The origin of the Mediterranean evaporite. *Initial Reports of the Deep Sea Drilling Project* (Washington, D.C.: U.S. Govt. Printing Office), vol. 13, pp. 1203–1231.

Kollman, K., 1963. Ostracoden aus der alpinen Trias; II—Weitere Bairdiidae. *Geol. Bundesanst. Jahrb.*, 106:121–203. (German with English summary.)

Kozur, H., 1969. Die Gattung *Speluncella* Schneider 1956 (Ostracoda) in der germanischen Trias. *Freiberger Forschungs.*, C 245:47–67.

————, 1971. Die Bairdiacea der Trias. *Geol. Pal. Mitt. Jahrb.*, 1 (3):1–27.

————, 1972. Die Bedeutung triassischer Ostracoden für stratigraphische und palaoökologische Untersuchungen. *Ges. Geol. Berghaustud. Mitt.*, 21:623–660.

Kristan-Tollman, E., 1969. Zur stratigraphischen Reichweite der Ptychobairdien und Anisobairdien (Ostracoda) in der alpinen Trias. *Geol. Pal.*, 3:81–95.

Lantzy, R. J., M. F. Dacey, and F. T. Mackensie, 1977. Catastrophe theory: application to the Permian mass extinction. *Geology*, 5:724–728.

McAlester, A. L., 1973. Phanerozoic biotic crisis. In: A. Logan, and L. V. Hills (eds.), The Permian and Triassic systems and their mutual boundary. *Canadian Soc. Petrol. Geol. Mem.*, 2:11–15.

Mörner, N.-A., 1978. Low sea levels, droughts and mammalian extinctions. *Nature*, 271:738–739.

Newell, N. D., 1963. Crises in the history of life. *Sci. Amer.*, 208 (2):76–92.

Peypouquet, J. P., and R. H. Benson (in press). Les Ostracodes et l'Évolution des Paléoenvironments de la Walvis Ridge depuis le Crétace. *Soc. Géol. France Séance Spec. Bull.*

Rhodes, F. H. T., 1967. Permo-Triassic extinctions. In: W. B. Harland, et al. (eds.), *The Fossil Record* (London: Geol. Soc. London), pp. 57–76.

Ruggieri, G., 1967. The Miocene and later evolution of the Mediterranean Sea. In: C. G. Adams and D. V. Ager (eds.), *Aspects of Tethyan Biogeography*. Systematics Assoc., Spec. Paper 7:283–290.

Ryan, W. B. F., 1973. Geodynamic implications of the Messinian crisis of salinity. In: C. W. Drooger (ed.), *Messinian Events in the Mediterranean* (Amsterdam: North-Holland Publ. Co.), pp. 26–38.

Schopf, T. J. M., 1974. Steady-state vs. empirical views of biological history. *Science*, 183:945–946.

Sohn, I. G., 1968. Triassic ostracodes from Makhtesh Ramon, Israel. *Geol. Surv. Israel Bull.*, 44:1–71.

Stevens, C. H., 1977. Was development of brackish oceans a factor in Permian extinctions. *Geol. Soc. Amer. Bull.*, 88:133–138.

Urlichs, M., 1971. Variability of some ostracods from the Cassian Beds (Alpine Triassic) depending on the ecology. In: H. J. Oertli (ed.), Paleoecologie d'Ostracodes, Pau, 1970. *Soc. Natl. Petrol. Aquitaine, Cent. Rech. Pau, Bull.*, 5:695–715.

Valentine, J. W., 1969. Patterns of taxonomic and ecologic structure of the shelf benthos during phanerozoic time. *Paleontology*, 2 (4):684–709.

————, 1973. *Evolutionary Paleocecology of the Marine Biosphere* (Englewood Cliffs, N. J.: Prentice-Hall), 511 pp.

Waterhouse, J. B., 1973. The Permian-triassic boundary in New Zealand and New Caledonia and its relationship to world climatic changes and extinction of Permian life. In: A. Logan, and L. V. Hills (eds.), The Permian and Triassic systems and their mutual boundary. *Canadian Soc. Petrol. Geol. Mem.,* 2:445–464.

Part IV

# CATASTROPHES AND THE REAL WORLD

Chapter 18

# MARINE MINERAL RESOURCES AND UNIFORMITARIANISM

KENNETH O. EMERY

Woods Hole Oceanographic Institution

INTRODUCTION

James Hutton introduced the concept of uniformitarianism near the end of the eighteenth cenntury in his "Theory of the Earth." He included such statements as: Earth phenomena should be explicable by powers that accord with the earth's composition and are natural to it, and whose principles are known—with no appeal to extraordinary events. The processes must be consistent with the propagation of plants and animals.

Hutton's preoccupation with the importance of animals as the objective of geological processes would not now be accepted, especially in view of the absence of animals on other planets of the solar system. Hutton and his disciple John Playfair admonished against concern with unreasonable precipitates, Noachian floods, igneous rocks formed through melting by coal fires, and other processes that now seem unrealistic or even magical. However, some "geology books" of the nineteenth and twentieth centuries are based upon unwarranted attempts to fit geology into creation myths invented several millenia ago. Entrancement with magic pseudoscience even in modern times is illustrated by popular beliefs in water witching, the Bermuda Triangle, astrology, pyramid power, UFOs, and such.

Our knowledge of oceanic processes has advanced over that of land processes a century or so ago, largely because the methods of reasoning that were developed for the land were easily transferred to the ocean to guide the new measurements and data collections. Nevertheless, the ocean floor *is* different from the continents in terms of age (mostly much younger), origin (sea-floor spreading), dimensions (changing with time), and many processes (turbidity currents, bottom currents, thermal circulation in spreading belts, among others). Most of these facts and processes

449

are unique to the ocean. They also were beyond the observation potential of men in Hutton's time, and they were observed or inferred only during the 1960s or 1970s (with consequent probability of additional new surprises during the present decade). Nevertheless, the lesson is to interpret new observations in the light of known physical, chemical, biological, and geological principles applied only to a new environment on earth.

The newness of many ocean-floor observations means that some phenomena may not unanimously be recognized as clear examples of uniformitarianism or catastrophism on the basis of scale, duration, and frequency. How widespread, how long, and how frequent must a process be in order to be termed unusual enough to be catastrophic? Can one man's catastrophe be just another man's nuisance or interruption? Attention is directed in this report to some ocean-floor minerals as examples of uniformitarian deposition by nature and their catastrophic mining and use by man.

## MINERAL DEPOSITION

More than 1500 minerals have been identified and analyzed. Ten are the dominant rock-making minerals, and 60 comprise about 99 percent of all igneous, sedimentary, and metamorphic rocks in the earth's crust. Among the remaining minerals that are far less common are those that are prized for the metals that can be extracted from them. Not all such minerals are ores, however, for ores are only those deposits that can be mined at a profit. Ores may occur in igneous, sedimentary, or metamorphic rocks, either as primary deposits or as secondarily enriched ones.

Minerals in sedimentary rocks can be most easily conceived as having been formed by uniformitarian processes, through slow but steady weathering and erosion of the parent rocks, followed by transportation and selective deposition. Deposition is controlled by common physical factors, such as speed of transport relative to grain size and shape; chemical factors, like temperature and salinity; and biological factors, that include organic withdrawal from solution as skeletal material or as precipitates caused by biochemical modifications of the environment.

Minerals in igneous rocks are deposited during cooling or by reactions with enclosing rocks. In metamorphic rocks, many minerals require high temperatures and high pressures that lead to the conversion of less dense minerals into more dense ones. In one sense, both igneous and metamorphic minerals can be considered to be catastrophic in origin, because the processes that form primary deposits are discontinuous in time and space. On the other hand, these processes are natural, probably occurring some place on earth at all times, and the secondary enrichment that is required for most ore deposits is a long-term process. For these reasons I consider both the primary and secondary mineral deposits to be examples of uniformitarianism.


MINERAL USE

Man's first use of rocks was probably as fist-strengtheners that later evolved into hammers and axes. With the advent of fire came hearth-stones, and later the use of pyrite and flint. In time, stones were employed as hard tips for spears and arrows, and were later replaced by points made of strong, hard and nearly amorphous flint and obsidian. The development of weaponry probably was accompanied by the use of white chalk, yellow clay, and red ochre to go with black charcoal for personal adornment. Control over the sources of the materials used for weapons must have been as important as later need for control over sources of salt needed by the body.

As technology advanced, the superiority of copper weapons to those of stone was recognized. Then bronze was found to be better even than copper, with the attendant need for tin, arsenic, or other alloy hardeners. When iron weapons broke or dulled those of bronze, the competition at first was less for low-grade iron sources than for knowledge of the necessary technology (I Samuel, 13: 19–21). Later, the increasing need for iron, and thus for the wood and coal to work it, produced some of the world's major wars. In fact, nearly all wars appear basically to have been due to competition for mineral sources, either for direct mining, for trading, or for the wherewithal to trade. Similarly, the possessors of these minerals were "king of the mountain" for a time, able to repel the have-nots. Among the metals and minerals for which wars have been fought and with which some of them have been won are copper, tin, iron, coal, potash and sulfur (for gunpowder), halite, silver, and gold. Wars over oil and uranium perhaps have not yet occurred, but they are highly likely to take place because of their great strategic and economic importance.

At present, the world demand for minerals is enormous, particularly among the industrially advanced nations that have high levels of technology. For example, the annual production of all minerals (fuels, metals, and nonmetals) in the United States is about 5% of the nation's GNP (Gross National Product), with almost as much additional minerals imported. Similar small percentages of total GNP in mineral production obtains in other industrially advanced nations. In contrast, large percentages are present in many underdeveloped nations which produce them for export (Emery, 1974, 1976a). Use of minerals by the world is increasing logarithmically, so that critical shortages of many of the rarer metals threatens. As a result, conferences have been held and many books and articles written during the preceding decades. Subjects treated in the books can be grouped into several major categories: assessment of United States and world stocks of metals and energy minerals, methods of conserving them, effects of their use on the environment, and the future of mankind bereft or short of minerals. For those who wish to follow this interest, more than a score of these recent books are listed in the bibliography. A small effort could double or triple the length of this list.

The main point is that man's dependence upon and his use of minerals are increasing at great speed, at the same time that the stock of minerals is diminishing. That is, the demand is roughly inversely proportional to the remaining supply. The critical shortages of the present and immediate future are the result of less than a century of intense mining of deposits that had accumulated during several billions of years of earth history. I see this as an example of the catastrophic use of a uniformitarian accumulation and natural enrichment.

OCEAN-FLOOR MINERALS

Just as for minerals in general, there have been many conferences held, books produced, and articles written on minerals of the ocean floor. I have included several books and papers in the bibliography for those who are particularly interested in the subject. These works discuss mineral assessments, mining methods, the economics, and the politics that ensue. In large part, the conferences and writings are so repetitive that the later books contain little that is new, and the narrow audience has become surfeited.

To put the picture into perspective, the annual total world oil production and offshore oil production (in millions of tons or tons X $10^6$) for a few recent years has been, respectively: 1973: 2690 and 490; 1974: 2760 and 452; 1975: 2620 and 403; 1976: 2780 and 459; 1977: 2760 and 557 (Anonymous, 1977). By 1980 these productions were 2910 and 670 million tons. The decrease in production between 1973 and 1976 was due to the rapid price increases engineered by OPEC (Organization of Petroleum Exporting Countries), and it most affected the higher cost of offshore production. If we take $90/ton ($12/barrel) as the unsubsidized price of oil in 1977, the total gross income to offshore producers of the world has been $50 X $10^9$. To this amount must be added about $25 X $10^9$ for the natural gas produced offshore, which yielded a total of about $75

X $10^9$ for the expected 1977 gross income to producers of all offshore petroleum (oil and gas) before taxes and expenses. This was about three times the then current income to the fishermen of all fish that were caught (about 60 X $10^6$ tons times about \$300/ton). All other ocean-floor mineral production had and has relatively negligible value: sand and gravel—\$500 X $10^6$; heavy minerals (tin, ilmenite, rutile, titanite, magnetite)—\$50 X $10^6$; and shells, coral, and pearls—\$20 X $10^6$, all for estimated 1977 production value. Contrary to many popular articles on the subject, neither gold nor diamonds are produced from the ocean floor.

Income from possible new mineral production in the immediate future is also negligible compared with that from petroleum. For example, the Red Sea hot brine deposits of zinc, copper, lead, and silver had a total estimated value of \$2.8 X $10^9$ in 1967 (Bischoff and Manheim, 1969), but increased to \$4 X $10^9$ by 1977 through higher mineral prices, which were largely due to monetary inflation. Even if the metals can be produced at a profit (after costs for dredging, shipping, and refining), the total amount is relatively small and the annual production is even smaller. Similar deposits elsewhere in the World Ocean are unlikely, because they are believed to form only during the earliest stages of sea-floor spreading when thermal circulation of ocean water includes passage through continental as well as oceanic crust (Emery and Skinner, 1977). Similar deposits probably did form during the early stages of the opening of the Atlantic Ocean, but they now are buried under thick prisms of continental rise and slope sediments.

The ocean-floor deposit that is most widely acclaimed by underdeveloped countries and yet is poorly understood is the vast expanse of manganese nodules. The valuable metals in these nodules are copper, nickel, and cobalt (1.1, 1.3, and 0.3%, respectively, in the richest deposits of the Pacific Ocean). Present annual production of these three metals on land totals about \$10 X $10^9$, and if offshore production should prove to be so profitable that offshore products penetrate the copper market 10% (a large amount) the annual gross value from the offshore metals would be \$1 X $10^9$. This amount again is small compared with the value of offshore petroleum, but it is enough to be interesting to mining companies. Incidentally, the nodules are being deposited in accordance with the principles of uniformitarianism and apparently at a rate that is faster than man's use of the important metal components.

A last mineral from the ocean floor that is much mentioned is phosphorite. The richest deposits occur on outer continental shelves and bank tops off southern California-Mexico, Peru-Chile, South Africa, and Morocco, but in none of these places is the concentration of phosphate so high in the nodules as in many deposits that are mined on land—32%.

Accordingly, offshore production is likely to be long delayed, especially in view of the fact that the land deposits have reserves of several hundred years at the present rate of mining.

## OCEAN ENERGY SOURCES

### Fossil Fuels

Petroleum began to be won from the ocean floor almost a century ago, but its production began in earnest only after World War II, when the technology of marine exploration and drilling began sustained advances. From 1971 to 1976 about 17% of total oil production (less than for that of natural gas) has come from offshore drilling. Because the main environment of petroleum deposition is the ocean, we can expect increasing percentages of it to come from the ocean floor in the future; most of that from land sites comes from uplifted former ocean floors.

Far more fossil energy than that in petroleum is locked into coal, oil shale, and tar sands (Cloud and Skinner, 1975), which mainly occur on land. Most coal was deposited in ancient fresh-water swamps; probably most rich oil shales were deposited in ancient lakes; tar sands are residues of oil reservoirs breached by erosion on land. Similarly, most of the exhausted oil fields are on land, and they still contain about two-thirds of their original trapped oil. All these deposits contain enormous fossil energy, hundreds of times more than has so far been extracted by man. Unfortunately, however, the deeper and poorer deposits may require more energy to mine than they can yield, thus providing no net energy (Huettner, 1976). A possible remedy is to partially burn them *in situ* and to collect the gases that are produced. This practice is opposed because of the real dangers of runaway subterranean fires, settlement of overlying strata, and leakage of poisonous fumes. Other rich coal deposits that are near the surface can be strip mined and used to fuel large, nearby electric generating plants or crushed for shipment via pipelines. In opposition are the high electric transmission losses from generators distant from industrial centers and lack of sufficient water to carry crushed coal in pipelines to the industrial centers. Low-grade deposits near the surface can be strip mined and pyrolized so that some of their fuel content would serve to heat the rest enough to produce liquid or gaseous fuel. Objections to this method have centered around lack of sufficient available water for the process plus the problems of huge waste shale disposal.

Whether from the land or the ocean floor, the fossil fuels have the advantage of being highly concentrated sources of energy that are more accessible than other kinds of energy. Oil is in particular demand, because it is a liquid rather than a solid or gas and thus is more easily transported and handled, making it ideal for small mobile uses such as automobiles, tractors, and small ships. Coal is more suitable for large stationary uses such as municipal and industrial electric generating plants. All these uses, however, have reached the point at which the carbon dioxide and perhaps the heat itself, after long storage in fossil form, are released so rapidly as to observably increase the temperature of the earth's atmosphere. The effect of carbon dioxide is probably more important than the direct heat production, because it appears to form a blanket in the atmosphere that permits transit of sunlight but blocks the heat into which the sunlight is converted—greenhouse effect (Wilson, 1970). The result can be great changes in climate zones of the earth's surface. It is still uncertain whether the increased temperatures would melt the remaining glacial ice to cause a 60-m rise in sea level or would increase air circulation enough to initiate a new ice advance. Either extreme would cause problems for mankind, his food supply, and his habitations which have become more or less adapted to present climate zonation.

Some idea of the rapidity at which man is using the fossil energy that required such a long time to accumulate and concentrate is given by Table 18–1. Considerable energy losses occur during transitions from sunlight to fixation of carbon by photosynthesis, to burial of organic carbon, to formation and migration of petroleum, to exploitation of petroleum by man. The annual accumulation of petroleum was obtained by dividing the total amount of cumulative production, proved reserves, and estimated undiscovered petroleum that is producible with present technologies by the time span that petroleum accumulated. This suggests a rate of about 1000 tons per year; it would have been higher, had not most petroleum escaped by natural erosion of traps during geologic time. The present annual rate of exploitation by man is about $5 \times 10^6$ times the rate of petroleum accumulation. Most interesting, however, is the fact that sunlight reaching the earth is about $7 \times 10^{10}$ greater than the annual rate of natural accumulation of petroleum, and about 14,000 times the present rate of use of petroleum. Another way of looking at the matter is that the energy from present annual combustion of oil, gas, and coal is about the same as that from only 52 minutes of direct solar energy striking the entire earth. These ratios in favor of sunlight are so large that many scientists believe that direct solar energy must be tapped in the future; moreover, this source would avoid the changes in atmospheric temperature caused by the burning of fossil fuels.

|  | OIL<br>(X 10⁹ tons) | GAS<br>(X 10¹² m³) |
|---|---|---|
| Annual Production (1977) | 3 | 2 |
| Cumulative Production (to January 1977) | 48 | 25 |
| Proved Reserves | 80 | 60 |
| Undiscovered Resources | 142 | 135 |
| Total Original Producible Oil and Gas | 270 | 220 |

|  | ENERGY EQUIVALENT IN TONS OF OIL (X 10⁹) |
|---|---|
| Same Expressed as Oil Equivalent | 470 |
| Annual Solar Energy Reaching Earth | 70,000 |
| Annual Fixation of Carbon by Photosynthesis | 60 |
| Annual Burial of Carbon in Marine Sediments | .07 |
| Annual Petroleum Formed and Trapped until<br>Exploited by Man | .000001 |
| Annual Petroleum Exploited by Man | 5 |

Table 18-1. Some world data for petroleum.*
* partly from Emery (in press)

## PRESENT SOLAR ENERGY

Capture of solar energy can be accomplished in many ways, both on land and in the ocean, but several of the ways are more effective in the ocean. Direct conversion of sunlight into electrical energy can be accomplished by selenium and silicon cells, but it is very inefficient and requires an enormous collection area. For example, at 100% efficiency a total of 13,000 km² of cells would be needed at the equator to equal the world's present annual energy production from oil, gas, and coal. At a more reasonable efficiency of 5 percent, the area would be 260,000 km², or a square 510 km on a side. Such an area is large whether on land or ocean, and the capital investment in terms of money or energy for the enterprise would be huge compared with the expected rate of energy production.

Perhaps a better way of utilizing solar energy is to let nature do much of the concentrating cheaply, though wastefully (as in fossil fuels), by collecting the energy in winds, waves, temperature differences, salinity differences, currents, and tides of the ocean. Each of these points of energy collection has been proposed, but each is only locally enhanced and each requires huge capital costs per unit of energy recovered. Largely for these reasons, much attention has been directed toward nuclear sources, primarily fission, because fusion seems very far in the future. Problems of high construction costs, danger from sabotage and terrorists, and environmental contamination have delayed progress, but the major problem that commonly is not faced is that of safe disposal of wastes during the tens of thousands of years required for radioactivity to die to acceptable levels. The ocean may become the chief receptacle, as it already has for many other wastes produced by man.

## ECONOMIC, SOCIAL, AND POLITICAL IMPACT OF MINERAL CATASTROPHE

The mining of fossil fuels and metals at a catastrophic rate is likely to produce equally catastrophic changes in some of man's affairs—economics, politics, and relations between countries. Some of the mined minerals are completely dissipated by use, such as fuels; others are so widely dispersed as lost and worn-out manufactured items that they cannot be recovered economically. A few are intrinsically so valuable and durable that they are almost completely recycled; for example, 95% of all the gold that has ever been mined is believed to be still in use.

The rapid use of fuels and many metals is due to logarithmically increasing levels of technology and logarithmically increasing world populations, whereby a large percentage of the people who ever lived are alive today. In the past, population size was controlled by the Four Horsemen of the Apocalypse (War, Famine, Anarchy, and Pestilence). For better (short term) or worse (long term), modern technology distributed to underdeveloped countries has reduced the present roles of the Four Horsemen and raised the specter of their later, even more catastrophic, roles when the supples of fuels and metals fail. The means of delaying the moment of truth taken by industrially advanced countries are completely different from those taken by underdeveloped countries.

Industrially advanced countries are beginning to realize that shortages of some fuels and metals are imminent, and their attempts to solve the problems take many forms. Simplest is that of search for new supplies in poorly studied regions, particularly including the ocean floor. This means development of new exploration techniques such as geophysical tools and methods and deep-ocean drilling capabilities. New demands for access to scarce materials are likely to occur, as they have in past millenia. Many governments have taken over oil companies on the theory that availability of petroleum and revenue derived therefrom are too important to be left to private enterprise. Conservation by rationing, high costs (by punitive taxes), and recycling is increasing, as is the search for substitute materials (such as aluminum for copper, and new organic compounds for aluminum in electrical transmission lines). Temporary smoothings of irregularities in supplies are provided through stockpiling and subsidies by government orders.

Underdeveloped countries that depend upon revenues from raw materials have a very different set of approaches. First are attempts at cartelization to increase prices for their raw materials. Unsuccessful cartels have been tried for rubber, tin, copper, aluminum, sugar, and other materials. The OPEC cartel has been successful for five years, but it has hastened efforts by industrially advanced countries to find alternate sources for oil and oil substitutes; and cracks in the cartel are beginning to appear. Concurrent is confiscation (in the name of nationalization) of mines, smelters, refineries, and other installations built by companies and governments of industrially advanced countries. This method is effective only in the short term, because the process eliminates most access to technological know-how for exploration and improvement of methods, and it results in the complete loss of new capital inflow from private enterprise. The next stage, usually, is heavy demand for technology (Haskins, 1964; Myrdal, 1974; Ross and Smith, 1974; Revelle, 1976) and low-interest loans of capital from governmental or international sources. Neither new technology nor loans are feasible for many countries, because their levels of education are too low for receiving the technology (Wade, 1975; Emery, 1976b), and their cash flow is too small to repay the loans. Next are demands in the name of justice and necessity for forgiveness of the loans coupled with claims for the transfer of payments from rich to poor countries as foreign aid.

Worst situated of all are the underdeveloped countries that have few or no raw materials to export. Nevertheless, United Nations membership has trebled since the organization was founded through the addition of such countries, largely newly developed from former colonial lands.

In the recent past, two new methods of gaining revenue have evolved. One is confrontation at the United Nations General Assembly, where votes are equal regardless of the size of populations or power of countries. Thus, large blocs of several political and economic persuasions take adversary positions with respect to the industrially advanced countries. The other method is that of acceptance of the concept of the former Malta ambassador to the United Nations, Arvid Pardo, and of President Lyndon Johnson's 1966 statement that the ocean resources are the "common heritage of mankind" (Friedmann, 1971, p. 21, 28). Argument has been based mainly upon access to manganese nodules, over whose mining and revenues the United Nations Sea Bed authority claims jurisdiction. This claim appears rather silly, because not one of the underdeveloped countries or any combination of those, has the technical ability, capital, or interest to do the mining. Moreover, if the mining were to be extensive enough to make a 10% penetration of the world market for copper with a 10% profit on the copper, nickel, and cobalt, the income would be only $0.03 per year per capita of the population in the world's underdeveloped countries. In addition, of course, the ocean-floor mining, if successful, would displace markets for the same minerals now produced by some underdeveloped countries (UNCTAD secretariat, 1974), requiring some form of entitlement payments and large associated bureaucracy—with an attendant high cost of operation.

Transfer of technology and funds from industrially advanced to underdeveloped countries has risen during the recent past, especially after the end of colonialism. Moreover, crises in the underdeveloped countries have been averted by charitable shipments of food in time of and to places of famine. Growth of food on the land and capture of fish at sea consumes about 10 calories of fossil energy for every calorie of food that reaches the consumer (Steinhart and Steinhart, 1974), so the impending shortage of fossil energy means that food production and stockpiles are sure to diminish (Chancellor and Goss, 1976). Unfavorable changes in climate are adding to this effect. The same lack of fuel can reduce the availability of large-scale transport of food. Industrially advanced countries may soon not be able to respond to serious famines in either underdeveloped or other advanced countries, particularly when response means larger populations and greater demands will have to be met at the next crisis. An intelligent layman should keep in mind these close interrelationships between technology, science, politics, and cultures in order to understand the coming effects of mineral depletion.

Lest industrially advanced countries consider themselves safe, they must recognize that their envied position depends upon favorable climate (personal energy, and good rainfall, and growing temperatures), favorable standard of living (access to sufficient needed raw materials for food and industry), and favorable religion or government (incentive to work and think). Variations in these controls have caused shifts in dominant advanced countries during the past: Egypt-Mesopotamia to Greece to Rome to the Arab world to western Europe to the United States. Progressively shorter periods of dominance are observed. China to Japan appears to illustrate a separate evolution. Status quo cannot remain forever; where next? The Soviet Union, India, southeastern Asia? In a sense, the catastrophe of excessive mining on land and ocean floor can, and probably will, lead to catastrophes in human affairs in both underdeveloped countries and industrially advanced ones. New cultures with priorities different from those of the present world community may become dominant.


## POSTSCRIPT

During the six years since the date of the conference in July 1977 and the date of writing this article, several profound changes have occurred. These changes were powered by OPEC's intense pressure to increase the price of exported oil to the present $34/barrel, about three times the price in 1977. The predictable result was proportional increases in the price of all mined, agricultural, and manufactured products that depend upon fossil energy, soon followed by near-proportional increases for all products and all labor. These increases affected all populations in industrialized, OPEC, and underdeveloped nations. For industrialized nations, they signalled successive price increases of everything, inflation to cover higher costs, strikes to increase wages, recession in manufacturing and business, increased outlays to cover employment compensation, and, of course, decreased demand for oil. These changes in turn affected OPEC through increased prices for their own imports and decreased quantities and revenues for their oil exports, which resulted in decreased solidarity of OPEC countries. Worst off are the non-oil-exporting lesser developed nations whose costs rose for oil and everything else that they imported, accompanied by an inability to repay loans and even (for some) interest on loans. These difficulties arose at the same time that more aid was needed but less was available from either industrialized or OPEC nations.

The economic difficulties produced by political manipulation of the products of mineral extraction will not have run their full course by the date of publication of this book. For example, the sudden increase in the price of oil has led to new exploration and increased production from new sources (especially the North Sea and Mexico) with a higher percentage of increases from offshore than onshore fields. Total world production decreased 8%, but OPEC production decreased 29% between 1977 and 1981 in favor of the new and more stable sources closer to the users. Moreover, the new fields are young, annual production continues to increase, and other, new fields are coming on stream. This particular human-induced catastrophe shall be seen from the vantage of a half-century hence as representing only a minor perturbation on the long-term expansion curve of mineral extraction to satisfy the needs of increasing populations—thus in reality long-term uniformitarianism still continues for mineral extraction.

## ACKNOWLEDGMENTS

This study was funded by the Woods Hole Oceanographic Institution Ocean Industry Program.

## LITERATURE CITED AND RECOMMENDED READING

Albers, J. P., M. D. Carter, A. L. Clark, A. B. Coury, and S. P. Schweinfurth, 1973. Summary petroleum and selected mineral statistics for 120 countries, including offshore areas. *U. S. Geol. Surv. Prof. Paper*, 817, 149 pp.

Alexander, L. M. (ed.), 1967. *The Law of the Sea* (Columbus, Ohio: Ohio State Univ. Press), 321 pp.

Anderson, R. A., (ed.-in-chief), 1969. Ocean resources. *Stanford Jour. Int. Studies*, 4, 142 pp.

Anon., 1970. *Mineral Resources of the Sea* (New York: United Nations), 49 pp.

Anon., 1977. Subsea production systems. *Ocean Industry*, 12 (7):57–73.

Bhatt, J. J., 1974. *Environmentology: Earth's Environment and Energy Resources* (Buffalo, N.Y.: Modern Press), 412 pp.

Bischoff, J. L., and F. T. Manheim, 1969. Economic potential of the Red Sea heavy metal deposits. In: E. T. Degens and D. A. Ross (eds.) *Hot Brines and Recent Heavy Metal Deposits in the Red Sea* (New York: Springer-Verlag), pp. 535–541.

Borgese, E. M. (chairman), 1975. *Man in the Ocean, Symposis of Expo '75, Pacem in Maribus VI* (Malta: Int. Ocean Inst.), 848 pp.

Bregel, B., et al. (eds.), 1973. Public policy toward environment 1973: a review and appraisal. *N. Y. Acad. Sci. Ann.*, 216, 202 pp.

Brobst, D. A., and W. P. Pratt (eds.), 1973. United States mineral resources. *U. S. Geol. Surv. Prof. Paper*, 820, 722 pp.

Brockett, E. D., and H. D. Hedberg (chairmen), 1969. *Petroleum Resources Under the Ocean Floor* (Washington, D. C.: National Petroleum Council), 107 pp.

Cameron, E. N. (ed.), 1973. *The Mineral Position of the United States, 1975-2000* (Madison, Wisc.: Univ. Wisconsin Press), 159 pp.

Chancellor, W. J., and J. R. Goss, 1976. Balancing energy and food production, 1975-2000. *Science*, 192:213–218.

Cloud, P. (chairman), 1969. *Resources and Man* (San Francisco: W. H. Freeman & Co.), 259 pp.

Cloud, P., and B. J. Skinner, 1975. *Mineral Resources and the Environment* (Washington, D. C.: Nat. Acad. Sci.), 4 vols.

Committee on Geological Sciences, 1972. *The Earth and Human Affairs* (San Francisco: Canfield Press), 142 pp.

Daddario, E. Q., 1976. *An Assessment of Alternative Economic Stockpiling Policies* (Washington, D.C.: U. S. Congress Office of Technology Assessment), 327 pp.

Dean Witter & Co., 1969. *The Ocean and the Investor* (New York: Dean Witter Research Department), 132 pp.

Emery, K. O., 1974. Latitudinal aspects of the Law of the Sea and of petroleum production. *Ocean Develop. Int. Law Jour.,* 2:137–149.

————, 1976a. Some characteristics of nations. *Illinois Bus. Rev.,* 33 (1):6–8.

————, 1967b. Perspectives of shelf sedimentology. In: D. J. Stanley and D. J. P. Swift (eds.), *Marine Sediment Transport and Environmental Management* (New York: John Wiley & Sons), pp. 581–592.

———— (In Press). Potential for deep-ocean petroleum. *Wallenberg Symposium, Sweden,* August 1977.

Emery, K. O., and B. J. Skinner, 1977. Mineral deposits of the deep-ocean floor. *Marine Mining,* 1:1–71.

Fischer, A. G., and S. Judson (eds.), 1975. *Petroleum and Global Tectonics* (Princeton, N. J.: Princeton Univ. Press), 322 pp.

Freeman, S. D., 1974. *Exploring Energy Choices: A Preliminary Report* (Washington, D. C.: Ford Foundation Energy Policy Project), 81 pp.

Friedmann, W., 1971. *The Future of the Oceans* (New York: George Braziller), 132 pp.

Gordon, B. L. (ed.), 1974. *Marine Resources Perspectives* (Watch Hill, R. I.: Book & Tackle Shop), 366 pp.

Gullion, E. A. (ed.), 1968. *Uses of the Sea* (Englewood Cliffs, N. J.: Prentice-Hall), 202 pp.

Halbouty, M. T., J. C. Maher, and H. M. Lian (eds.), 1976. Circum-Pacific energy and mineral resources. *Amer. Assoc. Petrol. Geol., Mem.,* 25, 608 pp.

Haskins, C. P., 1964. *The Scientific Revolution and World Politics* (New York: Harper & Row), 115 pp.

Haun, J. D. (ed.), 1975. Methods of estimating the volume of undiscovered oil and gas resources. *Amer. Assoc. Petrol. Geol. Studies Geol., no. 1,* 206 pp.

Horn, D. R. (ed.), 1972. *Ferromanganese Deposits on the Ocean Floor: International Decade of Ocean Exploration* (Washington, D. C.: Nat. Sci. Found.), 293 pp.

Huddle, F. P. (ed.), 1974. *Requirements for Fulfilling a National Materials Policy* (Washington, D. C.: U. S. Cong. Off. Tech. Assess.), 194 pp.

Huettner, D. A., 1976. Net energy analysis: an economic assessment. *Science,* 192:101–104.

Jackson, H. M. (chairman), 1974. *U. S. Energy Resources, A Review as of 1972,* 93rd Congress, 2nd Session (Washington, D. C.: U. S. Govt. Printing Office), 267 pp.

Jaffee, R. I. (chairman), 1975. *National Materials Policy* (Washington, D. C.: National Academy of Sciences), 215 pp.

Kash, D. E., and I. L. White (chairmen), 1973. *Energy Under the Oceans: A Technology Assessment of Outer Continental Shelf Oil and Gas Operations* (Norman, Okla.: Univ. Oklahoma Press), 378 pp.

Kato, I. (chairman), 1977. *Utilization and Development of the Pacific Ocean: Proceedings International Pacific Symposium on the Pacific Ocean* (Tokyo, Japan: Ocean Association of Japan), 134 pp.

Klaff, J. L. (chairman), 1973. *Material Needs and the Environment Today and Tomorrow* (Washington, D. C.: U. S. Govt. Printing Office), 297 pp.

Kovach, E. G. (chairman), 1976. *Rational Use of Potentially Scarce Metals* (Brussels, Belgium: NATO Scientific Affairs Division), 129 pp.

Maddox, P., 1977. Offshore oil gains 6.5% over last year. *Offshore,* 37 (7):57–64; 66–68; 70–71; 77.

McLean, J. G. (chairman), 1972. *U. S. Energy Outlook: A Summary Report* (Washington, D. C.: National Petroleum Council), 134 pp.

Meadows, D. H., D. L. Meadows, J. Randers, and W. W. Behrens, Ill., 1972. *The Limits to Growth: A Report for the Club of Rome's Project on the Predicament of Mankind* (New York: New American Library), 207 pp.

Menard, H. W., 1974. *Geology, Resources, and Society* (San Francisco, Calif.: W. H. Freeman & Co.), 619 pp.

Meyer, R. F. (ed.), 1977. *The Future Supply of Nature-Made Petroleum and Gas, Technical Reports* (New York: Pergamon Press), 1046 pp.

Moore, J. R., 1975. *Mining in the Outer Continental Shelf and in the Deep Ocean* (Washington, D. C.: Natl. Acad. Sci.), 119 pp.

Myrdal, G., 1974. The transfer of technology to underdeveloped countries. *Sci. Amer.,* 231 (3):173–178; 180; 182.

Osgood, R. E. (director), 1974. *Perspectives on Ocean Policy* (Washington, D. C.: U. S. Govt. Printing Office), 435 pp.

Osgood, R. E., A. L. Hollick, C. S. Pearson, and J. C. Orr, 1975. *Toward a National Ocean Policy: 1976 and Beyond: Ocean Policy Project, John Hopkins University* (Washington, D. C.: U. S. Govt. Printing Office), 207 pp.

Padelford, N. J., 1968. *Public Policy and the Use of the Seas* (Cambridge, Mass.: Mass. Inst. Technology), 361 pp.

Park, C. F., Jr., 1968. *Affluence in Jeopardy: Minerals and the Political Economy* (San Francisco, Calif.: Freeman, Cooper & Co.), 368 pp.

———, 1975. *Earthbound: Minerals, Energy, and Man's Future* (San Francisco, Calif.: Freeman, Cooper & Co.), 279 pp.

Ravelle, R., 1976. The resources available for agriculture. *Sci. Amer.,* 235 (3):165–168; 170; 172–174; 177–178.

Ross, D. A., and L. J. Smith, 1974. Training and technical assistance in marine science—a viable transfer product. *Ocean Develop. Int. Law Jour.,* 2:219–253.

Spangler, M. B., 1970. *New Technology and Marine Resources Development: A Study in Government-Business Cooperation* (New York: Praeger), 607 pp.

Steinhart, J. S., and C. E. Steinhart, 1974. Energy use in the U. S. food system. *Science,* 184:307–316.

Trondsen, R., and W. J. Mead, 1977. California offshore phosphorite deposits, an economic evaluation. *Univ. Calif. Inst. Mar. Res. Ref.,* 77–106, 188 pp.

UNCTAD secretariat, 1974. The effects of production of manganese from the sea-bed, with particular reference to effects on developing country producers of manganese ore. *U. N. Conf. Trade Develop. 14th Session, Geneva,* 24 pp.

Wade, N., 1975. Third world: science and technology contribute feebly to development. *Science,* 189:770–771; 774–776.

Wilson, C. L. (ed.), 1970. *Man's Impact on the Global Environment* (Cambridge, Mass.: Mass. Inst. Technology), 319 pp.

Wilson, E. B., and J. M. Hunt (chairmen), 1975. *Petroleum in the Marine Environment* (Washington, D. C.: Natl. Acad. Sci.), 107 pp.

World Bank, 1974. *Atlas, Population, Per Capita Product, and Growth Rates* (New York: World Bank), 22 pp.

Library of Congress Cataloging in Publication Data

Main entry under title:
Catastrophes and earth history.

Papers from a symposium held in the Woods Hole
Oceanographic Institution, June 6-10, 1977, and from
a symposium held in connection with the Second North
American Paleontological Convention in Aug. 1977 at
Lawrence, Kan.

Bibliography: p.

1. Catastrophes (Geology)—Congresses.
I. Berggren, William A.   II. Van Couvering, John A.
QE506.C37   1984        551.7'001        83-11026
ISBN 0-691-08328-2
ISBN 0-691-08329-0 (pbk.)